T0178054

Kratom and Other Mitragynines

The Chemistry and Pharmacology of Opioids from a Non-Opium Source

Kratom and Other Mitragynines

The Chemistry and Pharmacology of Opioids from a Non-Opium Source

Edited by
Robert B. Raffa

Assistant Editors
Jaclyn R. Beckett, Vivek N. Brahmbhatt, Theresa M. Ebinger,
Chrisjon A. Fabian, Justin R. Nixon, Steven T. Orlando,
Chintan A. Rana, Ali H. Tejani and Robert J. Tomazic

CRC Press
Taylor & Francis Group
Boca Raton London New York

CRC Press is an imprint of the
Taylor & Francis Group, an **informa** business

CRC Press
Taylor & Francis Group
6000 Broken Sound Parkway NW, Suite 300
Boca Raton, FL 33487-2742

First issued in paperback 2019

ISBN-13: 978-0-4822-2518-1 (hbk)
ISBN-13: 978-0-367-86924-3 (pbk)

Library of Congress Cataloging-in-Publication Data

Kratom and other mitragynines : the chemistry and pharmacology of opioids from a non-opium source / editor, Robert B. Raffa.
    p. ; cm.
    Includes bibliographical references and index.
    ISBN 978-1-4822-2518-1 (hardcover : alk. paper)
    I. Raffa, Robert B., editor.
    [DNLM: 1. Analgesics, Opioid--chemistry. 2. Mitragyna--chemistry. 3. Analgesics, Opioid--pharmacology. 4. Opioid-Related Disorders--prevention & control. 5. Plant Preparations--chemistry. 6. Plant Preparations--pharmacology. QV 89]

    RM666.O6
    615.7'822--dc23   2014033470

Visit the Taylor & Francis Web site at
http://www.taylorandfrancis.com

and the CRC Press Web site at
http://www.crcpress.com

Contents

Preface

Morphine and other opioid drugs are extracts or analogs of compounds isolated from a single source, the opium poppy (*Papaver somniferum*). However, it seems that biological diversity has given rise to an alternative source. Specifically, at least two alkaloids isolated from the plant *Mitragyna speciosa*—and their synthetic analogs—display pain-relieving activity. Several characteristics of these compounds suggest a classic opioid mechanism of action: nM affinity for opioid receptors, competitive interaction with the opioid receptor antagonist naloxone, and two-way analgesic cross-tolerance with morphine. However, other characteristics suggest novelty, particularly chemical structures that are unlike that of morphine and with a possible greater separation from side effects. This book reviews the chemical and pharmacological properties of these compounds, as well as their therapeutic and abuse potential.

The *M. speciosa* plant is known as "kratom" (Thailand) or "biak-biak" (Malaysia) in Southeast Asia. The plant has been used for its distinctive psychoactive properties, primarily by laborers as a stimulant to counteract work fatigue. At high doses, morphine-like effects prevail. It has also been chewed or dissolved in teas and used as treatment for opiate withdrawal, and for fever reduction, analgesia, diarrhea, coughing, hypertension, and even depression.

More than twenty alkaloids, several of which are biologically active, have been isolated from the *M. speciosa* plant, with mitragynine being the major one. The content varies from plant to plant and from young plants to older ones, and a likely pathway for the biogenesis of these compounds within the plant has been established. MG and 7-hydroxy-MG also exhibit opioid-like activity; 7-OH-MG is more than 10 times more potent than morphine. Tolerance and cross-tolerance to morphine, withdrawal signs (a hallmark indication of physical dependence), and conditioned place-preference have also been demonstrated.

The opioid-like receptor binding and in vivo activity raise the question of abuse potential. For many decades, laborers in Southeast Asia have used mitragynine in low doses as a stimulant and energy supplement. In higher doses, kratom has been used as an opium substitute or to manage opium withdrawal symptoms. Abuse has led to the drug being made illegal in Thailand, Malaysia, Myanmar, and Australia. It remains a popular street drug in these areas, and reports of abuse in other countries have appeared (with street names such as Thang, Kakuam, Thom, Ketum, and Biak in the

United States). Mitragynine is widely available on the internet, and several websites sell it in the form of leaves (whole or crushed), as an extract, a powder or encapsulated powder, and extract-resin "pies." Seeds and whole plants for cultivation are also available. Several Internet websites provide detailed information about mitragynine and its use. Many of these sites include users' experiences with it. It is not controlled in the United States under the Controlled Substances Act, but the DEA (Drug Enforcement Administration) has placed it on their list of drugs and chemicals of concern. This book also reviews the evidence for current and possible more serious future abuse of these substances.

This book evolved from an elective course taught at Temple University School of Pharmacy (by R.B. Raffa). It has benefited greatly from helpful discussions with several faculty members at the school, including Drs. Michael Borenstein, Scott Gothe, and Wayne Childers, whose contributions are acknowledged and greatly appreciated. The authors for each chapter were selected based on their expertise and contributions to their field. Each one has provided an account of his or her subject matter in an authoritative and readable way. A particular strength of the book is the breadth of experience and the wide geographical distribution of the authors. Another is the varied perspective that each provides.

The book is designed so that it can be read either straight through or as individual chapters, depending on the level of background and interest of the reader. There is thus some overlap of material throughout the chapters, but this is intentional so that individual chapters can be read without loss of continuity. We hope that this book—the first comprehensive one on the basic science and clinical use/abuse of kratom—provides the reader with a concise yet comprehensive introduction to and understanding of Nature's "other opioid."

Robert B. Raffa, PhD
Philadelphia, 2014

About the Editors

Robert B. Raffa, PhD, Senior Editor, is Professor of Pharmacology in the Department of Pharmaceutical Sciences at Temple University School of Pharmacy and a research professor in the Department of Pharmacology Temple University School of Medicine. He has bachelor's degrees in chemical engineering and in physiological psychology (both from the University of Delaware), master's degrees in biomedical engineering (Drexel University) and toxicology (Thomas Jefferson University), and a doctorate in pharmacology (Temple University School of Medicine). He was a Research Fellow and team co-leader for analgesics discovery at Johnson & Johnson and was pivotal in the elucidation of the mechanism of action and development of the analgesic drug tramadol (Ultram™). He is co-holder of several patents, including the combination of tramadol plus acetaminophen (Ultracet™). He has published over 280 papers in refereed journals, is the co-author or editor of several books on pharmacology, chemotherapy-induced adverse effects, and thermodynamics, and is a co-editor of the *Journal of Clinical Pharmacy and Therapeutics*. He is a past president of the Mid-Atlantic Pharmacology Society and is the recipient of research and teaching awards. He maintains an active research effort and lectures and consults worldwide on analgesics and analgesic combinations.

ASSISTANT EDITORS

Jaclyn R. Beckett, PharmD
Vivek N. Brahmbhatt, PharmD
Theresa M. Ebinger, PharmD
Chrisjon A. Fabian, PharmD
Justin R. Nixon, PharmD
Steven T. Orlando, PharmD
Chintan A. Rana, PharmD
Ali H. Tejani, PharmD
Robert J. Tomazic, PharmD

Drs. Beckett, Brahmbhatt, Ebinger, Fabian, Nixon, Orlando, Rana, Tejani, and Tomazic were classmates at Temple University School of Pharmacy. They participated in an elective course that resulted in the publication of a review article on the subject matter of this book, "Orally Active

Opioid Compounds from a Non-Poppy Source," which was published in the American Chemical Society's *Journal of Medicinal Chemistry* (56 4840–4848, 2013). They maintain an active interest in the subject, present posters, and have related manuscripts in preparation. They are pursuing their professional careers in pharmacy practice.

Contributors

Ines A. Ackerman
Department of Forensic Toxicology
Institute of Forensic Medicine
St. Gallen, Switzerland

Bonnie A. Avery, PhD
Department of Pharmaceutics
University of Mississippi School
  of Pharmacy
University, Mississippi

Zoriah Aziz, PhD
Department of Pharmacy
Faculty of Medicine
University of Malaya
Kuala Lumpur, Malaysia

Jaclyn R. Beckett, PharmD
Temple University School
  of Pharmacy
Philadelphia, Pennsylvania

Jochen Beyer, PhD
Department of Forensic Toxicology
Institute of Forensic Medicine
St. Gallen, Switzerland

Vivek N. Brahmbhatt, PharmD
Temple University School
  of Pharmacy
Philadelphia, Pennsylvania

Annette Cronin, PhD
Department of Forensic Toxicology
Institute of Forensic Medicine
St. Gallen, Switzerland

Stephen J. Cutler, PhD
Department of Medicinal Chemistry
University of Mississippi School
  of Pharmacy
University, Mississippi

Theresa M. Ebinger, PharmD
Temple University School
  of Pharmacy
Philadelphia, Pennsylvania

Sasha W. Eisenman, PhD
Department of Landscape
  Architecture and Horticulture
Temple University
Ambler, Pennsylvania

Earth Erowid
Erowid Center
Grass Valley, California

Fire Erowid
Erowid Center
Grass Valley, California

Chrisjon A. Fabian, PharmD
Temple University School
  of Pharmacy
Philadelphia, Pennsylvania

Vedanjali Gogineni, MS
Department of Medicinal
  Chemistry
University of Mississippi School
  of Pharmacy
University, Mississippi

Ruri Kikura-Hanajiri, PhD
Section Chief of Narcotic
  Section
Division of Pharmacognosy,
  Phytochemistry and Narcotics
National Institute of Health
  Sciences
Tokyo, Japan

Syunji Horie, PhD
Laboratory of Pharmacology
Faculty of Pharmaceutical
  Sciences
Josai International University
Togane, Japan

Jennifer L. Ingram-Ross, PhD
Janssen Research and
  Development
Raritan, New Jersey

Mariko Kitajima, PhD
Graduate School of Pharmaceutical
  Sciences
Chiba University
Chiba, Japan

Francisco Leon, PhD
Department of Medicinal
  Chemistry
University of Mississippi School
  of Pharmacy
University, Mississippi

Allan Patrick G. Macabeo, PhD
Department of Chemistry, College
  of Science
The University of Santo Tomas
Manila, Philippines

Sharif M. Mansor, PhD
Centre for Drug Research
Universiti Sains Malaysia
Penang, Malaysia

Kenjiro Matsumoto, PhD
Laboratory of Pharmacology
Faculty of Pharmaceutical Sciences
Josai International University
Togane, Japan
and
Division of Pathological Sciences
Department of Pharmacology and
  Experimental Therapeutics
Kyoto Pharmaceutical University
Kyoto, Japan

Christopher McCurdy, PhD
Department of Medicinal
  Chemistry
University of Mississippi School
  of Pharmacy
University, Mississippi

Justin R. Nixon, PharmD
Temple University School
  of Pharmacy
Philadelphia, Pennsylvania

Steven T. Orlando, PharmD
Temple University School
  of Pharmacy
Philadelphia, Pennsylvania

Oné R. Pagán, PhD
Department of Biology
West Chester University
West Chester, Pennsylvania

Joseph V. Pergolizzi, Jr, MD
Department of Medicine
Johns Hopkins University School
   of Medicine
Baltimore, Maryland
and
Department of Anesthesiology
Georgetown University School
   of Medicine
Washington, D.C.

Robert B. Raffa, PhD
Department of Pharmaceutical
   Sciences
Temple University School
   of Pharmacy
Philadelphia, Pennsylvania

Surash Ramanathan, PhD
Centre for Drug Research
Universiti Sains Malaysia
Penang, Malaysia

Chintan A. Rana, PharmD
Temple University School
   of Pharmacy
Philadelphia, Pennsylvania

Scott M. Rawls, PhD
Temple University School of
   Medicine and
Center for Substance Abuse
   Research
Temple University
Philadelphia, Pennsylvania

Hiromitsu Takayama, PhD
Graduate School of Pharmaceutical
   Sciences
Chiba University
Chiba, Japan

Ali H. Tejani, PharmD
Temple University School
   of Pharmacy
Philadelphia, Pennsylvania

Robert J. Tomazic, PharmD
Temple University School
   of Pharmacy
Philadelphia, Pennsylvania

Jordan K. Zjawiony, PhD
Department of BioMolecular
   Sciences
University of Mississippi School
   of Pharmacy
University, Mississippi

1 Why Do Plants Contain Biologically Active Compounds?

Oné R. Pagán

CONTENTS

1.1 PRESSURE FOR SURVIVAL

From our admittedly biased perspective as members of the animal kingdom, at first glance it seems remarkable that plants are able to survive at all. Plants lack a basic defensive response that almost every animal has—the ability to flee from the danger. However, plants are anything but defenseless. It is evident that plants are some of the most successful types of organisms on Earth. This success is due in part to the ability of many plants to produce compounds (secondary metabolites) that help maximize their chances of survival—and therefore reproduction and perpetuation.

1.2   CHEMICALS AS BENEFICIAL FOR SURVIVAL

There are at least three mechanisms that illustrate how the actions of chemical substances can be beneficial to a plant. One is that some chemical substances deter the growth of competing plants, or are even capable of killing predators, which is a straightforward defensive mechanism. Another is that some plants produce compounds that attract beneficial animals, which in turn serve as pollinators as well as many other fascinating mechanisms. The study of the interrelationship of chemicals and living organisms is an aspect of chemical ecology (Eisner and Meinwald 1995). Chemicals as toxins or as attractive molecules for reproduction are widely discussed in the scientific literature (Vereecken et al. 2010; Stökl et al. 2011; Ibanez et al. 2012; Friedman and Rasooly 2013). A third, generally lesser known, aspect of the chemical survival mechanism used by plants is the effect they have on the behavior of the target organisms. A remarkable consequence of this phenomenon includes the fact that many types of plants have co-evolved with humans.

1.3   COEVOLUTION WITH HUMANS

Nature is the best chemical engineer. During the millions of years through which evolution has worked, representatives of every class of organisms have developed substances to help them capture prey or to avoid becoming prey. As part of this evolutionary process, chemicals (molecular structures) that are beneficial for the survival of the organism are conserved; many of these molecules include small organic toxins (Speijers 1995; Mebs 2001; Wittstock and Gershenzon 2002; Brenner et al. 2003; Pagán 2005).

1.3.1   Evolution of Chemicals

From a biological point of view, the main mechanism by which plants are capable of producing such substances is through evolutionary change. The scientific evidence is without dispute. As part of this process, chemical configurations that prove to be beneficial for the survival of the organism are conserved and passed along to the next generation.

1.3.2   Complementary Targets

The production of defensive chemical compounds (toxins) is an aspect of the so-called "evolutionary arms race" (Dawkins and Krebs 1979), where these toxins are structurally optimized for their interaction with molecular targets, usually macromolecules such as receptors, transporters, and enzymes. At the organismal level, many plants display the phenomenon of *allelopathy*, which involves the effect of plants on other plants within their environment, a phenomenon that has been known for thousands of years (Li et al. 2010). We now know that most of these interactions involve the release of plant-derived chemicals into the environment, which can have either deleterious or beneficial effects on other plants in their immediate vicinity (Tharayil 2009).

1.4 PSYCHOACTIVE EFFECTS

Not surprisingly, many plant-derived substances are capable of inducing psychoactive responses. Animal life overwhelmingly displays behavioral responses, generated by the nervous system and optimized by evolution, that aid survival. This presents an apparent paradox, because psychoactive chemicals induce changes in mood, perception, and related subjective states in animals, which undoubtedly influences their behavior. Yet it is common knowledge that many species of animals, including humans, purposively consume plants that generate psychoactive effects. This paradox can be stated as follows: why would a plant make psychoactive substances at all?

1.4.1 PLANT PERSPECTIVE

Plants do not seem to use psychoactive substances for any of their intrinsic physiological processes, as these compounds do not seem to play any bioenergetic role in the plants that make them. In addition, many of these compounds are metabolically "expensive" in terms of the energy-requiring specialized enzymes that are often required to synthesize specific structural features, including any "correct" stereochemical arrangement. Thus, on the face of it, it does not make much sense for a plant to dedicate valuable and probably scarce metabolic resources to synthesize a compound that does not appear to be immediately required for survival.

Also, plants do not possess anything like a proper nervous system, and as far as we know, a nervous system is an indispensable requirement for producing the phenomenon of "mind"; however, there is talk of a new field called *plant neurobiology*, at this point rather controversial (Brenner et al. 2006; Alpi et al. 2007; Baluska and Mancuso 2007). Seen from this perspective, once again plants do not seem to exhibit any "psycho" aspect that may be affected by psychoactive chemicals. Until relatively recently this has been a mystery.

It is easy to forget that the natural history of multicellular animals is built upon the physiology of much simpler unicellular organisms. The appearance of nervous systems in nature did not occur in a vacuum; many of its components (in terms of chemistry, morphology, and physiology) were already present in unicellular organisms. For instance, paramecia use neurotransmitter-like molecules, and even *bona fide* neurotransmitters, as signaling molecules. Paramecia also make use of electrophysiological phenomena in their physiology that are reminiscent of action potentials in higher organisms. Many other types of microorganisms, from bacteria to protists, as well as many types of multicellular organisms—including plants and fungi—use these kinds of molecules for cell-to-cell signaling purposes (Eckert and Brehm 1979; Bonner 2008; Roshchina 2010). Thus it is not surprising that plants possess chemicals capable of interacting with animal nervous systems.

1.4.2 ANIMAL PERSPECTIVE

Many abused drugs of natural origin are secondary metabolites of plants, including the well-known compounds *nicotine* and *cocaine*, produced primarily by tobacco and coca plants, respectively. Both compounds have been described as natural insecticides in the sense that they often fatally interfere with the normal physiology of the nervous system of insects. For example, an insect feeding on a tobacco plant is exposed to nicotine, and this exposure will repel the insect and can even kill it if the insect ingests a high enough amount. On the other hand, a human, a much larger organism, exposed to the same amount of nicotine as an insect will likely experience psychoactive, rather than lethal, effects. There is abundant evidence suggesting that plants use chemicals as defense from insects (Hartmann 2004; Pieterse and Dicke 2007). Many proteins that are important to nervous system function in insects have counterparts in vertebrates; this is why invertebrate animal models can provide insights into human nervous system functions. Therefore the ingestion of these plant-derived chemicals by humans can have psychoactive effects. The reason for this is again related to evolution. In the same way that many defensive toxins are conserved for their positive effects on survival, their molecular targets are conserved as well. It is a general rule that once life has a set of molecular tools, these tools may be reused or co-opted for alternative functions.

1.4.3 PLANT–ANIMAL INTERACTIONS

Going back to the question of why psychoactive substances are beneficial to the plants that produce them, one proposed interpretation is that many types of plants are engaged in a "chemopsychological war" against animals (Elrich and Raven 1964, 1967). From this point of view, plants that produce such substances may have an evolutionary edge at several levels. We have already mentioned their possible insecticidal effects. Additionally, owing to the presence of molecular targets to these substances in humans, a person may experience desirable sensations upon ingestion of a plant species containing specific molecules. When people identify a plant species with a desirable trait, an artificial selection process may start. Through this process, desirable plants are protected from insects or other predators such as grazing animals. Eventually these plants will be actively cultivated. In due course, a specific plant breed is obtained, which can be further selected for potency.

Some schools of thought propose that plants that produce psychotropic substances (including currently abused drugs) coevolved with humans; for example, the ingestion of neurotransmitter-like molecules derived from plants could act as substitutes for bioenergetically expensive neurotransmitters under states of nutritional deprivation (Sullivan and Hagen 2002; Kennedy and Wightman 2011). In other words, we could say that psychoactive plants acted as the first nutritional supplements ever used by humans. However, there is the concept of the paradox of drug reward (Hagen et al. 2009; Sullivan et al. 2008), which essentially asks: If plant defensive substances tend to cause unpleasant sensation in animals, why do humans seek such substances?

It is undeniable and not surprising that human–plant coevolution has sociological implications. Traditionally, people who possessed knowledge about the effects of plants on humans were valued and revered as sacred persons. There is archeological evidence indicating that this practice has apparently been around for quite a long time.

1.5 MUTUAL BENEFITS

A very interesting question is how were psychoactive plants discovered? We will probably never know for sure; this surely happened thousands of years ago, before historical records were made. And it is certain that the story will be subtly different depending on the specific plant. Nonetheless, there are certain distinct possibilities, such as when a sick human, say with a fever, ate a plant just because of hunger and then, lo and behold, the fever disappeared faster than it would have if it had been left to run its course. An alert human may have noticed that. Also, it is easy to imagine our ancestors gathering dry leaves for kindling to start a fire, using tobacco or marijuana plants for this purpose. The psychoactive effects of breathing such smoke would be immediately evident.

1.6 SUMMARY

Plants produce biologically active substances that influence the behavior of potential predators or organisms that can assist in their reproduction. Humans are not the "intended" target of such substances, but biologically and eventually culturally we have learned to take advantage of such substances, for good or bad. There is certainly much more work to be done in this interesting field, especially in light of the promise of the chemistry of natural products as a source of potential medications.

REFERENCES

Alpi, A., Amrhein, N., Bertl, A. et al. 2007. Plant neurobiology: No brain, no gain? *Trends Plant Sci* 12(4): 135–36.

Baluska, F. and Mancuso, S. 2007. Plant neurobiology as a paradigm shift not only in the plant sciences. *Plant Signal Behav* 2(4): 205–207.

Bonner, J.T. 2008. *The Social Amoebae: The Biology of Cellular Slime Molds*. Princeton: Princeton University Press.

Brenner, E.D., Stevenson, D.W., and Twigg, R.W. 2003. Cycads: Evolutionary innovations and the role of plant-derived neurotoxins. *Trends Plant Sci* 8(9): 446–52.

Brenner, E.D., Stahlberg, R., Mancuso, S., Vivanco, J., Baluska, F., and Van Volkenburgh, E. 2006. Plant neurobiology: An integrated view of plant signaling. *Trends Plant Sci* 8: 413–19.

Dawkins, R., and Krebs, J.R. 1979. Arm races between and within species. *Proc R Soc Lond B* 205: 489–511.

Eckert, R., and Brehm, P. 1979. Ionic mechanisms of excitation in *Paramecium*. *Ann Rev Biophys Bioeng* 8: 353–83.

Eisner, T. and Meinwald, J. (eds). 1995. *Chemical Ecology: The Chemistry of Biotic Interaction*. Washington D.C.: National Academy Press.

Elrich, P.R. and Raven, P.H. 1964. Butterflies and plants: A study on coevolution. *Evolution* 18(4): 586–608.

Elrich, P.R. and Raven, P.H. 1967. Butterflies and plants. *Sci Amer* 216: 104–13.

Friedman, M., and Rasooly, R. 2013. Review of the inhibition of biological activities of food-related selected toxins by natural compounds. *Toxins* 5(4): 743–75.

Hagen, E.H., Sullivan, R.J., Schmidt, R., Morris, G., Kempter, R., and Hammerstein, P. 2009. Ecology and neurobiology of toxin avoidance and the paradox of drug reward. *Neuroscience* 160(1): 69–84.

Hartmann, T. 2004. Plant-derived secondary metabolites as defensive chemicals in herbivorous insects: A case study in chemical ecology. *Planta* 219(1): 1–4.

Ibanez, S., Gallet, C., and Despres, L. 2012. Plant insecticidal toxins in ecological networks. *Toxins* 4(4): 228–43.

Kennedy, D.O., Wightman, E.L. 2011. Herbal extracts and phytochemicals: Plant secondary metabolites and the enhancement of human brain function. *Adv Nutr* 2(1): 32–50.

Li, Z.H., Wang, Q., Ruan, X., Pan, C.D., and Jiang, D.A. 2010. Phenolics and plant allelopathy. *Molecules* 15(12): 8933–52.

Mebs, D. 2001. Toxicity in animals: Trends in evolution? *Toxicon* 39: 87–96.

Pagán, O.R. 2005. Synthetic Local Anesthetics as Alleviators of Cocaine Inhibition of the Human Dopamine Transporter. PhD dissertation. p. 1. Cornell University, Ithaca, NY.

Pieterse, C.M., and Dicke, M. 2007. Plant interactions with microbes and insects: From molecular mechanisms to ecology. *Trends Plant Sci* 12(12): 564–69.

Roshchina, V.V. 2010. Evolutionary Considerations of Neurotransmitters in Microbial, Plant, and Animal Cells. Chapter 2 in: *Microbial Endocrinology: Interkingdom Signaling in Infectious Disease and Health*, Lyte M., and Freestone P.P.E. (eds.). New York: Springer.

Speijers, G.J. 1995. Toxicological data needed for safety evaluation and regulation on inherent plant toxins. *Nat Toxins* 3(4): 222–26.

Stökl, J., Brodmann, J., Dafni, A., Ayasse, M., and Hansson, B.S. 2011. Smells like aphids: Orchid flowers mimic aphid alarm pheromones to attract hoverflies for pollination. *Proc Biol Sci* 278(1709): 1216–22.

Sullivan, R.J., and Hagen, E.H. 2002. Psychotropic substance-seeking: Evolutionary pathology or adaptation? *Addiction* 97(4): 389–400.

Sullivan, R.J., Hagen, E.H., and Hammerstein, P. 2008. Revealing the paradox of drug reward in human evolution. *Proc Biol Sci* 275(1640): 1231–41.

Tharayil, N. 2009. To survive or to slay: Resource-foraging role of metabolites implicated in allelopathy. *Plant Signal Behav* 4(7): 580–83.

Vereecken, N.J., Cozzolino, S., and Schiestl, F.P. 2010. Hybrid floral scent novelty drives pollinator shift in sexually deceptive orchids. *BMC Evol Biol* 10: 103.

Wittstock, U. and Gershenzon, J. 2002. Constitutive plant toxins and their role in defense against herbivores and pathogens. *Curr Opin Plant Biol* 5: 1–8.

2 Short Overview of Mitragynines

Robert B. Raffa, Jaclyn R. Beckett,
Vivek N. Brahmbhatt, Theresa M. Ebinger,
Chrisjon A. Fabian, Justin R. Nixon,
Steven T. Orlando, Chintan A. Rana,
Ali H. Tejani, and Robert J. Tomazic

CONTENTS

2.1 INTRODUCTION

The opioid drugs, such as morphine, codeine, methadone, oxycodone, and many others, are derived from or are analogs of extracts of compounds isolated from the opium poppy (*Papaver somniferum*). The naturally occurring exogenous opioids (e.g., morphine and codeine) and endogenous opioids (e.g., the endorphins and enkephalins) co-evolved (see Chapter 1) the ability to produce a great variety of biological effects, some beneficial and some not. For a long time, it was thought that the poppy was nature's sole source of opioid compounds. However, it now appears that biological diversity has evolved an alternative source of opioid compounds, and that is the subject of this book.

2.1.1 MITRAGYNINE (MG) AND 7-HYDROXYMITRAGYNINE (7-OH-MG)

At least two alkaloids isolated from the plant *Mitragyna speciosa*, mitragynine (MG) and 7-hydroxymitragynine (7-OH-MG), and several synthetic analogs (mitragynines) display opioid characteristics: high and selective affinity for one or more of the opioid receptors, competitive interaction with the selective opioid receptor antagonist naloxone, and two-way analgesic cross-tolerance with morphine. They also display features that might offer therapeutic benefit compared with morphine and other opioids. We recently reviewed this topic (Raffa et al. 2013) and present an abridged version here.

M. speciosa has long been used for its distinctive properties, in particular by laborers as a stimulant to alleviate work fatigue (Jansen and Prast 1988a,b). At higher doses, opium-like effects prevail, which has led to its use as treatment for pain, diarrhea, cough, hypertension, and depression and in opiate withdrawal and fever reduction (Jansen and Prast 1988a,b; Khor et al. 2011).

2.1.2 OTHER ALKALOIDS

More than twenty alkaloids have been isolated from the *M. speciosa* plant (León et al. 2009), with mitragynine accounting for 66.2% of the crude base (Takayama 2004) and 6% by weight of the dried plant. Others include paynantheine, speciogynine, 7α-hydroxy-7*H*-mitragynine (Matsumoto et al. 2004), speciociliatine, mitragynaline, mitragynalinic acid, corynantheidaline, and corynantheidalinic acid (Takayama 2004). Ajmalicine and mitraphylline are structurally related alkaloids isolated from *M. speciosa*. Mitragynine is a minor constituent in young plants and the dominant indole alkaloid in older plants (Shellard et al. 1978; León et al. 2009). A likely pathway for the biogenesis of alkaloid compounds within the plant has been suggested (Shellard et al. 1978).

MG, 7-OH-MG, and mitragynine pseudoindoxly have high affinity (nM range) for opioid receptors (Takayama 2004). MG and 7-OH-MG also exhibit opioid-mediated (naloxone-sensitive) antinociceptive activity: 7-OH-MG is 40-fold more potent than mitragynine and 10-fold more potent than morphine (Matsumoto et al. 2004).

2.2 EVIDENCE FOR CLASSIFICATION AS AN OPIOID

Substances are typically classified as opioid based on the following criteria: binding affinity for one or more of the major 7TM-GPCR (seven transmembrane G-protein–coupled receptor) subtypes of opioid receptor (MOR, DOR, KOR, ORL-1), morphine-like in vivo effects such as pain relief (called analgesia in humans, antinociception in animals), miosis, constipation, respiratory depression, tolerance, cross-tolerance to a known opioid, development of physical dependence (manifested by abstinence-induced and/or antagonist-induced withdrawal signs), and similarity of chemical structure to morphine or another known opioid.

2.2.1 In Vitro

The two studies that have reported the affinities of MG and/or 7-OH-MG for opioid receptors (in guinea pig brain homogenates) used the MOR-selective ligand DAMGO (D-Ala^2, N-MePhe^4, Gly-ol]-enkephalin), the DOR-selective ligand DPDPE (D-Pen^2-D-Pen^5-enkephalin), and the KOR-selective ligand U-69,593 (N-methyl-2-phenyl-N-[(5R,7S,8S)-7-(pyrrolidin-1-yl)-1-oxaspiro[4.5]dec-8-yl]acetamide). The affinity of MG at MOR, DOR, and KOR is 7.2 nM, 60 nM, and >1,000 nM, respectively. It therefore displays a nearly 10-fold selectivity for MOR over DOR sites and greater than 1,000-fold selectivity for MOR over KOR (Takayama et al. 2002). The affinity of morphine in the same study was 3.5 nM, 417 nM, and 468 nM at MOR, DOR, and KOR, respectively. The affinity of 7-OH-MG in the same study was 13 nM at MOR, 155 nM at DOR, and 123 nM at KOR. MG and 7-OH-MG thus have affinity comparable with morphine for MOR. The relative receptor-binding profile of MG is MOR > DOR >> KOR, and of 7-OH-MG it is MOR > DOR ≅ κ (compared to morphine's profile of MOR >> DOR ≅ KOR). A subsequent study (Matsumoto et al. 2004) reported similar results for 7-OH-MG (9.8 nM, 145 nM, and 195 nM at MOR, DOR, and KOR, respectively). To our knowledge, there are no reported measures of affinity at the ORL-1 (opioid receptor-like; nociceptin; orphanin FQ) receptor.

2.2.2 Isolated Tissue

MG has been reported to have activity (Takayama et al. 2002) in the classic electrically stimulated guinea pig ileum isolated tissue preparation (Kosterlitz and Robinson 1957). Both MG and 7-OH-Mg displayed nearly the same intrinsic activity as morphine. MG was about one-fourth as potent as morphine, and 7-OH-MG was about 10-fold more potent than morphine (Kitajima et al. 2006). The effect of MG could not be due to activity at ORL-1, since it was not blocked by an ORL-1 antagonist (Horie et al. 2005).

MG-pseudoindoxyl, speciogynine, speciociliatine, 7-hydroxyspeciocillatine, and paynantheine also exhibit agonist action in electrically stimulated guinea pig ilium (Takayama et al. 2002; Takayama 2004) and an antagonist, corynantheidaline, reverses morphine-inhibited twitch contraction in guinea pig ileum (Takayama 2004).

2.2.3 In Vivo

2.2.3.1 Antinociception: Mice, Rats, and Dogs

The subcutaneous (s.c.) administration of 7-OH-MG produces a full and long-lasting antinociceptive effect in the mouse tail-flick test (Thongpradichote et al. 1998). Oral administration of 7-OH-MG also produces a full antinociceptive effect in this test. It has been demonstrated that the antinociceptive effect of MG is due to action in the central nervous system (CNS), because this effect was produced when MG was injected directly into the brain by intracerebroventricular administration. The maximum antinociceptive effect by this route was greater than 50%, but less than the full effect produced by morphine. In the mouse hot-plate test (54.5°C), MG produced 100% antinociception by oral administration, but the same dose was ineffective by s.c. administration (Macko et al.1972).

MG is also orally active in the rat tail-flick test and the Randall and Selitto rat paw-pressure test (Macko et al. 1972), and in an antinociceptive test in dogs (hindleg flick). Its potency in the rat tests was about one-half that of codeine; it was about equipotent to codeine in the dog test (Macko et al. 1972).

2.2.3.2 Antinociception: In Vivo Receptor Profile

The ability of the selective opioid receptor antagonist naloxone to attenuate the effect of MG has been demonstrated and the opioid receptor subtype selectivity of MG-induced intracerebroventricular antinociception has been assessed using the receptor-selective compounds cyprodime (MOR), naltrindole (DOR), and nor-binaltorphimine (nor-BNI) (KOR) (Thongpradichote et al. 1998). The antinociceptive effects produced by either peripheral (intraperitoneal) or central (intracerebroventricular) administration of MG to mice was antagonized by central (intracerebroventricular) administration of naloxone (Matsumoto et al. 1996b); and, consistent with the binding studies in vitro, morphine- and 7-OH-MG-induced antinociception in the mouse tail-flick and hot-plate (55°C) were antagonized by naloxone and cyprodime (MOR), but less so, or not at all, by naltrindole (DOR) and nor-BNI (KOR).

2.2.3.3 Naloxone-Precipitated Withdrawal in Mice

Another indication of opioid-like activity is precipitation of withdrawal by an opioid receptor antagonist. When morphine or 7-OH-MG was injected s.c. to mice over a period of 5 days following standard protocols (Suzuki et al. 1995; Kamei and Ohsawa 1997; Tsuji et al. 2000) the injection of naloxone elicited classic opioid withdrawal signs (jumping, rearing, urination, and forepaw tremor). Naloxone challenge did not elicit as much diarrhea in 7-OH-MG-treated mice as it did in morphine-treated mice.

2.2.3.4 Naloxone-Precipitated Withdrawal in Zebrafish

In a recent study (Khor et al. 2011), zebrafish were exposed to morphine for two weeks, and then tested for signs of abstinence-induced withdrawal, in the absence of MG and in the presence of MG. In the absence of MG, withdrawal from morphine was pronounced, but was much less so in the presence of MG. Although there are a number of possible alternative explanations (Khor et al. 2011), one is that MG substitutes for morphine in this test, consistent with an agonist action at MOR.

2.2.3.5 Antinociceptive Tolerance in Mice

Tolerance and cross-tolerance are classic signs of poppy-derived opioids and analogs. Tolerance to the antinociceptive effect of opioids develops over the course of repeated administrations. For example, the antinociceptive effect of morphine administered s.c. twice daily declines from nearly 100% on day 1 in the mouse tail-flick test to nearly no effect on day 5 (Matsumoto et al. 2005). A similar pattern of development of tolerance was observed for 7-OH-MG, and two-way cross-tolerance between morphine and 7-OH-MG was observed (that is, a diminished 7-OH-MG effect in morphine-tolerant mice and a diminished morphine effect in 7-OH-MG-tolerant mice). The development of tolerance by itself is not indicative of opioid activity, but two-way cross-tolerance with morphine is.

2.2.4 OTHER EFFECTS

Receptor-mediated effects should include not only the therapeutic effects, but also the side effects associated with that receptor. Opioids produce characteristic side effects, such as emesis, inhibition of cough (antitussive action), constipation (or antidiarrheal action), respiratory depression, and Straub tail in rodents (a vertical or nearly vertical position of the tail, sometimes with a slight forward pointing of the tip). Summarizing the results of a study in which several of these effects were examined (Macko et al. 1972), oral MG dose-dependently inhibits cough reflex in unanaesthetized dogs; MG does not produce emesis in dogs at doses that produce full oral antinociceptive effect; codeine appears to be more active than MG in depressing respiration in anesthetized dogs; MG does not induce Straub tail; oral MG at the high end of its antinociceptive dose-response curve produces < 20% inhibition of gastrointestinal transit in rats (compared to 70–75% inhibition by morphine and codeine); and intraperitoneally administered 7-OH-MG produced no significant effect. In addition, to our knowledge, there have been no reports that either MG or 7-OH-MG produce diuresis (a characteristic effect of several KOR agonists).

2.2.4.1 Inhibition of Gastrointestinal Transit

Morphine and related opioids inhibit GI transit in animals and humans. The dose needed for pain relief often equals or exceeds the dose that inhibits transit, so constipation is a common problem associated with opioid use. The relative separation between MG-induced antinociception and constipation was examined further in a recent study using the charcoal meal test in mice (Matsumoto et al. 2006). 7-OH-MG inhibited transit in a dose-dependent manner, but was an estimated 4.9–6.4 times less constipating than was morphine at antinociceptive doses. Taken together, the two studies (Macko et al. 1972; Matsumoto et al. 2006) suggest that MG and 7-OH-MG might have a greater separation between pain relief and constipation.

2.2.4.2 Conditioned Place-Preference in Mice

In this measure of drug "likeability," animals learn to associate one location ("place") with a drug being administered and a different place with no drug being administered. At a pre-set time after the final conditioning session, the animals are allowed to choose which place they prefer, that is, to display a place-preference. Mice display a place-preference for opioids. In such a test, 7-OH-MG induced dose-related place-preference compared to vehicle (Matsumoto et al. 2008), which was about equal to that of morphine at similar antinociceptive doses. Although not in any way proof of an opioid action, the results are nevertheless consistent with the other findings with 7-OH-MG.

2.3 CHEMICAL STRUCTURES

A cursory comparison suggests some similarities in the chemical structures of MG, 7-OH-MG, and morphine. Each has the three functional groups deemed important for opioid receptor binding, namely, a tertiary nitrogen atom, a benzene residue, and a phenolic hydroxyl (Dhawan et al. 1996). However, a comparison suggests that the groups in MG and morphine do not overlay well in 3-D space (Takayama et al. 2002), and it is believed that MG and morphine bind in different ways to (or subsites on) opioid receptors (Matsumoto et al. 2005). The hydroxyl group at C7 of 7-OH-MG (absent in MG) increases potency significantly. For example, 7-OH-MG is nearly 50-fold more potent than MG and more than 10-fold more potent than morphine in the guinea pig ileum preparation (Takayama 2004). This has been postulated (Matsumoto et al. 2006) to be due to its greater lipophilicity than morphine and faster distribution across the blood–brain barrier.

2.4 RECEPTOR SUBTYPE SELECTIVITY

MOR-selective structures display a large degree of diversity. DOR-selective compounds usually contain a hydrophobic region (e.g., indole or spiroindane) that fits a hydrophobic receptor pocket and might also contribute to selectivity by providing added bulk that inhibits binding to other sites (e.g., a tryptophan within the binding cavity of MOR appears to sterically hinder the added bulk associated with DOR-selective compounds), as evidenced by the fact that when tryptophan is switched to alanine, leucine, or lysine (amino acids smaller than tryptophan), DOR-selective compounds can bind (Kane et al. 2006). 7-OH-MG does not contain this hydrophobic region, possibly accounting for its greater affinity for MOR than DOR. KOR-selective compounds typically contain a basic moiety that binds to a glutamate within the KOR binding pocket (Kane et al. 2006).

2.4.1 STRUCTURE–ACTIVITY RELATIONSHIP (SAR)

The SAR of MG-related compounds is extensively covered in other chapters, which are authored by the researchers who carried out the work. The results of their extensive and comprehensive work are only very briefly summarized here. The results of these studies have identified the C9 position, C7 position, N_b lone pair, and β-methoxyacrylate moiety as the four major sites (Takayama 2004; Matsumoto et al. 2006) (Table 2.1). Note that in the IUPAC nomenclature the methoxy group is attached at the C8 carbon atom of the octahydroindoloquinazoline scaffold, whereas the same carbon is referred to as C9 position in the corynantheidine scaffold: the C7 position is important for binding affinity; the C9 position on the corynantheidine scaffold is important for the intrinsic activity of the compounds; a C10-fluorinated, ethylene-glycol–bridged derivative of MG displays nM binding affinity at MOR and KOR, oral antinociceptive potency greater than that of

TABLE 2.1
Mitragynine (MG) and Analogs

	R_1	R_2	Name
	OCH_3	H	Mitragynine (MG)
	H	H	Corynantheidine
	OH	H	9-Hydroxycorynantheidine
	$OCOCH_3$	H	9-Acethoxycorynantheidine
	OCH_2OCH_3	H	9-Methoxymethylcorynantheidine
	OCH_3	OH	7-Hydroxy-MG (7-OH-MG)
	OCH_3	OCH_3	7-Methoxy-MG
	OCH_3	OCH_2CH_3	7-Methoxy-MG
	OCH_3	$OCOCH_3$	7-Acethoxy-MG

morphine, and weaker inhibition of gastrointestinal transit than morphine at equi-analgesic doses; the β-methoxyacrylate residue ($-C-CO_2-CH_3$) and the N_b lone pair appear to be essential for opioid agonist activity.

2.4.2   MG-PSEUDOINDOXYL

MG-pseudoindoxyl has high affinity for MOR and DOR (0.01 and 3.0 nM, respectively) and negligible affinity for KOR (79 nM) (Takayama 2004). It has a naloxone-sensitive inhibitory effect on guinea pig ileum preparation, but much less antinociceptive activity than morphine (possibly owing to instability in brain) (Matsumoto et al. 2005, 2006).

2.5   NON-OPIOID PROPERTIES?

2.5.1   In Vitro

Noradrenergic and serotoninergic (5-hydroxytryptamine, or 5-HT) systems are involved in analgesic pathways of the central nervous system (Raffa

2006; Raffa et al. 2012). MG displays some affinity for α_2-adrenoceptors (Boyer et al. 2008), but MG does not act directly on these receptors to produce antinociceptive activity (Tohda et al. 1997). MG has been reported to display some affinity for 5-HT_{2C}, 5-HT_7, and dopamine D_2 receptors (Boyer et al. 2008), but the details were not reported.

2.5.2 In Vivo

Similar to morphine, MG suppresses 5-HT–induced head-twitch response in mice, which suggests an action on 5-HT_{2A} receptors or α_2-adrenoceptors, or both (Matsumoto et al. 1997). Based on this, MG was tested for antidepressant activity using the mouse forced-swim and tail-suspension tests. MG was moderately active in these tests (Idayu et al. 2011), consistent with a previous study in which aqueous extract of *M. speciosa* had an antidepressant activity (Kumarnsit et al. 2007). In addition, a role for the descending noradrenergic and 5-HT inhibitory systems in MG-induced antinociception has been postulated (Matsumoto et al. 1996a); a role for cannabinoid CB1 receptors appears to have been ruled out (Shamima et al. 2012).

2.6 METABOLISM

MG undergoes Phase I and II hepatic metabolism in rodents and humans and is excreted primarily in the urine. The metabolism of MG in humans has been investigated (Philipp et al. 2011), revealing a number of Phase I metabolites and at least three glucuronides. The extent to which any of these contributes to MG's biological effects is unknown, although some contribution is suggested by the greater oral compared with s.c. activity of MG in some antinociceptive tests (Macko et al. 1972).

2.7 TOXICOLOGY

The toxicology of MG has been studied in single- and multiple-dosing regimens in rats and dogs (Macko et al. 1972). In rats, there was no observed toxicity following single oral doses as high as 806 mg/kg, and thirty multiple oral doses of up to 50 mg/kg/d did not produce any observable side-effects. In dogs, no side effects were observed during five daily oral doses of 16 mg/kg/d and two additional days of oral 32 mg/kg/d. Transient clinical findings (primarily blood dyscrasias) were observed at higher doses and longer exposures.

To our knowledge, there are no well-defined studies of toxicity of MG in humans, and in the case report of the identification of MG, among other substances, at autopsy, the medical examiner did not include mitragynine in the cause of death (Holler et al. 2011).

2.8 ABUSE POTENTIAL

Laborers in Southeast Asia used MG in low doses as a stimulant and energy supplement. In higher doses, it has been used as an opium substitute or to manage opium withdrawal symptoms (Shellard 1989). Abuse has led to the drug being legislated illegal in several Southeast Asian countries (León et al. 2009), and reports of abuse in other countries have appeared (the subject of a subsequent chapter) (DEA 2013). Mitragynine is readily and widely available on the Internet, and users' experiences with it are available and quite informative (see Chapter 16). Although it is not currently controlled in the United States under the Controlled Substances Act, it is on the DEA's (Drug Enforcement Administration) list of drugs and chemicals of concern (DEA 2013).

2.9 SUMMARY

The *M. speciosa* plant contains more than twenty alkaloids, at least three of which display significant affinity for opioid receptors. They therefore satisfy the modern in vitro criteria for classification as an opioid. The binding assays suggest a relatively greater selectivity for MOR, with some DOR and KOR activity. Two of the alkaloids, MG and 7-OH-MG, have been characterized as agonists (their effects are naloxone-reversible); and one, corynantheidine, has been characterized as an antagonist. MG inhibits (naloxone-reversibly) electrically induced guinea pig ileum contraction. In vivo, the compounds produce opioid receptor-mediated antinociceptive effect via a central site of action involving opioid receptors.

Whether or not MG and analogs bind to opioid receptors in the exact same site or in the same manner as morphine has not been completely elucidated. The actual details are still a matter of study. Slight differences might account for the observed differences in the in vivo profile of these compounds. For example, the compounds appear to produce less emesis and respiratory depression than does codeine, although much more detailed investigation is needed. The evidence for a greater antinociception to constipation ratio is more compelling. In at least two antinociception tests, 7-OH-MG was approximately 10-fold more potent than morphine, but it was equally potent in inhibiting GI transit—thus it had a 10-fold greater separation between antinociception and constipation than did morphine.

M. speciosa compounds have been subjected to abuse. They are widely available on the Internet, which also provides information about their use and psychotropic effects. Not currently scheduled substances in the United States, they are listed as chemicals of concern by the DEA.

Taken together, the data appear to be sufficient to conclude that compounds from *M. speciosa*, such as MG and 7-OH-MG, and synthetic analogs of these alkaloids are a novel (or at least distinct) class of opioids. It appears that biodiversity has provided an alternative natural source of opioid compounds.

ACKNOWLEDGEMENTS

The authors wish to thank Wayne Childers, PhD, and Michael R. Borenstein, PhD, Temple University School of Pharmacy, Philadelphia, PA, for helpful discussion and suggestions.

REFERENCES

Boyer, E.W., Babu, K.M., Adkins, J.E., McCurdy, C.R. and Halpern, J.H. 2008. Self-treatment of opioid withdrawal using kratom (*Mitragynia speciosa* Korth). *Addiction* 103: 1048–50.

DEA. 2013. Kratom (*Mitragyna speciosa* Korth.). January 2013.

Dhawan, B.N., Cesselin, F., Raghubir, R., Reisine, T., Bradley, P.B., Portoghese, P.S. and Hamon, M. 1996. International Union of Pharmacology. XII. Classification of opioid receptors. *Pharmacol Rev* 48: 567–92.

Holler, J.M., Vorce, S.P., McDonough-Bender, P.C., Magluilo, J., Jr., Solomon, C.J. and Levine, B. 2011. A drug toxicity death involving propylhexedrine and mitragynine. *J Anal Toxicol* 35: 54–59.

Horie, S., Koyama, F., Takayama, H., et al. 2005. Indole alkaloids of a Thai medicinal herb, *Mitragyna speciosa*, that has opioid agonist effect in guinea-pig ileum. *Planta Med* 71: 231–36.

Idayu, N.F., Hidayat, M.T., Moklas, M.A., Sharida, F., Raudzah, A.R., Shamima, A.R. and Apryani, E. 2011. Antidepressant-like effect of mitragynine isolated from *Mitragyna speciosa* Korth in mice model of depression. *Phytomedicine* 18: 402–07.

Jansen, K.L. and Prast, C.J. 1988a. Ethnopharmacology of kratom and the *Mitragyna* alkaloids. *J Ethnopharmacol* 23: 115–19.

Jansen, K.L. and Prast, C.J. 1988b. Psychoactive properties of *Mitragynine* (kratom). *J Psychoactive Drugs* 20: 455–57.

Kamei, J. and Ohsawa, M. 1997. Role of noradrenergic functions in the modification of naloxone-precipitated withdrawal jumping in morphine-dependent mice by diabetes. *Life Sci* 60: PL223–28.

Kane, B.E., Svensson, B. and Ferguson, D.M. 2006. Molecular recognition of opioid receptor ligands. *AAPSJ* 8:e126–37.

Khor, B.S., Jamil, M.F., Adenan, M.I. and Shu-Chien, A.C. 2011. Mitragynine attenuates withdrawal syndrome in morphine-withdrawn zebrafish. *PLoS One* 6: e28340.

Kitajima, M., Misawa, K., Kogure, N., et al. 2006. A new indole alkaloid, 7-hydroxyspeciociliatine, from the fruits of Malaysian *Mitragyna speciosa* and its opid agonistic activity. *Natural Medicine* 60: 28–35.

Kosterlitz, H.W. and Robinson, J.A. 1957. Inhibition of the peristaltic reflex of the isolated guinea-pig ileum. *J Physiol* 136: 249–62.

Kumarnsit, E., Keawpradub, N. and Nuankaew, W. 2007. Effect of *Mitragyna speciosa* aqueous extract on ethanol withdrawal symptoms in mice. *Fitoterapia* 78: 182–85.

León, F., Habib, E., Adkins, J.E., Furr, E.B., McCurdy, C.R. and Cutler, S.J. 2009. Phytochemical characterization of the leaves of *Mitragyna speciosa* grown in U.S.A. *Nat Prod Commun* 4: 907–10.

Macko, E., Weisbach, J.A. and Douglas, B. 1972. Some observations on the pharmacology of mitragynine. *Arch Int Pharmacodyn Ther* 198: 145–61.

Matsumoto, K., Hatori, Y., Murayama, T., et al. 2006. Involvement of mu-opioid receptors in antinociception and inhibition of gastrointestinal transit induced by 7-hydroxymitragynine, isolated from Thai herbal medicine *Mitragyna speciosa*. *Eur J Pharmacol* 549: 63–70.

Matsumoto, K., Horie, S., Ishikawa, H., Takayama, H., Aimi, N., Ponglux, D. and Watanabe, K. 2004. Antinociceptive effect of 7-hydroxymitragynine in mice: Discovery of an orally active opioid analgesic from the Thai medicinal herb *Mitragyna speciosa*. *Life Sci* 74: 2143–55.

Matsumoto, K., Horie, S., Takayama, H., et al. 2005. Antinociception, tolerance and withdrawal symptoms induced by 7-hydroxymitragynine, an alkaloid from the Thai medicinal herb *Mitragyna speciosa*. *Life Sci* 78: 2–7.

Matsumoto, K., Mizowaki, M., Suchitra, T., et al. 1996a. Central antinociceptive effects of mitragynine in mice: Contribution of descending noradrenergic and serotonergic systems. *Eur J Pharmacol* 317: 75–81.

Matsumoto, K., Mizowaki, M., Suchitra, T., Takayama, H., Sakai, S., Aimi, N. and Watanabe, H. 1996b. Antinociceptive action of mitragynine in mice: Evidence for the involvement of supraspinal opioid receptors. *Life Sci* 59: 1149–55.

Matsumoto, K., Mizowaki, M., Takayama, H., Sakai, S., Aimi, N. and Watanabe, H. 1997. Suppressive effect of mitragynine on the 5-methoxy-N,N-dimethyltryptamine-induced head-twitch response in mice. *Pharmacol Biochem Behav* 57: 319–23.

Matsumoto, K., Takayama, H., Narita, M., et al. 2008. MGM-9 [(E)-methyl 2-(3-ethyl-7a,12a-(epoxyethanoxy)-9-fluoro-1,2,3,4,6,7,12,12b-octahydro-8-methoxy indolo[2,3-a]quinolizin-2-yl)-3-methoxyacrylate], a derivative of the indole alkaloid mitragynine: A novel dual-acting mu- and kappa-opioid agonist with potent antinociceptive and weak rewarding effects in mice. *Neuropharmacology* 55: 154–65.

Philipp, A.A., Wissenbach, D.K., Weber, A.A., Zapp, J. and Maurer, H.H. 2011. Metabolism studies of the Kratom alkaloid speciociliatine, a diastereomer of the main alkaloid mitragynine, in rat and human urine using liquid chromatography-linear ion trap mass spectrometry. *Anal Bioanal Chem* 399: 2747–53.

Raffa, R. 2006. Pharmacological aspects of successful long-term analgesia. *Clin Rheumatol* 25 Suppl 1: S9–15.

Raffa, R.B., Beckett, J.R., Brahmbhatt, V.N., et al. 2013. Orally active opioid compounds from a non-poppy source. *J Med Chem* 56: 4840–48.

Raffa, R.B., Buschmann, H., Christoph, T., et al. 2012. Mechanistic and functional differentiation of tapentadol and tramadol. *Expert Opin Pharmacother* 13: 1437–49.

Shamima, A.R., Fakurazi, S., Hidayat, M.T., Hairuszah, I., Moklas, M.A. and Arulselvan, P. 2012. Antinociceptive action of isolated mitragynine from *Mitragyna speciosa* through activation of opioid receptor system. *Int J Mol Sci* 13: 11427–42.

Shellard, E.J. 1989. Ethnopharmacology of kratom and the *Mitragyna* alkaloids. *J Ethnopharmacol* 25: 123–24.

Shellard, E.J., Houghton, P.J. and Resha, M. 1978. The *Mitragyna* species of Asia: Part XXXII. The distribution of alkaloids in young plants of *Mitragyna speciosa* Korth grown from seed obtained from Thailand. *Planta Medica* 34: 253–63.

Suzuki, T., Tsuji, M., Mori, T., Misawa, M. and Nagase, H. 1995. Effect of naltrindole on the development of physical dependence on morphine in mice: A behavioral and biochemical study. *Life Sci* 57: PL247–52.

Takayama, H. 2004. Chemistry and pharmacology of analgesic indole alkaloids from the rubiaceous plant, *Mitragyna speciosa*. *Chem Pharm Bull (Tokyo)* 52: 916–28.

Takayama, H., Ishikawa, H., Kurihara, M., et al. 2002. Studies on the synthesis and opioid agonistic activities of mitragynine-related indole alkaloids: discovery of opioid agonists structurally different from other opioid ligands. *J Med Chem* 45: 1949–56.

Thongpradichote, S., Matsumoto, K., Tohda, M., Takayama, H., Aimi, N., Sakai, S. and Watanabe, H. 1998. Identification of opioid receptor subtypes in antinociceptive actions of supraspinally administered mitragynine in mice. *Life Sci* 62: 1371–78.

Tohda, M., Thongpraditchote, S., Matsumoto, K., et al. 1997. Effects of mitragynine on cAMP formation mediated by delta-opiate receptors in NG108-15 cells. *Biol Pharm Bull* 20: 338–40.

Tsuji, M., Takeda, H., Matsumiya, T., Nagase, H., Yamazaki, M., Narita, M. and Suzuki, T. 2000. A novel kappa-opioid receptor agonist, TRK-820, blocks the development of physical dependence on morphine in mice. *Life Sci* 66: PL353–58.

3 Psychopharmacological Indole Alkaloids from Terrestrial Plants

Allan Patrick G. Macabeo

CONTENTS

3.1 INTRODUCTION

A number of plants are traditionally used to produce psychoactivity and other mind-altering effects. Many plant-derived compounds can affect perception of reality and may invoke feelings of tranquility, invigoration, or "otherworldliness." Many use these substances to escape reality. Physiologically, these compounds are absorbed into the blood stream and transported to various sites where they exert their psychoactive effects. Generally, they influence the performance of the central and peripheral nervous systems, especially the brain, by affecting signal transduction in the neurons. Most psychoactive drugs alter or mimic the behavior of four kinds of natural neurotransmitters: acetylcholine, norepinephrine, serotonin, and neuropeptides. In most cases, such CNS-related biochemical pathways are impacted by psychoactive secondary metabolites such as alkaloids and a number of low-molecular-weight terpenoids. Often, these compounds exert mind-stimulating or -altering properties at low concentrations. Studies categorize psychoactive ingredients as sedatives, stimulants, or hallucinogens. The drug culture of the 1960s referred to these compounds as downers, uppers, or psychedelics.

An extensive survey of phytochemical efforts related to the discovery of psychopharmacological compounds from plants with purported mind-altering and -stimulating effects reveals the alkaloids to be one of the most active phytoingredients. Alkaloids are nitrogenous, complex plant constituents that serve as protective agents or toxic deterrents against herbivores. While many of these compounds are highly toxic to humans, in small amounts they exert interesting pharmacological activities, including psychoactivity. Morphine (1) (an isoquinoline alkaloid from *Papaver somniferum*), nicotine (2) (a tropane alkaloid from *Nicotiana tabacum*), atropine (3) (a tropane alkaloid from *Atropa belladonna*), and caffeine (4) (a xanthine alkaloid from coffee and tea leaves) are some of the known effective psychoactive plant alkaloids (Schultes 1976) (Figure 3.1).

The indole alkaloids comprise a myriad of compounds identified in many plant species today known to display molecular complexities and a number of beneficial properties for treating a wide range of human-related diseases. Their occurrence is well identified in plant families such as Apocynaceae, Rubiaceae, Loganiaceae, Rutaceae, and Nyssaceae (O'Connor and Maresh 2006). The biosynthetic cascades leading to several types of terpene-based indole alkaloids are well documented (Verpoorte et al. 1998). In old and recent literatures, indole alkaloids are frequently reported common alkaloid types that possess psychopharmacological activity—hallucinogenic activity, for example. While most derivatives are biochemically produced in plants, some low-molecular-weight psychoactive indole alkaloids are also found in other sources. Examples are psilocybin from sacred mushrooms; *N,N*-dimethyltryptamines from vinho de jurema and snuffs known as yopo, huilca and epena; bufotenine (5-hydroxy-*N,N*-dimethyltryptamine) from the skin of certain toads; and serotonin (5-hydroxytryptamine) present in minute amounts in the central nervous tissue of warm-blooded animals. Psychoactive indole alkaloids can be grouped into four types: β-carbolines, ergot alkaloids, iboga alkaloids, and tryptamines (Schultes 1976).

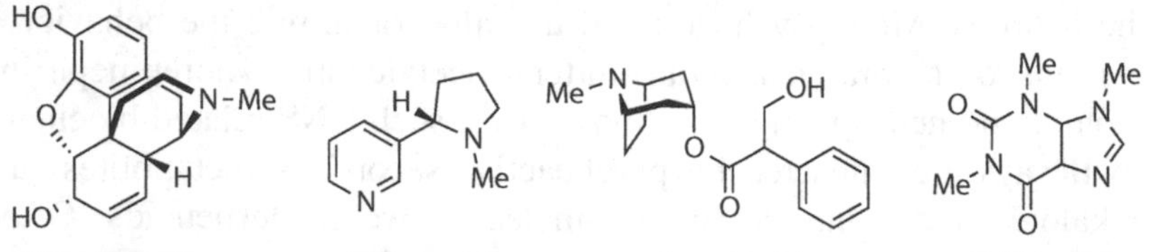

Morphine (1)   Nicotine (2)   Atropine (3)   Caffeine (4)

FIGURE 3.1 Common alkaloidal compounds with known psychoactivity.

3.2 PRIVILEGED PSYCHOACTIVE INDOLE ALKALOIDS

Intense exploration of plants with known ethnopsychoactivity has led to the isolation and identification of indole alkaloids **5–8** (Figure 3.2). Pharmacological studies indicate their vast potential in treating neurological and psychological disorders. The following discloses important findings observed for the psychopharmacological profile of indole alkaloids **5–8**.

3.2.1 IBOGAINE

The indole alkaloid ibogaine (**5**) is one of the naturally occurring basic constituents with a wide spectrum of psychoactivity in the apocynaceous plants *Tabernaemontana iboga* and *Voacanga africana* of Central Africa. In earlier studies, **5** has been identified as the active principle for the ethnohallucinogenic properties of *T. iboga* (Alper 2001; Quevauviller and Blanpin 1957). The intake of **5** can reduce the craving for heroin and cocaine, as demonstrated in a number of studies (Maciulaitis et al. 2008). The mode of action of ibogaine has been related to the regulation of the neuronal excitability and synaptic transmission in the parabrachial nucleus, as shown in a nystatin-perforated patch-recorded assay. Modulation is induced by depolarization of parabrachial neurons with increasing excitability and firing rate; by down-regulation of the non-NMDA receptor-mediated fast synaptic transmission; and last, through dopamine receptor activation for its actions (Malandrino et al. 2008). In a colorimetric assay for acetylcholinesterase (AChE) inhibition, **5** did not demonstrate inhibitory activity, which might suggest that it is not involved in pathways linked to muscarinic acetylcholine transmission in humans or animals (Alper et al. 2012).

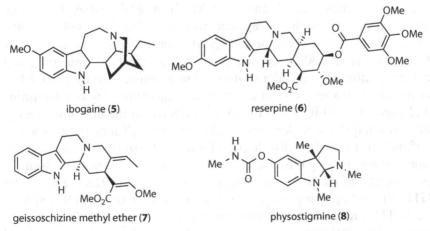

ibogaine (**5**)

reserpine (**6**)

geissoschizine methyl ether (**7**)

physostigmine (**8**)

FIGURE 3.2 Structures of well-studied psychoactive alkaloids.

3.2.2 RESERPINE

Reserpine (6) is an antipsychotic and antihypertensive monoterpene indole alkaloid from the African Apocynaceae plant *Rauwolfia serpentina* (Baumeister et al. 2003). It has been used to control high blood pressure and for the treatment of psychosis. The water infusion and teas extracted from the roots of *R. serpentina* are rich in 6 and have a calming, sedative action analogous to the action of typical antidepressants (Davis and Shepherd 1955). The antidepressant effect is attributed to the depletion of monoamine neurotransmitters such as catecholamines in the synapses, according to the monoamine hypothesis. Reserpine is the first alkaloid to demonstrate effective antidepressant action in a randomized placebo-controlled trial. The extended effect of 6 is due to a long period of replenishing the vesicular monoamine transmitter (VMAT). The irreversible block of VMAT results in the transport of free intracellular norepinephrine, serotonin, and dopamine in the presynaptic nerve terminal into presynaptic vesicles to ensure delivery into the synaptic cleft (Henry and Scherman 1989).

3.2.3 GEISSOSCHIZINE METHYL ETHER

Yokukansan, a traditional Japanese herb preparation used to treat aggressiveness in patients with dementia, contains seven medicinal herbs, including *Uncaria* hook, which is the hook or the hook-bearing stem of *Uncaria rhynchophylla* Miquel, *Uncaria sinensis* Haviland, and *Uncaria macrophylla* Wallich (Rubiaceae) (Iwasaki et al. 2005; Shinno et al. 2007; Mizukami et al. 2009). This cocktail has been ethnomedicinally used to treat headache, dizziness, seizure, and epilepsy. Among the psychoactive constituents identified from this plant is the indole alkaloid geissoschizine methyl ether (7, GM). In several studies, it plays an important function in the therapeutic efficacy of *Uncaria* hook to treat a variety of physiological and neurological disorders (Kanatani et al. 1985; Pengsuparp et al. 2001). The aripiprazole-like effect of 7 was observed in a study after it exhibited a relatively low intrinsic activity and evoked partial activation response in a subset of cells expressing the dopamine D(2L) receptor. While 7 was found less effective at the dopamine receptor than aripiprazole at dopamine D(2L) receptors, 7 may serve as a new scaffold for atypical antipsychotics (Ueda et al. 2011). GM also showed agonistic property against the serotonin 1A receptor when assessed for its competitive binding and [^{35}S] guanosine 5'-O-(3-thiotriphosphate) (GTPγS) binding using artificial Chinese hamster ovary cells expressing 5-HT1A receptors in a follow-up study. In socially isolated mice, 7 improved the isolation-induced increased aggressiveness and decreased sociality, the effect of which was offset by co-treatment with WAY-100635. Thus, GM validates the pharmacological effect of yokukansan on aggressiveness and sociality in socially isolated mice (Nishi et al. 2012).

Remyelination, a regenerative strategy to prevent the abnormal accumulation of myelin and oligodendrocyte in medial prefrontal cortex, has been an attractive area for discovering drugs with which to prevent the psychotic signs of schizophrenia. Using a cuprizone (CPZ)-induced demyelination model and evaluation of cellular changes in response to GM administration during the remyelination phase in adult mice revealed that newly formed oligodendrocytes were increased by GM treatment after CPZ exposure. GM also offset a decrease in myelin basic protein immunoreactivity caused by CPZ administration. Thus, GM therapy may improve myelin deficiency by mature oligodendrocyte formation and remyelination in the mPFC of CPZ-treated mice and may be a possible drug candidate for eradicating schizophrenic symptoms (Morita et al. 2014). Interestingly, GM also inhibited reversibly and competitively 50% of acetylcholinesterase activity, thus being a potential anti-Alzheimer agent (Yang et al. 2012).

The agonistic/antagonistic properties of **7** on the 5-HT7 receptor were also assessed and validated by measuring intracellular cAMP levels in HEK239 cells. GM prevented 5-HT-induced cAMP production in a concentration-dependent pattern, as well as the specific 5-HT7 receptor antagonist SB-269970. However, **7** did not stimulate intracellular cAMP assembly as does 5-HT. These results suggest that GM has an antagonistic effect on the 5-HT7 receptor (Ueki et al. 2013).

3.2.4 PHYSOSTIGMINE

Physostigmine (**8**) is a parasympathomimetic indole alkaloid isolated from *Physostigma venenosum* known for its inhibitory potency against AChE. It is a short-acting reversible AChE inhibitor and has demonstrated significant cognitive effect in both normal and Alzheimer's (AD) patients (Da-Yuan et al. 1996; Mukherjee 2001). Mechanistically, it inhibits the hydrolysis of acetylcholine by acetylcholinesterase at the transmitted sites of acetylcholine. This inhibitory property boosts the effect of acetylcholine, making it more potent for treating cholinergic-associated diseases and myasthenia gravis. As a model for discovering more potent AChE inhibitors, the structure of **8** has paved the way for the development of rivastigmine, a licensed derivative used in the United Kingdom for the symptomatic cure of mild to moderately severe AD. Thus it is apparent that plant-derived alkaloid AChE inhibitors may be important for the development of more appropriate drug candidates in treating AD (Foye et al. 1995).

Physostigmine has also been observed to counter the undesired side effects of diazepam, a benzodiazepine, by relieving the feeling of anxiety and tension. Additionally, it is also thought to reverse the effects of barbiturates, which may suggest its use as a sedative or hypnotic.

3.3 CHOLINESTERASE AND MONOAMINE OXIDASE (MAO) INHIBITORS

Alzheimer's disease (AD) is a neurodegenerative disorder characterized by a progressive loss of memory, decline in language skills, and other cognitive impairments (Goedert and Spillantini 2006). The loss of cholinergic neurons in AD results in a deficiency of acetylcholine in specific brain regions that mediate memory and learning abilities, according to the cholinergic theory (Cummings 2004). Observations that the ratio of butyrylcholinesterase (BChE) to acetylcholinesterase (AChE) gradually elevates in the AD brain, partially as a consequence of the progressive loss of the cholinergic synapses where AChE activity is localized, may favor the use of BChE inhibitors in treating moderate to severe forms of AD (Talea 2001). An attractive approach to discover potential anti-AD drugs is the multi-target-directed ligand approach based on the "one-molecule, multiple-target" paradigm. Thus another AD-related enzyme target is monoamine oxidase (MAO), which catalyzes the oxidative deamination of a variety of biogenic and xenobiotic amines with the concurrent production of hydrogen peroxide (Giacobini 2003).

Assessment of anti-AChE, anti-BChE, and anti-monoamine oxidases A and B (MAO-A and MAO-B) activity of thirteen *Psychotria* alkaloids showed the two quaternary β-carboline alkaloids prunifoleine (9) and 14-oxoprunifoleine (10) to be potent inhibitory agents (Figure 3.3). Both exhibited noncompetitive AChE and time-dependent MAO-A inhibitory activities. In addition, the monoterpene indole alkaloids angustine (11), vallesiachotamine lactone (12), *E*-vallesiachotamine (13), and Z-vallesiachotamine (14) inhibited BChE and MAO-A. Among the four alkaloids, alkaloid 11 was able to inhibit MAO-A in a reversible and competitive way, while the three vallesiachotamine congeners 12–14 displayed a time-dependent inhibition on this target (Passos et al. 2013).

The indole alkaloids coronaridine (15), voacangine (16), voacangine hydroxyindolenine (17), rupicoline (18), ibogamine, ibogaine (5), ibogaline, desethyl-voacangine, voachalotine, and affinisine were isolated from the chloroform extract of the stalk of *Tabernaemontana australis* and showed anti-cholinesterase activity with alkaloids 15–18 as the most inhibitory compounds exhibiting potency comparable to physostigmine

prunifoleine (9) 14-oxoprunifoleine (10) angustine (11)

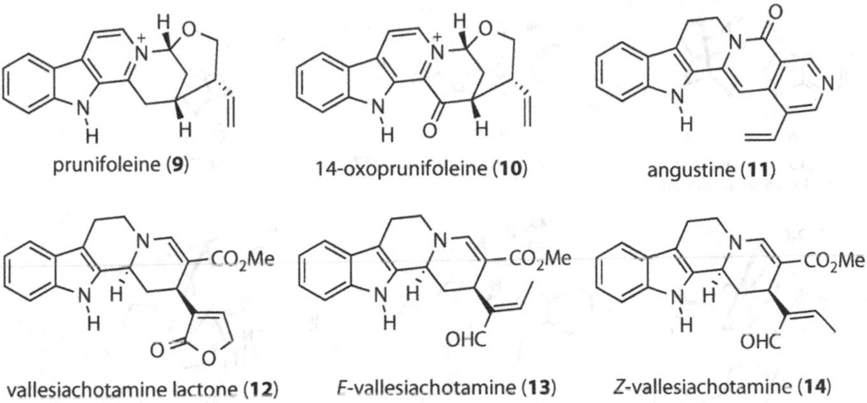

vallesiachotamine lactone (12) *E*-vallesiachotamine (13) *Z*-vallesiachotamine (14)

FIGURE 3.3 Psychoactive indole alkaloids from *Psychotria* species.

and galantamine in a TLC assay using the modified Ellman's method (Figure 3.4) (Andrade et al. 2006). The roots of *Tabernaemontana divaricata* afforded two bisindole alkaloids, 19,20-dihydrotabernamine (**19**) and 19,20-dihydroervahanine A (**20**) and have also shown activity against AChE. The inhibitory property of **20** was observed to be specific, reversible, and competitive. The AChE inhibitory results suggest that functionalization at carbons 11', 12', and 16' are important for activity against acetylcholinesterase (Ingkanina et al. 2013). In a separate study, the bisindole alkaloid 3'-R/S-hydroxyvoacamine (**21**) was also isolated from the stem extract of *T. divaricata*, a traditional Thai medicinal plant used for rejuvenation remedies and improving memory, through a bioassay-guided approach. Alkaloid **21** appeared to be a non-competitive inhibitor against AChE (Chaiyana et al. 2010). Bioassay-guided isolation of anticholinesterase agents from *Ervatamia hainanensis* led to the isolation of **15** and **16** with comparable inhibitory activity to galantamine in vitro (Zhan et al. 2010). Another Apocynaceae plant from Brazil, *Himatanthus lancifolius*, also yielded the alkaloid uleine (**22**) as an AChE inhibitory component.

In the search for AChE inhibitors from *Catharanthus roseus*, the monoterpenoid indole alkaloids catharanthine (**23**), serpentine (**24**), and ajmalicine (**25**) were isolated from the aqueous extract. Alkaloid **24** competitively blocked muscarinic receptors, whereas **25** was undifferentiated from the control, and **23** showed an insurmountable muscarinic antagonism at concentrations greater than 10 μM concentrations. Nicotinic-receptor–mediated diaphragm contractions were fully inhibited by **23** and **25** in a reversible but non-competitive manner, unlike the stronger nicotinic antagonist tubocurarine, whose competitive blockade was shut down by a physostigmine-accelerated increase in acetylcholine. At a concentration up to 100 μM, **24** did not change diaphragm contractions, indicating

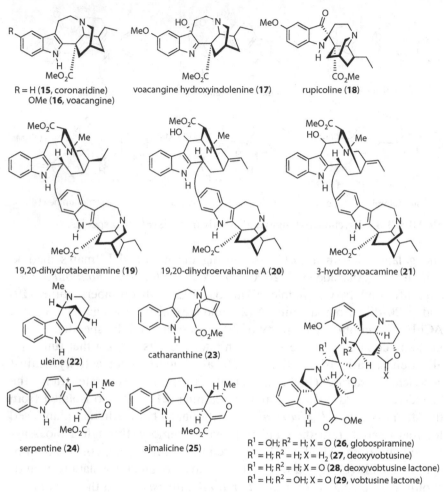

FIGURE 3.4 Anti-cholinesterase indole alkaloids from representative Apocynaceae plants.

reduced binding with neuromuscular nicotinic receptors. While alkaloid **24** exhibited strong AChE inhibition, it failed to reestablish diaphragm contractions upon submaximal tubocurarine blockade, indicating poor tissue penetration, which may prevent **24** from inhibiting AChE in deep neuromuscular synapses in the ex vivo preparation. With serpentine's potent in vitro AChE inhibitory activity and low cholinergic receptor affinity, it is assumed that minor structural variation may lead to a potent and selective AChE inhibitor, potentially useful for treating AD and/or myasthenia gravis (Pereira et al. 2010).

Globospiramine (**26**), a new spirobisindole alkaloid bearing an Aspidospermae–Aspidosperma skeleton, together with deoxyvobtusine (**27**), dihydroevodiamine (**30**)   turbinatine (**31**)   desoxycordifoline (**32**)

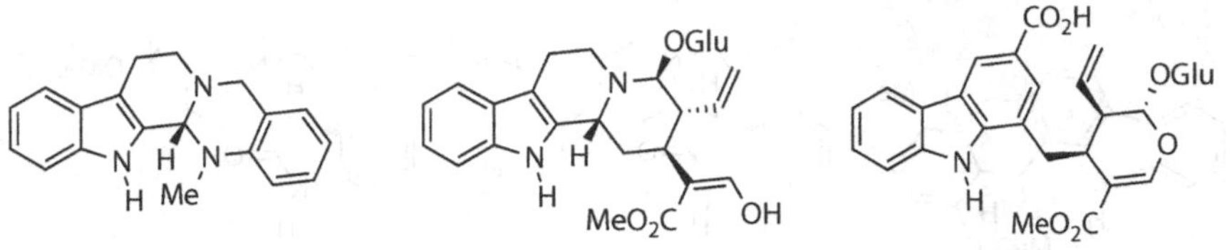

FIGURE 3.5   Indole alkaloids with anti-AChE inhibitory activity from other plant families.

deoxyvobtusine lactone (**28**), and vobtusine lactone (**29**) from the Philippine endemic Apocynaceae plant *Voacanga globose,* showed significant inhibitory activity on both AChE and BChE, with more pronounced activity against the latter enzyme using the colorimetric Ellman's assay (Decker 2005). All compounds turned out to be strong micromolar inhibitors. Bisindole alkaloid **27** displayed higher activity and revealed higher selectivity toward BChE over AChE. These results demonstrate for the first time the promise of spirocyclic Aspidospermae-Aspidosperma bisindole alkaloids as a new class of butyrylcholinesterase inhibitors and possibly as a new psychoactive compound type of the genus *Voacanga* (Macabeo et al. 2011).

From *Desmodium pulchellum* and *D. gangeticum* (Papilionaceae), four indole alkaloids were isolated, with leguein A being the most active against AChE (Ghosal et al. 1972). Dehydroevodiamine (**30**), from *Evodia rutaecarpa,* likewise inhibited AChE in vitro and reversed scopolamine-induced memory impairment in rats (Park et al. 1996). Turbinatine (**31**) and desoxycordifoline (**32**) from *Chimarrhis turbinata* (Rubiaceae) also demonstrated moderate inhibitory activity (Cardoso et al. 2004) (Figure 3.5).

Preliminary works describing the MAO inhibitory activity of alkaloid fractions of the neotropical *Psychotria suterella* and *Psychotria laciniata,* as well as two monoterpene indole alkaloids, were evaluated against monoamine oxidases (MAO-A and MAO-B) obtained from rat brain mitochondria. The alkaloids, *E-* (**13**) and *Z*-vallesiachotamine (**14**), also inhibited MAO-A and MAO-B, but in higher concentrations when compared with the fractions. These results suggest that species belonging to this genus are a potentially interesting source of new MAO inhibitors (Dos Santos Passos et al. 2013).

3.4   SEROTONIN AGONISTS AND ANTAGONISTS

The indole alkaloid alstonine (**33**) acts as an unusual antipsychotic agent in behavioral models, but differs in its dopamine and serotonin affinity profile (Figure 3.6). Alkaloid **33** produces anxiolytic activity in both

alstonine (**33**)   isorhynchophylline (**34**)   isocorynoxeine (**35**)

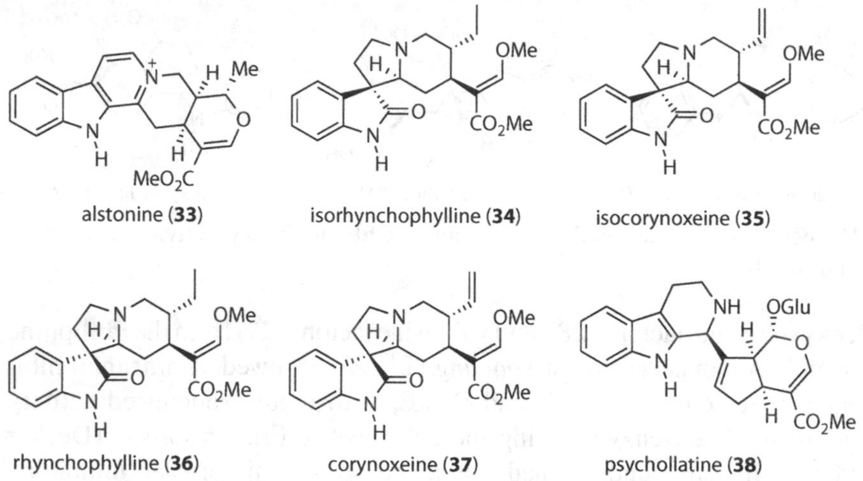

rhynchophylline (**36**)   corynoxeine (**37**)   psychollatine (**38**)

FIGURE 3.6   Indole alkaloids with activity against serotonin receptors.

hole-board and light/dark states. Pretreatment with the 5-HT2A/2C sero-tonin receptor antagonist ritanserin antagonized the effects of **33** in both the hole-board and light/dark models, suggesting the participation of these receptors in the mechanism of action of **33**. With alstonine partially revers-ing the increase in locomotion induced by MK-801 in the hole-board, as well as MK-801–induced hyperlocomotion in motor activity application, the association of glutamate NMDA receptors in this observed property should also be noted (Costa-Campos et al. 2004).

Isorhynchophylline (**34**), the major oxindole alkaloid from *Uncaria* species, dose-dependently inhibited 5-HT2A receptor-mediated head-twitch but not 5-HT1A receptor-mediated head-weaving responses evoked by 5-methoxy-*N,N*-dimethyltryptamine. Pretreatment with reserpine, a monoamine-depleting agent, enhances the head-twitching but does not influence the effect of isorhynchophylline on the behavioral response. Isocorynoxeine (**35**) also reduces the head-twitch response in reserpinized mice over the same dose range as alkaloid **34**, while both rhynchophylline (**36**) and corynoxeine (**37**) do not. None of the alkaloids tested had an effect on meta-chlorophenylpiperazine–induced hypolocomotion, a 5-HT2C-receptor–mediated behavioral response. In experiments in vitro, **34** and **35** dose-dependently and competitively inhibited 5-HT–evoked currents in *Xenopus* oocytes expressing 5-HT2A receptors, but had less of a suppres-sive effect on those in oocytes expressing 5-HT2C receptors. These results indicate that **34** and **35** preferentially suppress 5-HT2A receptor function in the brain probably via a competitive antagonism at 5-HT2A receptor sites and that the configuration of the oxindole moiety of **34** is essential for their antagonistic activity at the 5-HT2A receptor (Matsumoto et al. 2005).

Psychollatine (**38**), a glycoside indole monoterpene alkaloid from *Psychotria umbellata*, demonstrated anxiolytic-like property at concentrations that do not increase sleeping time nor modify spontaneous locomotor activity in the light/dark and hole-board models of anxiety. In the forced swimming model of depression, psychollatine effects were comparable to the known antidepressants imipramine and fluoxetine. Alkaloid **38** also controlled oxotremorine-induced tremors in all concentrations. In the step-down learning paradigm, psychollatine impaired the acquisition of learning and memory union, without obstruction with retrieval. Thus the effects of **38** on the central nervous system may involve serotonergic 5-HT2A/C receptors (Both et al. 2005). In a separate study, psychollatine prevented NMDA-induced seizures and mitigated MK-801–induced hyperlocomotion. It also prevented amphetamine-induced lethality and offset apomorphine-induced climbing behavior. Thus the psychopharmacological activity of psychollatine is attributed to the involvement of NMDA glutamate receptors (Both et al. 2006).

3.5 OTHER RELATED PSYCHOPHARMACOLOGICAL EFFECTS

Alstovenine (**39**) and venenatine (**40**), alkaloids from *Alstonia venenata*, had significant psychotropic activities in mice and rats (Figure 3.7). Alkaloid **39** inhibited monoamine oxidase, prevented reserpine syndrome, and potentiated the effects of dopa and serotonin. At high concentrations, **39** stimulated the central nervous system and induced stereotype behavior and convulsions. On the other hand, alkaloid **40** had a reserpine-like activity and induced sedation and ptosis. It stimulated hexobarbitone hypnosis and antagonized morphine-induced analgesia (Bhattacharya et al. 1975).

To prove the neuroactive efficacy of *Uncaria sinensis*, the indole alkaloids corynoxeine, rhynchophylline, isorhynchophylline, isocorynoxeine, hirsuteine (**41**), and hirsutine (**42**) were evaluated by 3-(4,5-dimethylthiazol-2-yl)-2,5-diphenyl-tetrazolium bromide (MTT) staining on their neuroprotective effects on glutamate-induced cell death and their inhibitory effects on $^{45}Ca^{2+}$ influx in cultured rat cerebellar granule cells.

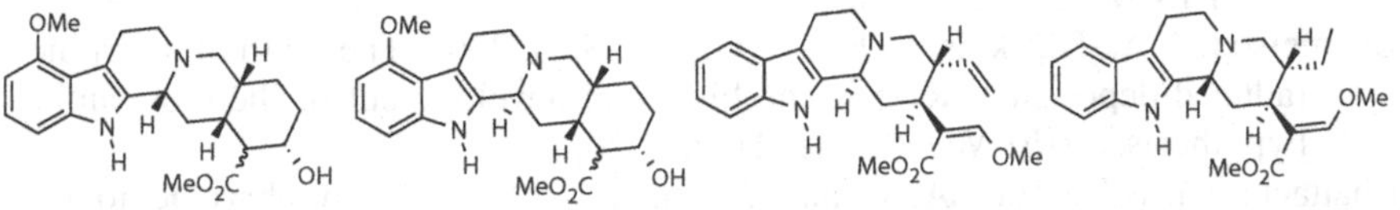

alstovenine (**39**)   venenatine (**40**)   hirsuteine (**41**)   hirsutine (**42**)

FIGURE 3.7 Indole alkaloids with other psychopharmacologic activities.

Rhynchophylline, isorhynchophylline, isocorynoxeine, hirsuteine, and hirsutine each significantly potentiated cell viability compared with exposure to glutamate only, with the effect of isorhynchophylline being the highest. The increased Ca^{2+} influx into cells induced by glutamate was significantly prevented by treatment with rhynchophylline, isorhynchophylline, isocorynoxeine, hirsuteine, or hirsutine. These results suggest that isorhynchophylline, isocorynoxeine, rhynchophylline, hirsuteine, and hirsutine are the active components of the hooks and stems of *Uncaria sinensis* that protect against glutamate-induced neuronal death in cultured cerebellar granule cells by inhibition of Ca^{2+} influx (Shimada et al. 1999).

3.6 CONCLUSION

This chapter presents results and the current status of pharmacological studies related to the exploration of indole alkaloids as psychoactive agents derived from well-known alkaloid-bearing plant families and genera. It describes some of the most interesting findings on the chemical and pharmacological researches on the various monoterpenoid indole alkaloids elaborated by member species with much recorded use in folk herbal medicine with known psychoactivity. Most of the identified indole alkaloids to date provide credence to the folkloric, ethnomedical uses of the source plants. Pharmacological studies on the alkaloids indicate the immense potential of the member plant species to be used for treating disorders linked to psychoactivity. Further research to discover new clinical applications of psychoactive alkaloids, clinical trials, and product development are warranted to fully exploit their psychopharmacological and therapeutic potential.

REFERENCES

Alper, K.R. 2001. Ibogaine: A review. *Alkaloids Chem Biol* 56: 1–38.
Alper, K., Reith, M.E., and Sershen, H. 2012. Ibogaine and the inhibition of acetylcholinesterase. *J Ethnopharmacol* 139(3): 879–82.
Andrade, M.T., Lima, J.A., Pinto, A.C., Rezende, C.M., Carvalho, M.P., and Epifanio, R.A. 2005. Indole alkaloids from *Tabernaemontana australis* (Muell. Arg) Miers that inhibit acetylcholinesterase enzyme. *Bioorg Med Chem* 13: 4092–95.
Baumeister, A.A., Hawkins, M.F., and Uzelac, S.M. 2003. The myth of reserpine-induced depression: Role in the historical development of the monoamine hypothesis. *J Hist Neurosci* 12(2): 207–20.
Bhattacharya, S.K., Ray, A.B., and Dutta, S.C. 1975. Psychopharmacological investigations of the 4-methoxyindole alkaloids of Alstonia venenata. *Planta Medica* 27(2): 164–70.

Both, F.L., Meneghini, L., Kerber, V.A., Henriques, A.T., and Elisabetsky, E. 2005. Psychopharmacological profile of the alkaloid psychollatine as a 5HT2A/C serotonin modulator. *J Nat Prod* 68(3): 374–80.

Both, F.L., Meneghini, L., Kerber, V.A., Henriques, A.T., and Elisabetsky, E. 2006. Role of glutamate and dopamine receptors in the psychopharmacological profile of the indole alkaloid psychollatine. *J Nat Prod* 69(3): 342–45.

Cardoso, I., Castro-Gamboa, D.H.S., Silva, M. et al. 2004. Indole glucoalkaloids from *Chimarrhis turbinata* and their evaluation as antioxidant agents and acetylcholinesterase inhibitors. *J Nat Prod* 67: 1882–85.

Chaiyana, W., Schripsema, J., Ingkaninan, K., and Okonogi, S. 2013. 3'-R/S-hydroxyvoacamine, a potent acetylcholinesterase inhibitor from *Tabernaemontana divaricata*. *Phytomedicine* 20(6): 543–48.

Costa-Campos, L., Dassoler, S.C., Rigo, A.P., Iwu, M., and Elisabetsky, E. 2004. Anxiolytic properties of the antipsychotic alkaloid alstonine. *Pharmacology, Biochemistry and Behavior* 77(3): 481–89.

Cummings, L. 2004. Treatment of Alzheimer's disease: Current and future therapeutic approaches. *J Rev Neurol Dis* 1: 60–69.

Davies, D.L., and Shepherd, M. 1955. Reserpine in the treatment of anxious and depressed patients. *The Lancet* 269: 117–20.

Da-Yuan, Z., Dong-Lu, B., and Xi-Can, T. 1996. Recent studies on traditional Chinese medicinal plants. *Drug Dev Res* 39: 147–57.

Decker, M. 2005. Novel inhibitors of acetyl- and butyrylcholinesterase derived from the alkaloids dehydroevodiamine and rutaecarpine. *Eur J Med Chem* 40: 305–13.

Dos Santos Passos, C., Soldi, T.C., Torres, A.R. et al. 2013. Monoamine oxidase inhibition by monoterpene indole alkaloids and fractions obtained from *Psychotria suterella* and *Psychotria laciniata*. *J Enzyme Inhib Med Chem* 28(3): 611–18.

Foye, W.O., Lemke, T.L., and Williams, D.A. 1995. *Principles of Medicinal Chemistry,* 4th edition. Baltimore: Williams and Wilkins.

Ghosal, S., Bhattacharya, S.K., and Mehta, R. 1972. Naturally occurring and synthetic carbolines as cholinesterase inhibitors. *J Pharm Sci* 61: 808–11.

Giacobini, E. 2003. Cholinergic function and Alzheimer's disease. *Int J Geriatr Psychiatr* 18: S1–S5.

Goedert, M., and Spillantini, M.G. 2006. A century of Alzheimer's disease. *Science* 314: 777–81.

Henry, J., and Scherman, D. 1989. Radioligands of the vesicular monoamine transporter and their use as markers of monoamine storage vesicles. *Biochem Pharmacol* 38(15): 2395–404.

Ingkaninan, K., Changwijit, K., and Suwanborirux, K. 2006. Vobasinyl-iboga bisindole alkaloids, potent acetylcholinesterase inhibitors from *Tabernaemontana divaricata* root. *J Pharm Pharmacol* 58(6): 847–52.

Iwasaki, K., Maruyama, M., Tomita, N. et al. 2005. Effects of the traditional Chinese herbal medicine Yi-Gan San for cholinesterase inhibitor-resistant visual hallucinations and neuropsychiatric symptoms in patients with dementia with Lewy bodies. *J Clin Psychiatry* 66: 1612–13.

Kanatani, H., Kohda, H., Yamasaki, K. et al. 1985. The active principles of the branchlet and hook of *Uncaria sinensis* Oliv. examined with a 5-hydroxy-tryptamine receptor binding assay. *J Pharm Pharmacol* 37: 401–404.

Macabeo, A.P.G., Vidar, W.S., Wan, B. et al. 2011. Mycobacterium tuberculosis $H_{37}Rv$ and cholinesterase inhibitors from *Voacanga globosa*. *Eur J Med Chem* 46: 3118–23.

Maciulaitis, R., Kontrimaviciute, V., Bressolle, F.M.M., and Briedis, V. 2008. Ibogaine, an anti-addictive drug: Pharmacology and time to go further in development: A narrative review. *Hum Exp Toxicol* 27: 181–94.

Malandrino, S., Cristoni, A., Ja Sakolov, S., Gabetta, B., and Pifferi, G. 1993. Antihypoxic activity of cuanzine and derivatives. *Pharmacol Res* 27(Sup.1): 121–22.

Matsumoto, K., Morishige, R., Murakami, Y., Tohda, M., Takayama, H., Sakakibara, I., and Watanabe, H. 2005. Suppressive effects of isorhynchophylline on 5-HT2A receptor function in the brain: Behavioural and electrophysiological studies. *Eur J Pharmacol* 17(3): 191–99.

Mizukami, K., Asada, T., Kinoshita, T. et al. 2009. A randomized cross-over study of a traditional Japanese medicine (kampo), yokukansan, in the treatment of the behavioural and psychological symptoms of dementia. *Int J Neuropsychopharmacol* 12: 191–99.

Morita, S., Tatsumi, K., Makinodan, M., Okuda, H., Kishimoto, T., and Wanaka, A. 2014. Geissoschizine methyl ether, an alkaloid from the Uncaria hook, improves remyelination after cuprizone-induced demyelination in medial prefrontal cortex of adult mice. *Neurochem Res* 39(1): 59–67.

Mukherjee, P.K. 2001. Evaluation of Indian traditional medicine. *Drug Inf J* 35: 620–23.

Mukherjee, P.K., Kumar, V., Mal, M., and Houghton, P.J. 2007. Acetylcholinesterase inhibitors from plants. *Phytomedicine* 14(4): 289–300.

Nishi, A., Yamaguchi, T., Sekiguchi, K. et al. 2012. Geissoschizine methyl ether, an alkaloid in *Uncaria* hook, is a potent serotonin $_1A$ receptor agonist and candidate for amelioration of aggressiveness and sociality by yokukansan. *Neuroscience* 207: 124–36.

O'Connor, S.E., and Maresh, J.J. 2006. Chemistry and biology of monoterpene indole alkaloid biosynthesis. *Nat Prod Rep* 23: 532–47.

Park, C.H., Kim, S., Choi, W., Lee, Y., Kim, J., and Kang, S.S. 1996. Novel anticholinesterase and antiamnesic activities of dehydroevodiamine, a constituent of *Evodia rutaecarpa*. *Planta Med* 62: 405–409.

Passos, C.S., Simões-Pires, C.A., Nurisso, A. et al. 2013. Indole alkaloids of *Psychotria* as multifunctional cholinesterases and monoamine oxidases inhibitors. *Phytochemistry*. 86: 8–20.

Pengsuparp, T., Indra, B., Nakagawasai, O. et al. 2001. Pharmacological studies of geissoschizine methyl ether, isolated from *Uncaria sinensis* Oliv., in the central nervous system. *Eur J Pharmacol* 425: 211–18.

Pereira, D.M., Ferreres, F., Oliveira, J.M. et al. 2010. Pharmacological effects of *Catharanthus roseus* root alkaloids in acetylcholinesterase inhibition and cholinergic neurotransmission. *Phytomedicine* 17(8–9): 646–52.

Quevauviller, A., and Blanpin, O. 1957. Pharmacodynamic study of voacamine, an alkaloid of *Voacanga africana*. *Therapie* 12: 636–47.

Schultes, R. 1976. Indole alkaloids in plant hallucinogens. *J Psychedelic Drugs* 8 (1): 7–25.

Seidl, C., Correia, B.L., Stinghen, A.E., and Santos, C.A. 2010. Acetylcholinesterase inhibitory activity of uleine from *Himatanthus lancifolius*. *Zeitschrift für Naturforschung C* 65(7–8): 440–44.

Shimada, Y., Goto, H., Itoh, T., Sakakibara, I., Kubo, M., Sasaki, H., and Terasawa, K. 1999. Evaluation of the protective effects of alkaloids isolated from the hooks and stems of *Uncaria sinensis* on glutamate-induced neuronal death in cultured cerebellar granule cells from rats. *J Pharm Pharmacol* 51(6): 715–22.

Shinno, H., Utani, E., Okazaki, S. et al. 2007. Successful treatment with Yi-Gan San for psychosis and sleep disturbance in a patient with dementia with Lewy bodies. *Prog Neuropsychopharmacol Biol Psychiatry* 31: 1543–45.

Talea, V. N. 2001. Acetylcholinesterase in Alzheimer's disease. *Mech Aging Dev* 122: 1961–69.

Ueda, T., Ugawa, S., Ishida, Y., and Shimada, S. 2011. Geissoschizine methyl ether has third-generation antipsychotic-like actions at the dopamine and serotonin receptors. *Eur J Pharmacol* 671(1–3): 79–86.

Ueki, T., Nishi, A., Imamura, S. et al. 2013. Effects of geissoschizine methyl ether, an indole alkaloid in *Uncaria* hook, a constituent of yokukansan, on human recombinant serotonin 7 receptor. *Cell Mol Neurobiol* 33(1): 129–35.

Verpoorte, R., Heijden, R.V.D., and Memelink, J. 1998. Plant biotechnology and the production of alkaloids: Prospects of metabolic engineering. *The Alkaloids* 50: 453-508.

Yang, Z.D., Duan, D.Z., Du, J., Yang, M.J., Li, S., and Yao, X.J. 2012. Geissoschizine methyl ether, a corynanthean-type indole alkaloid from *Uncaria rhynchophylla* as a potential acetylcholinesterase inhibitor. *Nat Prod Res* 26(1): 22–28.

Youdim, M.B.H., Finberg, J.P.M., and Tipton, K.F. 1988. Monoamine Oxidase. In: *Handbook of Experimental Pharmacology*, vol. 90. Trendelenburg, U., and Weiner, N. (eds.). 119–92. Berlin: Springer Verlag.

Zhan, Z.J., Yu, Q., Wang, Z.L., and Shan, W.G. 2010. Indole alkaloids from *Ervatamia hainanensis* with potent acetylcholinesterase inhibition activities. *Bioorg Med Chem Lett* 20(21): 6185–87.

4 Salvinorin A
Example of a Non-Alkaloidal Bioactive Opioid from a Plant Source

Jordan K. Zjawiony

CONTENTS

4.1 INTRODUCTION

Salvinorin A is a major active secondary metabolite isolated from the hallucinogenic plant *Salvia divinorum*. This neo-clerodane diterpenoid is the first known non-alkaloidal natural opioid with high affinity to and selectivity for the kappa-opioid receptor (KOR). As such, it serves as another interesting example, as does mitragynine, of a bioactive opioid from a non-poppy source.

Salvinorin A acts as an agonist to KOR and is the first example of a non-serotonergic hallucinogen. It also produces antinociceptive, antidepressant, anti-inflammatory, and antidiarrheal activities in vivo. The natural source of salvinorin A, *Salvia divinorum* is an endemic plant that grows wild in the Sierra Mazateca of the Oaxaca region of southern Mexico. The plant was used for centuries by shamans in religious ceremonies and in medical practices under several common names, such as *ska Maria Pastora*, *hierba Maria*, and *hojas de la Pastora* (Valdes et al. 1983). It is now recognized as the most potent hallucinogen of natural origin. *Salvia divinorum* is widely available over the Internet, where it is sold in various forms, from unprocessed leaves to fortified leaf preparations, tinctures, and even chewing gums. Despite its prohibition in many countries, it is increasingly used as a recreational drug. Semi-synthetic modifications of salvinorin A provide a number of derivatives that may serve as affinity labels for KOR to better understand the structure and dynamics of the binding site or as new drug leads for therapy of central nervous and gastrointestinal system disorders.

As of June 2014, research on *S. divinorum* and salvinorin A was the subject of nearly 400 publications, including 20 review articles (Benoni 2001; Sheffler and Roth 2003; Yan and Roth 2004; Prisinzano 2005, 2013; Prisinzano and Rothman 2008; Imanshahidi and Hosseinzadeh 2006; Vortherms and Roth 2006, 2011; Fichna et al. 2009; Lozama and Prisinzano 2009; Hanson 2010; Saric et al. 2010; Cunningham et al. 2011; Listos et al. 2011; Lovell et al. 2011, 2012; Piekielna et al. 2012; Diaz 2013; Zawilska and Wojcieszak 2013; Casselman et al. 2014). This chapter is a condensed summary of knowledge of this unique plant and its extraordinary metabolite.

4.2 BOTANICAL FEATURES OF *S. DIVINORUM*

Salvia divinorum Epling & Játiva-M. belongs to the family Lamiaceae (alt. Labiatae). Among the members of this family one can find common herbs such as mint, rosemary, basil, oregano, marjoram, thyme, sage, and lavender. The family consists of 238 genera and approximately 6500 species (Mabberley 2008). *Salvia* is the largest genus of the family, with approximately 900 species distributed all over the world. The most commonly known species of this genus is *Salvia officinalis* (common or culinary sage).

S. divinorum is a semitropical perennial and grows wild at altitudes of 700–1700 meters in the foggy highlands of the Sierra Mazateca northeast of Oaxaca, Mexico. In its natural habitat, it prefers ravines and other shadowy places. The plant grows to about 1 meter in height, has large, oval green leaves, hollow square stems, and white flowers with purple calyces. Unlike other salvias, *S. divinorum* produces few seeds that rarely germinate. It is not known why pollen fertility is reduced. In the wild, the plant reproduces asexually with stems reaching the ground, rooting, and developing new shoots. It grows easily indoors and can be propagated vegetatively from cuttings or layering. From personal experience it grows very well in hydroponic equipment in a greenhouse, but it can also grow outside in a semi-tropical climate. The plant is susceptible to insects, fungi, and viruses. The most common insects attacking *S. divinorum* are whitefly, aphids, slugs, caterpillars, thrips, spider mites, and scale insects.

Most plants of the Lamiaceae family have glands (glandular trichomes) on the leaf surface containing terpenes that are responsible for the aroma of culinary herbs from this family. In *S. divinorum* these glandular trichomes, specifically peltate glandular trichomes located mostly on the underside (abaxial) surface of leaves, contain the active metabolite diterpenoid salvinorin A. Anatomical and morphological studies with use of confocal and scanning electron microscopy further confirmed the presence and structure of peltate glandular trichomes and their role in biosynthesis, sequestration, and excretion of salvinorin A (Siebert 2004; Kowalczuk et al. 2014). The question why salvinorin A is produced and stored only in the abaxial surface of the leaves remains unanswered.

The content of salvinorin A in leaves varies from plant to plant depending on growing conditions and other environmental factors. Quantitative analysis of 20 leaf samples from separate plants gave a range 0.089–0.37% (Siebert 2004), with amounts of salvinorin A in stems being only about 4% of the level found in leaves.

4.3 SALVINORIN A

4.3.1 BIOSYNTHETIC PATHWAY

Incorporation experiments with [1-^{13}C]-glucose and [1-^{13}C; 3,4-^{2}H$_2$]-1-deoxy-D-xylulose performed in our laboratories (Kutrzeba et al. 2007) proved that salvinorin A is biosynthesized exclusively via the 1-deoxy-D-xylulose-5-phosphate (DOXP) pathway instead of the classic mevalonic acid pathway. Our work also provided evidence that, similarly to cannabinoids and other plant terpenoids, the site of salvinorin A biosynthesis is compartmentalized in glandular trichomes.

4.3.2 ISOLATION FROM *S. DIVINORUM*

There are several published methods of isolation of salvinorin A from *S. divinorum*, including exhaustive extraction with ethyl ether, chloroform, methanol, or acetone followed by flash column chromatography on activated carbon or silica gel (Ortega et al. 1982; Valdes et al. 1984; Munro et al. 2003; Tidgewell et al. 2004; Shirota et al. 2006) or liquid–liquid centrifugal partition chromatography (Shirota 2007). In our laboratory we utilize a cost-effective method of isolation by exhaustive extraction of dried and ground leaves with acetone followed by repeated crystalliza-tion from ethanol (Kutrzeba et al. 2009a). Taking into account the fact that glandular trichomes with salvinorin A are located on the surface of the leaves, we arrived at a rapid method of isolation of salvinorin A by fresh-leaf-surface extraction with chloroform followed by crystallization from ethanol (Kutrzeba et al. 2009a). This time- and cost-effective method provides virtually chlorophyll-free extract and allows for isolation of sal-vinorin A with an average 0.2% yield of dry weight, which is comparable with conventional methods.

4.3.3 Chemistry

Salvinorin A, a neoclerodane diterpenoid, is the major active secondary metabolite of *S. divinorum*. Altogether 37 metabolites were isolated from this plant (Ortega et al. 1982; Valdes et al. 1984, 2001; Valdes 1986; Giroud et al. 2000; Bigham et al. 2003; Munro et al. 2003; Harding et al. 2005b; Lee et al. 2005; Shirota et al. 2006; Kutrzeba et al. 2009b), including 10 salvinorins (A–J), 6 divinatorins (A–F), 4 salvidins (A–D), 2 salvinicins (A and B), hardwickiic acid, and 14 other structurally non-related com-pounds (Figure 4.1). Of all these, only salvinorin A has shown hallucino-genic activity.

The structure and absolute configuration of salvinorin A was deter-mined by NMR and X-ray crystallography (Ortega et al. 1982) and later by circular dichroism (Koreeda et al. 1990). In 2007 the first total synthesis of salvinorin A was reported with 29 steps (Scheerer et al. 2007), later improved to 20, and finally to 16 steps in 0.75% overall yield from R-(–)-Wieland–Miescher ketone 5-methyl analog (Nozawa et al. 2008; Hagiwara et al. 2009). Despite this progress the total synthesis of salvinorin A is not competitive with its isolation from the plant.

The relatively high yield (average 0.2% of dry weight), low cost (no chromatography), and simplicity of one-step isolation of salvinorin A from the plant have facilitated semisynthesis of a large number of salvinorin derivatives and analogs, generating extensive structure–affinity relation-ship data. As of the end of 2013, the semisynthesis of 233 salvinorin A analogs involving all functional groups in the molecule has been reported (Lovell et al. 2010, 2012; Ma et al. 2010; Cunningham et al. 2011; Fichna et al. 2011; Lozama et al. 2011; Prevatt-Smith et al. 2011; Polepally et al. 2013). Out of this number 27 compounds had affinity (K_i) lower than 10 nM (Figure 4.2). All of the 27 compounds are highly selective to KOR without affecting other opioid receptors. It is worth noting that most of them (22 out of 27) are modified analogs of salvinorin A at C-2, making

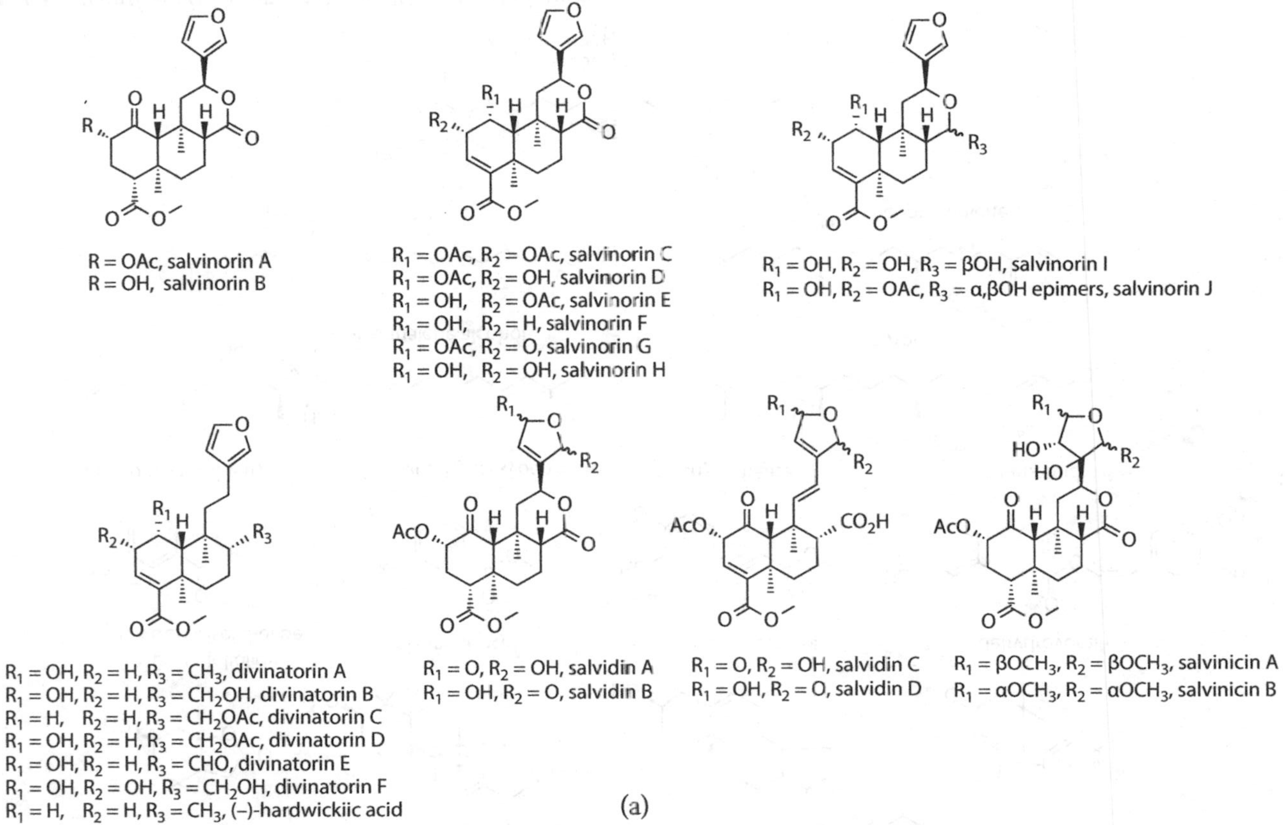

FIGURE 4.1 Compounds isolated from *Salvia divinorum*.

R₁ = OAc, salvinorin A
R = OH, salvinorin B

R₁ = OAc, R₂ = OAc, salvinorin C
R₁ = OAc, R₂ = OH, salvinorin D
R₁ = OH, R₂ = OAc, salvinorin E
R₁ = OH, R₂ = H, salvinorin F
R₁ = OAc, R₂ = O, salvinorin G
R₁ = OH, R₂ = OH, salvinorin H

R₁ = OH, R₂ = OH, R₃ = CH₃, divinatorin A
R₁ = OH, R₂ = H, R₃ = CH₂OH, divinatorin B
R₁ = H, R₂ = H, R₃ = CH₂OAc, divinatorin C
R₁ = OH, R₂ = H, R₃ = CH₂OAc, divinatorin D
R₁ = OH, R₂ = H, R₃ = CHO, divinatorin E
R₁ = OH, R₂ = OH, R₃ = CH₂OH, divinatorin F
R₁ = H, R₂ = H, R₃ = CH₃, (–)-hardwickiic acid

R₁ = OH, R₂ = OH, R₃ = βOH, salvinorin I
R₁ = OH, R₂ = OAc, R₃ = α,βOH epimers, salvinorin J

R₁ = O, R₂ = OH, salvidin A
R₁ = OH, R₂ = O, salvidin B

R₁ = O, R₂ = OH, salvidin C
R₁ = OH, R₂ = O, salvidin D

R₁ = βOCH₃, R₂ = βOCH₃, salvinicin A
R₁ = αOCH₃, R₂ = αOCH₃, salvinicin B

(a)

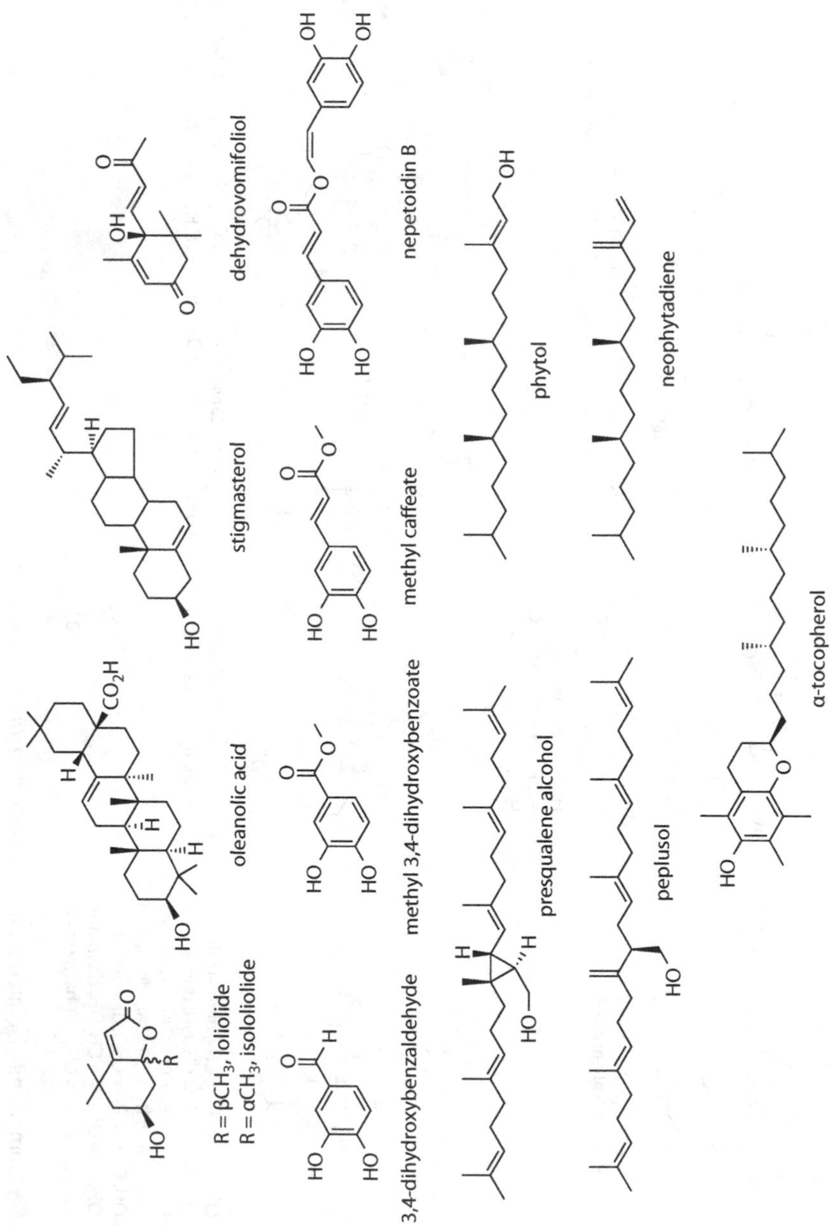

FIGURE 4.1 (*continued*) Compounds isolated from *Salvia divinorum*.

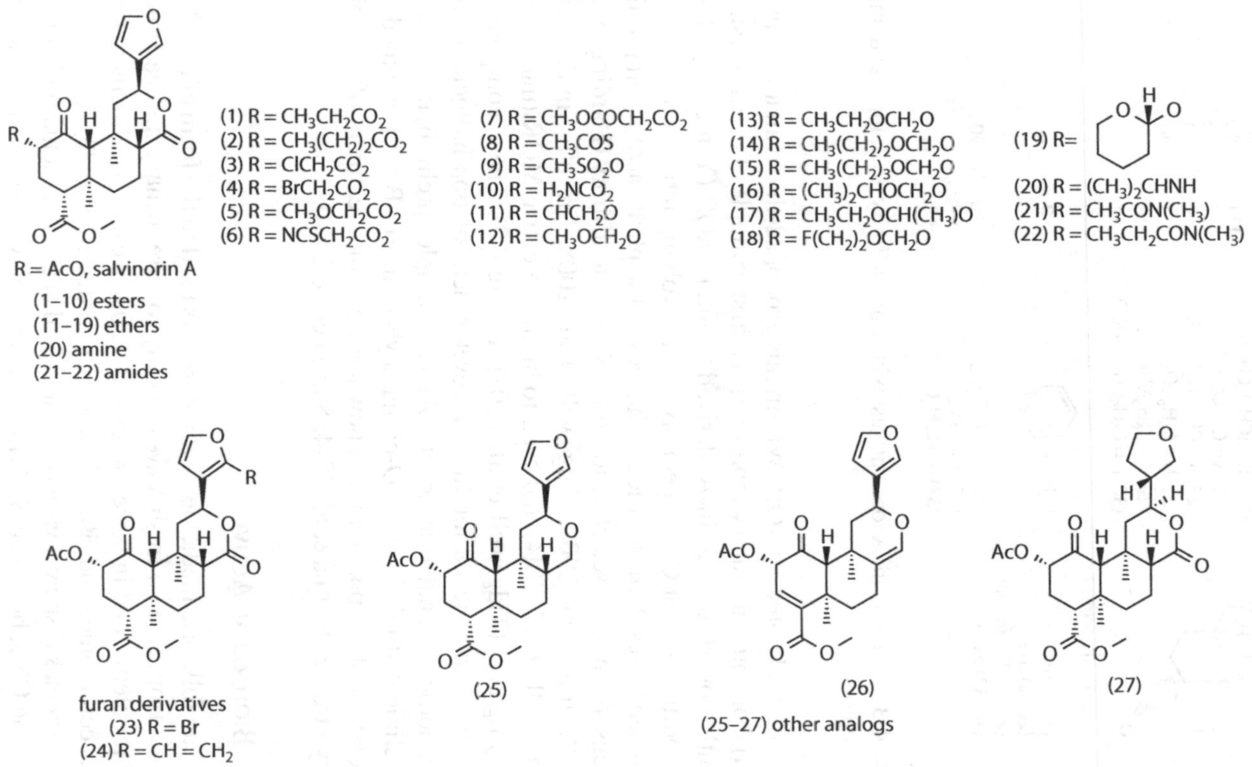

FIGURE 4.2 Salvinorin A derivatives with high affinity and selectivity for kappa opioid receptor (K_i < 10nM).

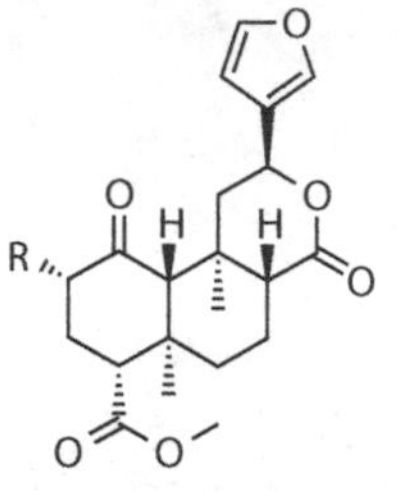

Affinity constants K_i [nM]

	MOR	KOR
(13) R = $CH_3CH_2OCH_2O$	41	3.13
(20) R = $(CH_3)_2CHNH$	111	4.5
(21) R = $CH_3CON(CH_3)$	135	0.37
(22) R = $CH_3CH_2CON(CH_3)$	15	0.11
(28) R = $C_6H_5CO_2$	12	90
(29) R = $C_6H_5NHCO_2$	16	93
(30) R = $C_6H_5CH=CHCO_2$	52	9.6
(31) R = 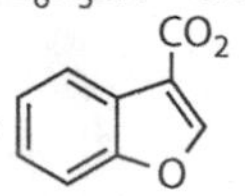	10	70
(32) R = 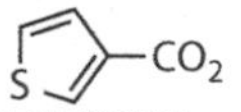	10	90
(33) R = C_6H_5CONH	3.1	7,430

R = AcO, salvinorin A

(13) ether
(20) amine
(21–22,33) amides
(28–32) esters

FIGURE 4.3 Salvinorin A derivatives with dual affinity to kappa and miu opioid receptors.

them the best tolerated by KOR. Within this group, 22-thiocyanatosalvino-rin A (**6**) from our laboratory represents the first example of an irreversible (covalently bound) salvinorin-A–derived ligand to KOR (Yan et al. 2009). Some modifications at C-2, especially those involving introduction of aromatic moieties, result in significant changes in the pharmacological profile of analogs, from KOR selective to MOR/KOR dual affinity (Harding et al. 2005a; Beguin et al. 2008; Tidgewell et al. 2008; Prevath-Smith et al. 2011; Polepally et al. 2014) and even to highly selective MOR (mu-opioid receptor) ligands (Tidgewell et al. 2008) (Figure 4.3). These compounds may serve as good leads for further development of non-hallucinogenic and non-addictive analgesic agents. As an example, preliminary in vivo and preclinical studies with 2-*O*-cinnamoylsalvinorin B (**30**) obtained in my laboratory show potential as a new agent for the treatment of irritable bowel syndrome (Fichna et al. 2014; Salaga et al. 2014).

4.3.4 BIOLOGICAL ACTIVITY

In Mexican folkloric medicine, *S. divinorum* is used in the form of an infusion made from 8 to 10 fresh leaves, mostly to cure urinary and gastrointestinal problems, and in larger doses (40–120 leaves) as a hallucinogen in religious ceremonies and ritual divination (Valdes et al. 1983).

   Salvia species display a variety of biological activities, including interaction with the CNS, but so far *S. divinorum* is the only one that causes strong hallucinogenic effects. The effective dose of salvinorin A is 200 µg when vaporized and inhaled, and this potency is comparable to LSD. The hallucinating effects are short in duration, with the peak after 30 seconds, lasting for about 5–10 minutes, and gradually subsiding over 20–30 minutes (Siebert 1994). The nature of the effects (visions, motions, distorted and overlapping realities, out of body experiences, revisiting past, etc.) depends on various factors, including dose, individual intellect of subject, and environment (Siebert 1994; Gonzales et al. 2006; Baggott et al. 2010; Cunningham et al. 2011; Johnson et al. 2011; Kelly 2011; Sumnall et al. 2011; Vohra et al. 2011; Addy 2012; Ranganathan et al. 2012; MacLean et al. 2013). Salvinorin A displays antinociceptive (Ansonoff et al. 2006; John et al. 2006; McCurdy et al. 2006), antipruritic (Wang et al. 2005), anti-inflammatory (Aviello et al. 2011), and antidiarrheal activities (Capasso et al. 2006, 2008a,b; Fichna et al. 2009). Controversial reports of depressive-like (Carlezon et al. 2006; Ebner et al. 2010) or antidepressive effects (Braida et al. 2009; Harden et al. 2012) in rats were recently explained based on dose-dependence. However, there are not enough data to claim antidepressive activity of salvinorin A in humans owing to only one case reported (Hanes et al. 2001). Similar phenomenon of dose-dependence in pharmacological effects is also observed in humans (MacLean et al. 2013). At lower doses of inhaled salvinorin A (0.375–4.5 µg/kg) strong hallucinogenic effects were reported, whereas at higher doses (15–19.5 µg/kg) dissociative effects were pronounced—no persisting adverse effects were observed, however. These observations are in accord with earlier studies in animal models, where rewarding effects were observed after administration of low doses of salvinorin A to zebrafish and rats (Braida et al. 2007, 2008). The rewarding effects of salvinorin A were also reported by other authors, implicating the involvement or "cross-talk" of the endocannabinoid system in the mechanism of this action (Capasso et al. 2008; Walentiny et al. 2010). It has been also shown that salvinorin A may decrease responses to cocaine and in this way attenuate cocaine-induced drug-seeking behavior (Chartoff et al. 2008; Morani et al. 2009, 2012).

In the past several years, reviews on the pharmacology of *S. divinorum*, salvinorin A, and its analogs have been published (Valdes et al. 1983; Siebert 1994; Prisinzano 2005; Prisinzano and Rothman 2008; Grundmann et al. 2007; Cunningham et al. 2011; Listos et al. 2011; Zawilska et al. 2013) and a recent article from the Johns Hopkins University group significantly adds to the current knowledge about the effects of salvinorin A on humans (MacLean et al. 2013).

4.3.5 MECHANISM OF ACTION

Salvinorin A is a highly selective and potent agonist of KOR (Roth et al. 2002). It is highly unique by having no affinity for any other receptors interacting with psychoactive compounds, including the most common for hallucinogens: serotonin receptor 5-HT$_{2A}$. Considerable effort was taken to identify the molecular mechanism by which salvinorin A and its analogs bind to KOR (Yan et al. 2005, 2008, 2009; Kane et al. 2006, 2008; Groer et al. 2007; Rothman et al. 2007; Vortherms and Roth 2007; White et al. 2014). It is interesting that, contrary to other KOR agonists that produce dysphoria and other adverse reactions, low doses of salvinorin A show some rewarding effects. This phenomenon implies the involvement of some other mechanism of action. There are reports indicating that in addition to stimulating KOR, salvinorin A triggers secondary effects in dopaminergic (Zhang et al. 2005; Beerepoot et al. 2008; Gehrke et al. 2008; Phipps and Butterweck 2010) and endocannabinoid systems (Capasso et al. 2008; Walentiny et al. 2010). Additional support of the hypothesis for an indirect action of salvinorin A comes from studies using positron emission tomography (PET) scans of the distribution of [^{11}C]salvinorin A in the brains of baboons (Hooker et al. 2008). It was found that administration of the opioid antagonist naloxone neither reduced the concentration of [^{11}C]salvinorin A nor changed its distribution in the brain. Moreover, the metabolism of labeled salvinorin A occurred not only in the parts of the brain with high KOR density, but also in brain regions with little or no KOR presence (Hooker et al. 2009). There is still a need for further studies to fully understand the mechanism of action of salvinorin A.

4.4   LEGAL STATUS

As of the end of 2013 S. divinorum and salvinorin A are regulated in 24 countries throughout the world and 35 states in the United States and territories (Siebert 2013). S. divinorum and in some cases also salvinorin A are illegal to possess or sell in 16 countries (Australia, Belgium, Croatia, Czech Republic, Denmark, Germany, Hong Kong, Italy, Japan, Latvia, Lithuania, Poland, Romania, South Korea, Sweden, and Switzerland). In Chile, France, Russia, and Spain S. divinorum and salvinorin A are legal to possess, but illegal to sell. Russia also prohibits growing the plant. In Estonia, Finland, Iceland and Norway, S. divinorum is considered to be a medicinal herb that requires a doctor's prescription.

In the United States, S. divinorum and salvinorin A are not regulated by the federal government. But laws in 31 states and territories (Alabama, Arkansas, Colorado, Connecticut, Delaware, Florida, Georgia, Guam, Hawaii, Illinois, Indiana, Kansas, Kentucky, Louisiana, Michigan, Minnesota, Mississippi, Missouri, Nebraska, North Carolina, North Dakota, Ohio, Oklahoma, Pennsylvania, South Dakota, Tennessee, Texas, Vermont, Virginia, West Virginia, and Wyoming) prohibit possessing S. divinorum. In California, Maine, and Maryland possession of S. divinorum is legal for adults, except for providing it or selling it to minors. In Wisconsin, manufacturing, delivering, or selling salvinorin A (and presumably *S. divinorum*) is illegal, but possession is not.

4.5 CONCLUSIONS AND PERSPECTIVES

Owing to its selective affinity for KOR, strong effects on human mood, and low toxicity, salvinorin A has become an attractive experimental pharmacological tool and a lead for CNS-drug discovery and development. The kappa-opioid receptor has been implicated in a wide variety of conditions, including dementia, mood disorders, schizophrenia, drug abuse, alcohol addiction, chronic pain, seizure disorders, congestive heart failure, augmentation of renal function, and diuresis. The KOR system is a pharmaceutical curiosity in that it has ability to modulate pain without causing euphoria—and hence without opioid-like addition. Thus long-acting agonists or antagonists could be very useful for potential treatment of a variety of disease states and for pharmacological studies of KOR structure and function. In particular, salvinorin-A–derived agonists, partial agonists, and KOR/MOR dual-agonists that do not cross the blood–brain barrier could be ideal in treating peripheral chronic pain conditions such as irritable bowel syndrome, osteo- and rheumatoid arthritis, degenerative joint disease, and sciatica. On the other hand, salvinorin-A–derived antagonists able to cross the blood–brain barrier could be useful in treating mood disorders, schizophrenia, dementia, seizures, HIV-related neuropsychiatric disorders, and brain and spinal cord tumors.

ACKNOWLEDGEMENTS

The author would like to thank all students and visiting scientists contributing to his research on *Salvia divinorum* at the University of Mississippi and all colleagues from other universities collaborating with him on this subject. Special thanks to Dr. Daneel Ferreira and Dr. William Day for their constructive comments and edits of this chapter.

REFERENCES

Addy, P.H. 2012. Acute and post-acute behavioral and psychological effects of salvinorin A in humans. *Psychopharmacology* 220: 195–204.
Baggott, M.J., Erowid, E., Erowid, F., Galloway, G.P., and Mendelson, J. 2010. Use patterns and self-reported effects of *Salvia divinorum*: An Internet-based survey. *Drug Alcohol Depend* 111: 250–56.

Beerepoot, P., Lam, V., Luu, A., Tsoi, B., Siebert, D., and Szechtman, H. 2008. Effects of salvinorin A on locomotor sensitization to D2/D3 dopamine agonist quinpirole. *Neuroscience Lett* 446: 101–104.

Beguin, C., Potter, D.N., DiNieri, J.A. et al. 2008. N-methylacetamide analog of salvinorin A: A highly potent and selective κ-opioid receptor agonist with oral efficacy. *J Pharmacol Exp Ther* 324: 188–95.

Benoni, H. 2001. Salvinorin A - a hallucinogen from Aztec sage. *Naturwissen Rundsch* 54: 575–578.

Bigham, A.K., Munro, T.A., Rizzacasa, M.A., and Robins-Browne, R.M. 2003. Divinatorins A-C, new neoclerodane diterpenoids from the controlled sage *Salvia divinorum*. *J Nat Prod* 66: 1242–44.

Braida, D., Limonta, V., Pegorini, S., Zani, A., Guerini-Ricco, C., Gori, E., and Sala, M. 2007. Hallucinatory and rewarding effect of salvinorin A in zebrafish: κ-opioid and CB_1-cannabinoid receptor involvement. *Psychopharmacology* 190: 441–48.

Braida, D., Limonta, V., Capurro, V. et al. 2008. Involvement of κ-opioid and endocannabinoid system on salvinorin A-induced reward. *Biol Psychiatry* 63: 286–92.

Braida, D., Capurro, V., Zani, A., Rubino, T., Vigano, D., Parolaro, D., and Sala, M. 2009. Potential anxiolytic and antidepressant-like effects of salvinorin A, the main active ingredient of *Salvia divinorum*, in rodents. *Brit J Pharmacol* 157: 844–53.

Capasso R., Borelli, F., Cascio, M.G. et al. 2008. Inhibitory effect of salvinorin A from *Salvia divinorum* on ileitis-induced hypermotility: cross-talk between κ-opioid and cannabinoid CB_1 receptors. *Brit J Pharmacol* 155: 681–89.

Carlezon, W. A., Beguin, C., DiNieri, J. A. et al. 2006. Depressive-like effects of the κ-opioid receptor agonist salvinorin A on behavior and neurochemistry in rats. *J Pharmacol Exp Ther* 316: 440–47.

Casselman, I., Nock, C. J.,Wohlmuth, H., Weatherby, R. P., Heinrich, M. 2014. From local to global: fifty years of research on *Salvia divinorum*. *J Ethnopharmacol* 151: 768–783.

Chartoff, E.H., Potter, D., Damez-Werno, D., Cohen, B.M., and Carlezon Jr., W.A. 2008. Exposure to the selective κ-opioid receptor agonist salvinorin A modulates the behavioral and molecular effects of cocaine in rats. *Neuropsychopharmacology* 33: 2676–87.

Cunningham, C.W., Rothman, R.B., and Prisinzano, T.E. 2011. Neuropharmacology of the naturally occurring κ-opioid hallucinogen salvinorin A. *Pharmacol Rev* 63: 316–47.

Diaz, J.-L. 2013. *Salvia divinorum*: a psychopharmacological riddle and a mind body prospect. *Curr Drug Abuse Rev* 6: 43–53.

Ebner, S.R., Roitman, R.F., Potter, D.N., Rachlin, A.B., and Chartoff, E.H. 2010. Depressive-like effects of the kappa opioid receptor agonist salvinorin A are associated with decreased phasic dopamine release in the nucleus accumbens. *Psychopharmacology* 210: 241–52.

Fichna J., Schicho, R., Janecka, A., Zjawiony, J.K., and Storr, M. 2009. Selective natural κ opioid and cannabinoid receptor agonists with a potential role in the treatment of gastrointestinal dysfunction. *Drug News Perspect*, 22(7): 383–92.

Fichna, J., Lewellyn, K., Yan, F., Roth, B.L., and Zjawiony, J.K. 2011. Synthesis and biological evaluation of new salvinorin A analogues incorporating natural amino acids. *Biorg Med Chem Lett* 21: 160–63.

Fichna, J., Zjawiony, J.K., Polepally, P.R., De Rego J.-C., and Roth, B.L. 2014. New salvinorin B derivatives, their new medical application of the pharmaceutically acceptable form of the drug. *Polish patent application* WIPO ST 10/C PL 406776, January 3, 2014.

Gehrke, B.J., Chefer, V.I., and Shippenberg, T.S. 2008. Effects of acute and repeated administration of salvinorin A on dopamine function in the rat dorsal striatum. *Psychopharmacology* 197: 509–17.

Giroud, C., Felber, F., Augsburger, M., Horisberger, B., Fivier, L., and Mangin, P. 2000. *Salvia divinorum*: An hallucinogenic mint which might become a new recreational drug in Switzerland. *Forensic Sci Int* 112: 143–50.

Gonzales, D., Riba, J., Bouso, J.C., Gomez-Jarabo, G., and Barbanoj, M.J. 2006. Pattern of use and subjective effects of *Salvia divinorum* among recreational users. *Drug Alcohol Depend* 85: 157–62.

Groer, C.E., Tidgewell, K., Moyer, R.A., Harding, W.W., Rothman, R.B., Prisinzano, T.E., and Bohn, L.M. 2007. An opioid agonist that does not induce μ-opioid receptor-arrestin interactions or receptor internalization. *Mol Pharmacol* 71: 549–57.

Grundmann, O., Phipps, S.M., Zadezensky, I., and Butterweck, V. 2007. *Salvia divinorum* and salvinorin A: An update on pharmacology and analytical methodology. *Planta Med* 73: 1039–46.

Hagiwara, H., Suka, Y., Nojima, T., Hoshi, T. and Suzuki, T. 2009. Second-generation synthesis of salvinorin A. *Tetrahedron* 65: 4820–25.

Hanson, J.R. 2010. Natural products from the hallucinogenic sage. *Sci Prog* 93: 171–180.

Harden, M.T., Smith, S.E., Niehoff, J.A., McCurdy, C.R., and Taylor, G.T. 2012. Antidepressive effects of the κ-opioid receptor agonist salvinorin A in a rat model of anhedonia. *Behav Pharmacol* 3: 710–15.

Harding, W.W., Tidgewell, K., Byrd, N. et al. 2005a. Neoclerodane diterpenes as a novel scaffold for μ-opioid receptor ligands. *J Med Chem* 48: 4765–71.

Harding, W.W., Tidgewell, K., Schmidt, M. et al. 2005b. Salvinicins A and B, new neoclerodane diterpenes from *Salvia divinorum*. *Org Lett* 7(14): 3017–20.

Hooker, J.M., Xu, Y., Schiffer, W., Shea, C., Carter, P., and Fowler, J.S. 2008. Pharmacokinetics of the potent hallucinogen, salvinorin A in primates parallels the rapid onset and short duration of effects in humans. *NeuroImage* 41: 1044–50.

Hooker, J.M., Patel, V., Kothari, S., and Schiffer, W.K. 2009. Metabolic changes in the rodent brain after acute administration of salvinorin A. *Mol Imaging Biol* 11: 137–43.

Imanshahidi, M. and Hosseinzadeh, H. 2006. The pharmacological effects of *Salvia* species on the central nervous system. *Phytother Res* 20: 427–37.

Johnson, M.W., MacLean, K.A., Reissig, C.J., Prisinzano, T.E., and Griffiths, R.R. 2011. Human psychopharmacology and dose-effects of salvinorin A, a kappa opioid agonist hallucinogen present in the plant *Salvia divinorum*. *Drug Alcohol Depend* 115: 150–55.

Kane B.E., Nieto, M.J., McCurdy, C.R., and Ferguson, D.M. 2006. A unique binding epitope for salvinorin A, a non-nitrogenous kappa opioid receptor agonist. *FEBS Journal* 273: 1966–74.

Kane B.E., McCurdy, C.R., and Ferguson, D.M. 2008. Toward a structure-based model of salvinorin A recognition of the kappa opioid receptor. *J Med Chem* 51: 1824–30.

Kelly, B.C. 2011. Legally tripping: A qualitative profile of *Salvia divinorum* use among young adults. *J Psychoactive Drugs* 43: 46–54.

Koreeda, M., Brown, L., and Valdes III, L.J. 1990. The absolute stereochemistry of salvinorins. *Chem Lett* 19(11): 2015–18.

Kowalczuk, A.P., Raman, V., Galal, A.M., Khan, I.A., Siebert, D.J., and Zjawiony, J.K. 2014. Vegetative anatomy and micromorphology of *Salvia divinorum* (Lamiaceae) from Mexico, combined with chromatographic analysis of salvinorin A. *J Nat Med* 68: 63–73.

Kutrzeba, L., Dayan, F.E., Howell, J'.L., Feng, J., Giner, J.-L., and Zjawiony, J. 2007. Biosynthesis of salvinorin A proceeds via the deoxyxylulose phosphate pathway. *Phytochemistry* 68: 1872–81.

Kutrzeba L.M., Karamyan, V.T., Speth, R.C., Williamson, J.S., and Zjawiony, J. K. 2009a. In vitro studies on metabolism of salvinorin A. *Pharm Biol* 47(11): 1078–84.

Kutrzeba, L.M., Ferreira, D., and Zjawiony, J.K. 2009b. Salvinorins J from *Salvia divinorum*: Mutarotation in the neoclerodane system. *J Nat Prod* 72: 1361–63.

Lee, D.Y.W., Ma, Z., Liu-Chen, L.-Y., Wang, Y., Chen, Y., Carlezon Jr., W.A., and Cohen, B. 2005. New neoclerodane diterpenoids isolated from the leaves of *Salvia divinorum* and their binding affinities for human κ opioid receptors. *Biorg Med Chem* 13: 5635–39.

Listos, J., Merska, A., and Fidecka, S. 2011. Pharmacological activity of salvinorin A, the major component of *Salvia divinorum*. *Pharmacol Rep* 631305–09.

Lovell, K.M., Prevatt-Smith, K.M., Lozama, A., and Prisinzano, T.E. 2010. Synthesis of neoclerodane diterpenes and their pharmacological effects. *Top Curr Chem (Chemistry of Opioids)* 299: 141–85.

Lovell, K.M., Vasiljevik, T., Araya, J.J. et al. 2012. Semisynthetic neoclerodanes as kappa opioid receptor probes. *Bioorg Med Chem* 20: 3100–10.

Lozama, A., Prisinzano, T.E. 2009. Chemical methods for the synthesis and modification of neoclerodane diterpenes. *Bioorg Med Chem Lett* 19: 5490–5495.

Lozama, A., Cunningham, C.W., Caspers, M.J., Douglas, J.T., Dersch, C.M., Rothman, R.B., and Prisinzano, T.E. 2011. Opioid receptor probes derived from cycloaddition of the hallucinogen natural product salvinoirin A. *J Nat Prod* 74: 718–26.

Ma, Z., Deng, G., and Lee, D.Y.W. 2010. Novel neoclerodane diterpene derivatives from the smoke of salvinorin A. *Tetrahedron Lett* 51: 5207–09.

Mabberley, D. J. 2008. *Mabberley's Plant Book: A Portable Dictionary of Vascular Plants, Their Classification and Uses.* 3rd edition. Cambridge, UK: Cambridge University Press.

MacLean, K.A., Johnson, M.W., Reissig, C.D., Prisinzano, T.E., and Griffiths, R.R. 2013. Dose-related effects of salvinorin A in humans: Dissociative, hallucinogenic and memory effects. *Psychopharmacology* 226: 381–92.

Morani, A.S., Kivell, B., Prisinzano, T.E., and Schenk, S. 2009. Effect of kappa-opioid receptor agonists U69593, U50488H, spiradoline and salvinorin A on cocaine-induced drug-seeking in rats. *Pharmacol Biochem Behav* 94: 244–49.

Morani, A.S., Schenk, S. Prisinzano, T.E., and Kivell, B. 2012. A single injection of a novel kappa opioid receptor agonist salvinorin A attenuates the expression of cocaine-induced behavioral sensitization in rats. *Behav Pharmacol* 23: 162–70.

Munro, T.A., and Rizzacasa, M.A. 2003. Salvinorins D-F, new neoclerodane diterpenoids from *Salvia divinorum*, and an improved method for the isolation of salvinorin A. *J Nat Prod* 66: 703–05.

Nozawa, M., Suka, Y., Hoshi, T., Suzuki, T., and Hagiwara, H. 2008. Total synthesis of the hallucinogenic neoclerodane diterpenoid salvinorin A. *Org Lett* 10(7): 1365–68.

Ortega, A., Blount, F.J., and Manchand, P.S. 1982. Salvinorin, a new-*trans* neoclerodane diterpene from *Salvia divinorum* (Labiatae). *J Chem Soc Perkin Trans I* 1982: 2505–508.

Phipps, S.M., and Butterweck, V. 2010. A new digitized method of the compulsive gnawing test revealed dopaminergic activity of salvinorin A in vivo. *Planta Med* 76: 1405–10.

Piekielna, J., Fichna, J., Janecka, A. 2012. Salvinorin A and related diterpenes: biological activity and potential therapeutic uses. *Postepy Biochem* 58: 485–491.

Polepally, P.R., White, K., Vardy, E., Roth, B.L., Ferreira, D., and Zjawiony, J.K. 2013. Kappa-opioid receptor-selective dicarboxylic ester-derived salvinorin A ligands. *Biorg Med Chem Lett* 23: 2860–62.

Polepally, P.R., Huben, K., Vardy, E., Setola, V., Mosier, P.D., Roth, B.L. and Zjawiony, J.K. 2014. Michael acceptor approach to the design of new salvinorin A-based high affinity ligands for the kappa-opioid receptor. *Eur J Med Chem in press*.

Prevatt-Smith, K.M., Lovell, K.M., Simpson, D.S. et al. 2011. Potential drug abuse therapeutics derived from the hallucinogenic natural product salvinorin A. *Med Chem Commun* 2: 1217–22.

Prisinzano, T.E., 2005. Psychopharmacology of the hallucinogenic sage *Salvia divinorum*. *Life Sci* 78: 527–31.

Prisinzano, T.E., and Rothman, R.B. 2008. Salvinorin A analogs as probes in opioid pharmacology. *Chem Rev* 108: 1732–43.

Prisinzano, T.E. 2013. Neoclerodanes as atypical opioid receptor ligands. *J Med Chem* 56: 3435–43.

Ranganathan, M., Schnakenberg, A., Skosnik, P.D., Cohen, B.M., Pittman, B., Sewell, A., and D'Souza, D.D. 2012. Dose-related behavioral, subjective, endocrine, and psychophysiological effects of the κ-opioid agonist salvinorin A in humans. *Biol Psychiatry* 72: 871–79.

Roth, B.L., Banor, K., Westkaemper, R., Siebert, D., Rice, K.C., and Steinberg, S. 2002. Salvinorin A: A potent naturally occurring non-nitrogenous κ-opioid selective agonist. *Proc Natl Acad Sci USA* 99: 11934–39.

Rothman R.B., Murphy, D.L., Xu, H. et al. 2007. Salvinorin A: Allosteric interactions at the μ-opioid receptor. *J Pharmacol Exp Ther* 320: 801–10.

Salaga M., Polepally, P.R., Sobczak, M. et al. 2014. Novel orally available salvinorin A analog PR-38 inhibits gastrointestinal motility and reduces abdominal pain in mouse models mimicking irritable bowel syndrome. *J Pharmacol Exp Ther* 350: 69–78.

Saric, D., Kalodera, Z., Lackovic, Z. 2010. Psychotropic plant *Salvia divinorum* Epl. & Jativa - the source of the most potent natural hallucinogen. *Farm Glas* 66: 523–541.

Scheerer, J.R., Lawrence, J.F., Wang, G.C., and Evans, D.A. 2007. Asymmetric synthesis of salvinorin A, a potent κ opioid receptor agonist. *J Am Chem Soc* 129: 8968–69.

Sheffler, D.J., Roth, B.L. 2003. Salvinorin A: the "magic mint" hallucinogen finds a molecular target in the kappa opioid receptor. *Trends Pharmacol Sci* 24: 107–109.

Shirota, O., Nagamatsu, K., and Sekita, S. 2006. Neo-clerodane diterpenes from the hallucinogenic sage *Salvia divinorum*. *J Nat Prod* 69: 1782–86.

Shirota, O., Nagamatsu, K., and Sekita, S. 2007. Simple preparative isolation of salvinorin A from the hallucinogenic sage, *Salvia divinorum*, by centrifugal partition chromatography. *J Liq Chromatogr R T* 30: 1105–14.

Siebert, D.J. 1994. *Salvia divinorum* and salvinorin A: New pharmacologic findings. *J Ethnopharmacol* 43: 53–56.

Siebert, D.J. 2004. Localization of salvinorin A and related compounds in glandular trichomes of the psychoactive sage, *Salvia divinorum*. *Ann Bot* 93: 763–71.

Siebert, D.J. 2013. The legal status of *Salvia divinorum*. http://www.sagewisdom.org/legalstatus.html.

Sumnall, H.R., Measham, F., Brandt, S.D., and Cole, J.C. 2011. *Salvia divinorum* use and phenomenology: Results from an online survey. *J Psychopharmacol* 25: 1496–1507.

Tidgewell, K., Harding, W.W., Schmidt, M., Holden, K.G., Murry, D.J., and Prisinzano, T.E. 2004. A facile method for the preparation of deuterium labeled salvinorin A: Synthesis of [2,2,2-^{2}H$_3$]-salvinorin A. *Biorg Med Chem Lett* 14: 5099–102.

Tidgewell, K., Groer, C.E., Harding, W.W. et al. 2008. Herkinorin analogues with differential β-arrestin 2 interaction. *J Med Chem* 51: 2421–31.

Valdes III, L.J., 1986. Loliolide from *Salvia divinorum*. *J Nat Prod* 49(1): 171.

Valdes III, L.J., Diaz, J.L., and Paul, A.G. 1983. Ethnopharmacology of Ska Maria Pastora (*Salvia divinorum*, Epling and Jativa-M). *J Ethnopharmacol* 7: 287–312.

Valdes III, L.J., Butler, W.M., Hatfield, G.M., Paul, A.G., and Koreeda, M. 1984. Divinorin A, a psychotropic terpenoid, and Divinorin B from the hallucinogenic Mexican mint *Salvia divinorum*. *J Org Chem* 49: 4716–20.

Valdes III, L.J., Chang, H.-M., Visger, D.C., and Koreeda, M. 2001. Salvinorin C, a new neoclerodane diterpene from a bioactive fraction of the hallucinogenic Mexican mint *Salvia divinorum*. *Org Lett* 3(24): 3935–37.

Vohra, R., Seefeld, A., Cantrell, F.L., and Clark, R.F. 2011. *Salvia divinorum*: Exposures reported to a statewide poison control system over 10 years. *J Emerg Med* 40(6): 643–50.

Vortherms, T.A. and Roth, B.L. 2006. Salvinorin A from natural product to human therapeutics. *Mol Interv* 6(5): 259–67.

Vortherms T.A., Mosier, P.D., Westkaemper R.B., and Roth, B.L. 2007. Differential helical orientation among related G-protein coupled receptors provide a novel mechanism for selectivity. *J Biol Chem* 282: 3146–56.

Walentiny, D. M., Vann, R. E., Warner, J. A. et al. 2010. Kappa opioid mediation of cannabinoid effects of the potent hallucinogen, salvinorin A in rodents. *Psychopharmacology* 210: 275–84.

White, K.L., Scopton, A.P., Rives, M-L. et al. 2014. Identification of novel functionally selective kappa opioid receptor scaffolds. *Mol Pharmacol* 85(1): 83–90.

Yan, F., Roth, B.L. 2004. Salvinorin A: A novel and highly selective k-opioid receptor agonist. *Life Sci* 75: 2615–2619.

Yan, F., Mosier, P.D., Westkaemper, R.B. et al. 2005. Identification of the molecular mechanism by which the diterpenoid salvinorin A binds to κ-opioid receptors. *Biochemistry* 44: 8643–51.

Yan, F., Mosier, P.D., Westkaemper, R.B., and Roth, B.L. 2008. Gα-subunits differentially alter the conformation and agonist affinity of κ-opioid receptors. *Biochemistry* 47: 1567–78.

Yan, F., Bikbulatov, R. V., Dicheva, N. et al. 2009. Structure-based design, synthesis and biochemical and pharmacological characterization of novel salvinorin A analogues as active probes of the κ-opioid receptor. *Biochemistry* 48(29): 6898–908.

Zawilska, J.B., and Wojcieszak, J. 2013. *Salvia divinorum*: From Mazatec medicinal and hallucinogenic plant to emerging recreational drug. *Hum Psychopharmacol Clin Exp* 28: 403–12.

Zhang, Y., Butelman, E.R., Schlussman, S.D., Ho, A., and Kreek, M.J. 2005. Effects of the plant-derived hallucinogen salvinorin A on basal dopamine levels in the caudate putamen and in a conditioned place aversion assay in mice: Agonist actions at kappa opioid receptors. *Psychopharmacology* 179: 551–58.

5 The Botany of *Mitragyna speciosa* (Korth.) Havil. and Related Species

Sasha W. Eisenman

CONTENTS

5.1 INTRODUCTION

Mitragyna speciosa (Korth.) Havil., most commonly referred to as kratom (see Table 5.1 for a list of additional common names), is a tropical tree that is known to grow in peninsular Thailand, southeastern Myanmar, Malaysia, Borneo, Sumatra, the Philippines, and New Guinea. The alkaloid-containing leaves of *M. speciosa* have historically been used for a variety of purposes. As a traditional medicine, preparations of the leaves have been used for treating various illnesses. The fresh leaves have commonly been chewed by farmers, laborers, and fishermen in Malaysia, Thailand, and elsewhere to reduce fatigue while working in hot conditions. The dried leaves or concentrated leaf extracts have been ingested or smoked as an opium substitute and intoxicant. Over the last decade, the leaves of this plant have become increasingly available to a global audience via e-commerce. *M. speciosa*

TABLE 5.1

Common Names Used for *Mitragyna speciosa* Korth. (Havil.)

Country	Common Name
Indonesia	kadamba (Kelantan), puri (Batak Toba, Sumatra), keton
Malaysia	biak, biak-biak, ketum, kutum, pokok biak, pokok ketum, sepat (Sabah)
Myanmar	beinsa, bein-sa-ywat
Philippines	mambog (Tagalog), lugub (Mandaya), polapupot (Ibanag)
Thailand	ithang (central), thom (peninsular), bai krathom, gratom, kakaum, katawn, krathawm, kratom, kraton
Vietnam	giam d[ef]p, giam l[as] nh[or]

Source: Burkill 1935; Chua and Schmelzer 2001; Lace and Roger 1922; Nyein 2013; Rätsch 2005.

is now taken by people looking for a natural alternative to pharmaceutical pain relievers and antidepressants, and as an aid in opioid withdrawal, as well as by those seeking relaxation, spiritual experiences, or a "legal high."

5.2 TAXONOMY AND MORPHOLOGY

Mitragyna Korth. is a small genus in the Rubiaceae (coffee or madder) family. The Rubiaceae is extremely diverse, and with over 13,000 species it is considered to be the fourth largest plant family (Goevarts et al. 2007, 2013). Members of the family vary greatly, ranging from small herbs, to vines, to towering trees. In general, plants in the Rubiaceae can be recognized as having a combination of the following characteristics: opposite or occasionally whorled simple leaves, interpetiolar stipules, inferior ovaries and sympetalous flowers (Razafimandimbison and Bremer 2001). The family is found primarily in tropical and subtropical areas, although there are some genera, such as *Cephalanthus* L., *Diodia* L., *Galium* L., and *Mitchella* L., that have temperate distributions.

The family's best-known genus, *Coffea*, contains one of the most economically important crops in the world, *Coffea arabica* L. (as well as the less abundantly produced *C. canephora* Pierre ex A. Froehner and *C. liberica* W. Bull ex Hiern.). Coffee generates approximately $15 billion in export revenues annually, and is second only to crude oil as the most traded primary commodity (Baffes et al. 2004). While coffee also contains the most well-known alkaloid, caffeine, other genera in the Rubiaceae also have significant alkaloids. The roots from *Carapichea ipecacuanha* (Brot.) L. Andersson are used in the preparation of ipecac syrup. *C. ipecacuanha* contains the alkaloids cephaeline, emetine, and psychotrine, among others, and various formulations have been used as an expectorant and emetic and for the treatment of amoebic dysentery. The bark from *Cinchona* species contains the alkaloid quinine, which has been used for centuries as an antimalarial (Evans 2002; Balick and Cox 1996). *Psychotria viridis* Ruiz & Pav. and related species contain *N,N,-*dimethyltryptamine, and are used in the preparation of the hallucinogenic beverage known as ayahuasca or yagé (Schultes and Hofmann 1973). The bark of *Pausinystalia johimbe* Pierre ex Beille contains yohimbine and has been used for erectile dysfunction, athletic performance, and weight loss, among other things. (Barceloux 2008; NMCD 2011). The ornamental genera *Gardenia, Pentas,* and *Ixora* are also in the Rubiaceae family.

Below the family level, the genus *Mitragyna* belongs to the tribe Naucleeae of the subfamily Cinchonoideae. This tribe is characterized by congested inflorescence heads, flowers with free or fused ovaries, and dehiscent free or indehiscent multiple fruits (Razafimandimbison and Bremer 2001). The genus *Mitragyna* is distributed across the Paleotropical biogeographical region and contains seven to ten species, depending on the treatment followed (Leroy 1975, Ridsdale 1978; Razafimandimbison and Bremer 2002). Four species occur in Africa, and six species are found in South and Southeast Asia, with distributions between India and New Guinea. As is the case with many plant genera and species, the nomenclatures of both *Mitragyna* and *M. speciosa* have been changed by various authors over the years. The genus was first described by the Dutch botanist Pieter Willem Korthals (1807–1892). Korthals made significant botanical collections and discoveries in the Malay Archipelago while working as a member of the Commission for Natural Sciences of the Dutch East India Company from 1830 to 1837 (Steenis-Kruseman 1950). Korthals published the new genus, and the species name *Mitragyna speciosa* in the same publication, but because he did not include a botanical description of the new species (which is a necessary requirement to be considered a valid species under the International Code of Botanical Nomenclature), this name is considered a nomen nudum (Korthals 1839; McNeill et al. 2012). Korthals later transferred the species to another genus, *Stephegyne* (Korthals 1842). Other authors also transferred and renamed the species (*Nauclea korthalsii* (Steudel 1841), *Nauclea luzoniensis* (Blanco 1845), *Nauclea speciosa* (Miquel 1856)). In his revision of the tribe *Naucleeae*, Haviland (1897) transferred the species back into the genus *Mitragyna*, and set the current accepted species name as *Mitragyna speciosa* (Korth.) Havil. According to Burkill (1935) the common name, kratom, is likely derived from the Sanskrit *kadam*, a name that is used for the widespread and related species *Neolamarckia cadamba* (Roxb.) Bosser, a sacred tree in the Hindu religion. Similar names are used for a number of related tree species that are indigenous to the region.

Based on anatomical and morphological variation, Leroy (1975) segregated out three of the four African species and placed them in a new genus *Hallea* J.-F.Leroy. This treatment was supported by some authors (Verdcourt 1985; Brummit 1992; Huysman et al. 1994), but more recently, based on morphological and molecular data, Razafimandimbison and Bremer (2002) merged *Hallea* into *Mitragyna* once again. *Hallea* (sensu J.-F.Leroy) has also been determined to be an illegitimate name (nomen illegitimum) because it had already been applied to a fossil plant by G.B. Mathews in 1948. Deng (2007) proposed that the three species be transferred to a new genus, *Fleroya* Y.F. Deng. Table 5.2 contains a list of the *Mitragyna* and *Fleroya* species, their synonyms, and their distributions.

In general, plants in the genus *Mitragyna* are trees or shrubs with opposite leaves and entire stipules that are more or less keel shaped. The flowers are sessile, and arranged in compact globose heads with interfloral pale-aceous, spathuloid bracts. The heads are terminally located on side shoots of laterally growing (plagiotropic) branches. Most typically these side shoots may be branched in simple or compound dichasia, with each branch terminating with a head. More rarely side shoots form thyrse-like structures or a terminal compound umbel-like structure. The flowers are attached to a densely hairy receptacle and the sepals are connate, forming a short or long tube with five lobes. The corollas are funnel-shaped and 5-lobed. There are five stamens with lanceolate, cordate anthers. The ovary is two-celled with a cylindric or mitriform stigma (Haviland 1897; Ridsdale 1978).

Mitragyna speciosa is an evergreen tree that reaches 25 meters tall and 2 to 3 feet in diameter. The trunk is usually straight and the outer bark is smooth and gray, and the inner bark is pinkish. The twigs are obtusely angulate, and the leaves are generally elliptic (ranging to ovate or obovate) in shape, and pet-iolate. Fully expanded, the leaves are 14–20 cm long and 7–12 cm wide, are chartaceous, and have a sub-acuminate apex and a rounded to sub-cordate base. The leaves typically have 12–15(–17) pairs of veins with pale-hairy domatia in the axils. The stipules are lanceolate, 2–4 cm long, sparsely pubescent, and 9-nerved. The flowering heads are arranged in threes, one short-peduncled head between two longer-peduncled heads, and are often subtended by leaf-like bracts. The heads are 1.5–2.5 cm in diameter with pale-hairy interfloral bracteoles that are 4–6 mm long. The receptacle is densely hirsute. The flower calyx is approximately 2 mm long and 5-lobed. The corolla is funnel-shaped, deep yellow in color, and glabrous on the exterior. The corolla tube is 3.5–5 mm, and the corolla lobes are 2.5–3 mm long and glabrous with a revolute margin and a distinct ring of hairs inside at the base of the lobes. The five stamens intersect with the lobes and are reflexed with lanceolate anthers that protrude from the corolla. The style is 13 mm and the rounded stigma 2 mm. The fruiting heads are 2–3 cm in diameter, with 10-ribbed fruits that are 7–9 mm long and 4–5 mm wide and contain numerous seeds. The seeds are around 1 mm long and have a 1–2 mm papery wing at each end (Ridley 1923; Haviland, 1897; Ng 1989; Chua and Schmelzer 2001). Figure 5.1 presents a botanical illustration of *M. speciosa* from Korthals (1842).

TABLE 5.2
Species of *Mitragyna* and Their Distribution

Accepted Species Name	Synonyms	Distribution
Fleroya ledermannii (K.Krause) Y.F.Deng	*Adina ledermannii* K.Krause *Hallea ciliata* (Aubrév. & Pellegr.) J.-F.Leroy *Hallea ledermannii* (K.Krause) Verdc. *Mitragyna ciliata* Aubrév. and Pellegr. *Mitragyna ledermannii* (K.Krause) Ridsdale	Africa: Liberia east to the Central African Republic and south to Gabon, Democratic Republic of Congo and Angola (Lemmens et al. 2012)
Fleroya rubrostipulata (K.Schum.) Y.F.Deng	*Adina rubrostipulata* K.Schum. *Adina rubrostipulata* var. *discolor* *Hallea rubrostipulata* (K.Schum.) J.-F.Leroy *Mitragyna rubrostipulata* (K.Schum.) Havil.	Africa: Democratic Republic of Congo, Ethiopia, Kenya, Tanzania, Malawi, Mozambique (Lemmens et al. 2012)
Fleroya stipulosa (DC.) Y.F.Deng	*Adina stipulosa* (DC.) Roberty *Hallea stipulosa* (DC.) J.-F.Leroy *Mamboga stipulosa* (DC.) Hiern *Mitragyna chevalieri* K.Krause *Mitragyna macrophylla* Hiern *Mitragyna stipulosa* (DC.) Kuntze *Nauclea bracteosa* Welw. *Nauclea stipulosa* DC.	Africa: widespread from Senegal east to Uganda and south to Zambia and Angola (Lemmens et al. 2012)
Mitragyna diversifolia (Wall. ex G.Don) Havil.	*Mitragyna javanica* Koord. & Valeton *Nauclea adina* Blanco *Nauclea diversifolia* Wall. ex G.Don *Stephegyne diversifolia* (Wall. ex G.Don) Brandis *Stephegyne diversifolia* (Wall. ex G.Don) Hook.f. *Stephegyne parvifolia* Vidal *Stephegyne tubulosa* Fern.-Vill.	Asia: Cambodia, China, Indonesia, Laos, Malaysia, Myanmar, Philippines, Thailand, Vietnam (eFloras 2008)

(Continued)

Species of *Mitragyna* and Their Distribution (*continued*)

Accepted Species Name	Synonyms	Distribution
Mitragyna hirsuta Havil.	*Paradina hirsuta* (Havil.) Pit.	Asia: Cambodia, China, Laos, Myanmar, Thailand, Vietnam (eFloras 2008)
Mitragyna inermis (Willd.) Kuntze	*Adina inermis* (Willd.) Roberty *Cephalanthus africanus* Rchb. ex DC. *Mitragyna africana* (Willd.) Korth. *Nauclea africana* Willd. *Nauclea africana* var. *luzoniensis* DC *Nauclea inermis* (Willd.) Baill. *Nauclea platanocarpa* Hook.f. *Platanocarpum africanum* (Willd.) Hook.f. *Stephegyne africana* (Willd.) Walp. *Uncaria inermis* Willd.	Africa: Mauritania east to Sudan (Lemmens et al. 2012)
Mitragyna parvifolia (Roxb.) Korth.	*Nauclea parvifolia* Roxb. *Nauclea parvifolia* Willd. *Stephegyne parvifolia* (Roxb.) Korth.	Asia: India, Sri Lanka, Myanmar and Bangladesh (India Biodiversity Portal 2013; Roskov 2013)
Mitragyna rotundifolia (Roxb.) Kuntze	*Bancalus rotundifolius* (Roxb.) Kuntze *Mitragyna brunonis* (Wall. ex G.Don) Craib *Nauclea brunonis* Wall. ex G.Don *Nauclea rotundifolia* Roxb.	Asia: Bangladesh, China, India, Laos, Myanmar, Thailand (eFloras 2008)
Mitragyna speciosa (Korth.) Havil.	*Nauclea korthalsii* Steud. *Nauclea luzoniensis* Blanco *Nauclea speciosa* (Korth.) Miq. *Stephegyne speciosa* Korth.	Indonesia (Borneo, Sumatra, Papua, West Papua) Malaysia (peninsular, Sabah), Myanmar, New Guinea, Papua New Guinea, the Philippines, Thailand, Vietnam
Mitragyna tubulosa (Arn.) Kuntze	*Nauclea tubulosa* Arn.	Endemic to peninsular India (India Biodiversity Portal 2013)

Source: Govaerts et al. 2013b.

FIGURE 5.1   Illustration of *Mitragyna speciosa*, from Korthals (1842).

Haviland (1897) described two unnamed varieties based on the number of leaf veins and area of distribution. Variety "*a*" was 15-nerved and found in Borneo, while variety "*b*" was 10-nerved and found in the Philippines and New Guinea. Valeton (1925) described *M. speciosa* var. *glabra* Valeton from plant material collected in New Guinea. None of these varieties seem to be currently in use.

5.3   DISTRIBUTION AND HABITAT

The distribution of *Mitragyna speciosa* is typically given as a list of countries in which the species is known to occur. For example, Ridsdale (1978) described the distribution of *M. speciosa* as "Thailand (Peninsular, Southwestern), S. Vietnam (cult.), Malay Peninsula, Sumatra, Borneo, Philippines, New Guinea." Chua and Schmeltzer (2001) stated that the species is cultivated in southern Vietnam, peninsular Thailand and Burma

(Myanmar). The majority of this distribution falls within an area known as the Malesian floristic region, the northwestern end of the distribution lying within the Indo-Chinese region (Takhtajan 1986; Johns 1995). The distribution of the species within these countries is significantly more difficult to determine and must be cobbled together from many varied sources. Many of the early works describing the flora of this region give scant information, and may only cite one or two localities where the species was seen.

In Peninsular Malaysia, Wray (1906) described the species as being "widely distributed in Perak, and there is a place near Salak, in the Kuala Kangsar district, named after it … It occurs in the jungle and is planted in the kampongs, and also has been preserved when other trees were felled and cleared away. Consequently, it is frequently seen in and around the villages." In the *Flora of the Malay Peninsula*, Ridley (1923) described the distribution as the "Malay isles" and the habitat as "open country," and describes the species as being "rather rare." In *A Dictionary of the Economic Products of the Malay Peninsula*, Burkill (1935) gives a similar account to that of Wray, stating "Open country in the northern half of the peninsula, cultivated or else preserved from felling when forest has been cleared; growing … through Lower Siam in the same way, and in cultivation at least northwards to Bangkok." More recent floristic works describe the habitat and distribution in peninsular Malaysia as fresh water swamps and river systems and stream habitats in the states of Selangor, Kelantan, Pahang, and Perak (Said and Zakaria 1992; FRIM 2013).

The island of Borneo is divided between three countries, Brunei, Indonesia, and Malaysia. Miquel (1856) provided the localities of "Borneo, on the Doeson-river, in Banjermassing [Banjarmasin], Mantalet [Mantalat], in swampy areas and in the dry regions of Martapoera [Martapura]." These areas are all in South Kalimantan, Indonesia. In a report on the natural vegetation of the lower Kinabatangan floodplain region of Sabah, Malaysia, Azmi (1998) stated that although few historical herbarium collections existed from Sabah, and all were confined to the Kinabatangan area, *M. speciosa* is occasionally in lowland forest and is a common riverside pioneer, particularly along the low, muddy banks of rivers or tributaries. More specifically, *Mitragyna speciosa*, along with *Nauclea orientalis*, *N. subdita*, *Duabanga moluccana*, *Neolamarckia cadamba*, and *Octomeles sumatrana*, were determined to be common pioneer species, "on the accreting banks of the Kinabatangan River or areas of recent alluvial deposits (low muddy riverbanks or banks of abandoned river channels)." *M. speciosa* is described as appearing to be "the most flood-tolerant species and forms dense stands in the new alluvial substrate of ox-bow lakes and low muddy river banks that are frequently inundated." Extensive groves of *Mitragyna* are stated to occur in the seasonal swamp forests of Danau Luagan. This forest is described as consisting of nearly pure stands of *M. speciosa*, particularly along the edge of the river, and this is thought to be the most extensive population of *M. speciosa* in Sabah. A more mature, heterogeneous forest was reported to occur at Tenegang lake (in Danau Pitas). Nilus et al. (2011) described the natural habitat for *M. speciosa* growing in Sabah, Malaysia, as "freshwater swamp forest" and stated that the species is "is often found growing gregariously in the backwater swamps, regenerating on shallow peat soils, and occasionally on deeper peat soils in larger floodplains." *M. speciosa* has been primarily produced from seed, but clonal propagation of the species has been studied and utilized for purposes of ecological restoration particularly in flood-prone areas of Kinabatangan (Ajik and Kimjus 2010; Nilus et al. 2011). Chey (2005) observed 90% infestation on seedlings of *Mitragyna speciosa* by the defoliating hawkmoth caterpillar *Cephonodes hylas* Linnaeus (Sphingidae) at the Forest Research Centre nursery in Sepilok, and Chung et al. (2012) documented significant damage to nursery stock by caterpillars of *Dysaethria quadricaudata* Walker.

In their *List of Trees, Shrubs, and Principal Climbers, etc. Recorded from Burma*, Lace and Rodger (1922) reported *beinsa* as being the common name for *M. speciosa* in Myanmar. The distribution was reported as being "South Tenasserim" (Southeastern Myanmar). Interestingly, around that time, *beinsa* was also the term for opium eater, and *bein* the term for opium (Ferrars and Ferrars 1900). In the Philippines there are records from lowland forests on Luzon (Cagayan), Mindoro, and Mindanao (Davao, Agusan) (Pelser et al. 2011). The island of New Guinea is divided between two countries: the western half is part of Indonesia (Papua and West Papua Provinces), while the eastern half is the Independent State of Papua New Guinea. The *Plants of Papua New Guinea* database has over 40 records for specimens of *M. speciosa*, documenting that it is distributed across much of the island (Conn et al. 2004).

5.4 USES OF *MITRAGYNA* SPECIES AND *M. SPECIOSA*

5.4.1 WOOD

Though *Mitragyna* species are mostly documented for their use in traditional medicine, some of the African species (primarily *M. ciliata*) are used as a source of wood. The wood is used for a wide array of purposes including construction, flooring, joinery, trim, furniture, sporting goods, toys, novelties, musical instruments, transmission poles, boxes, crates, canoes, oars, and tool handles. (Nyemb 2011). The wood of *M. speciosa* has also been described as suitable for veneer, as well as coating plywood, non-structural laminated veneer lumber, and profile molding (Ajik and Kimjus 2010).

5.4.2 TRADITIONAL MEDICINE

Many of the *Mitragyna* species have been used as traditional medicines. In Africa, preparations of the stem bark of *M. ciliata* have been used for the treatment of a variety of ailments, including amenorrhea, broncho-pulmonary diseases, colds, dysentery, fever, general weakness, gonorrhea, hypertension, inflammation, leprosy, malaria, and rheumatism, and to facilitate childbirth (Dongmo 2003; Bidie 2008; Nyemb 2011). Preparations of the bark and/or leaves of *M. inermis* and *M. stipulosa* have been used as a diuretic and analgesic, and to treat lactation failure, fever, diarrhea, diabetes, hypertension, and malaria (Adjanohoum et al. 1985; Toure et al. 1996; Oliver-Bever 1986; Betti 2002; Asase et al. 2005; Igoli 2005). *M. africana* (some synonymize this species with *M. inermis*) is used in West Africa for the treatment of bacterial infections, dysentery, gonorrhea, mental disorders, and epilepsy. It is used in combination with other plant species for treatment of sleeping sickness (human African trypanosomiasis), sterility, and mental illness (Aji et al. 2001).

In India, *Mitragyna parvifolia* bark and roots are used to treat a wide array of conditions, including burning sensations, colic, cough, edema, gynecological disorders, fever, muscular pain, and poisoning, and as an aphrodisiac. The leaves are applied to wounds and sores to alleviate pain and swelling, and to aid in healing (Badgujar and Surana 2010). In Bangladesh, the bark of *M. diversifolia* is used traditionally as a remedy for diarrhea (Uddin et al. 2009). In Thailand, a decoction or drink prepared from the stem bark, wood, and roots of *Mitragyna hirsuta* Hav. (common name *krathum bok* or *krathum khok*) and *Polyalthia cerasoides* is used as a galactogogue, and a decoction of the stems and roots of *M. rotundifolia* is used to treat rheumatoid arthritis (Chuakul et al. 2002; Junsongduang et al. 2014).

It has been said that the leaves of *Mitragyna speciosa* have been used medicinally in Southeast Asia "from time immemorial" (Assanangkornchai et al. 2007). In addition to being chewed, the plant has been in folk medicine for treating a variety of different ailments. The leaves have been applied to wounds and used as a vermifuge and local anesthetic, and extracts of the leaves have been used to treat coughs, diarrhea, and musculoskeletal pain as well as to increase sexual prowess (Burkill 1935; Suwanlert 1975; Chua and Schmelzer 2001; Chan et al. 2005; Reanmongkol et al. 2007; Ward et al. 2011; Ahmad and Aziz 2012; Sakaran et al. 2014).

5.4.3 PHARMACOLOGICAL EFFECT

Even before the genus and species were described botanically, Low (1836) briefly detailed its use: "The *Beah* is the leaf of a moderately high tree so named, which opium-smokers substitute for that drug when it is not procurable. The leaf is serrated, and is sold occasionally at ¼ rupee a cattie." Wray (1906) and Burkill (1935) give detailed accounts on the preparation and use of dried, ground *M. speciosa* leaves and the making of a concentrated extract that was ingested or rolled into a pellet (with finely shredded *Licuala paludosa* palm leaves) and smoked as a substitute for opium.

Ong and Nordiana (1999) reported that a decoction of the leaves is used in Malay ethnomedicine to overcome drug addiction.

Both the effects and use of *Mitragyna speciosa* have contradictory aspects. It has long been noted that small doses have a stimulating cocaine-like effect, while large doses have a more sedative and intoxicating effect (Grewal 1932a,b; Jansen and Prast 1988; Ward et al. 2011). Current perceptions of kratom use are also strongly divided; the plant is viewed as a narcotic intoxicant by some, while others laud the plant for providing a spiritual experience, or as a natural alternative to pharmaceutical pain relievers and antidepressants. Early documents confused the situation, with most referring to the use of *M. speciosa* as an opium "substitute" while other early reports spoke of its use as an anti-opium plant or an opium "remedy" (Wray 1906; Hooper 1907; J.M.H. 1907). Although early documentation focused on the opium-like use of kratom extracts, according to Sewanlert (1975) traditional users of kratom tend to be peasants, laborers, and farmers, who chew fresh leaves to reduce fatigue and hunger and to increase tolerance for extended periods of hard labor. According to Tanguay (2011), "In southern Thailand, traditional kratom use is not perceived as 'drug use' and does not lead to stigmatization or discrimination of users. Kratom is generally part of a way of life in the south, closely embedded in traditions and customs such as local ceremonies, traditional cultural performances and teashops, as well as in agricultural and manual labor in the context of rubber plantations and seafaring." Reanmongkol et al. (2007) reported that while new users only needed a few leaves to have an impact, heavy users may need to chew the leaves from 3 to 10 times a day, and some users find that over time they needed to increase the number of leaves chewed to 10–30 leaves or more per day. Recreational use is also well documented and kratom preparations have been reported to be available in local coffee and teashops in Thailand and northern Malaysia. A mix of kratom extract with codeine- or diphenhydramine-containing cough syrup, soda, and ice, and then laced with various pharmaceuticals, drugs, or chemicals, is referred to as "4x100" and has gained some degree of notoriety (Tungtananuwat and Lawanprasert 2010; Ahmad and Aziz 2012; DEA 2013).

In a survey of more than 550 kratom users in northern Malaysia, Ahmad and Aziz (2012) found that 88% reported daily usage, and that a variety of reasons were given for using kratom. The primary reasons were for "stamina and endurance, social and recreational use, and improving sexual performance," while other reasons for use included "help with sleeping, to reduce withdrawal symptoms and euphoric effects." They also found that medicinal use for treatment of diabetes, diarrhea, fever, hypertension, and pain relief constituted about one-third of the total use. In a survey of kratom users in Thailand, Assanangkornchai et al. (2007), found that kratom leaves were usually chewed without swallowing, that men were more likely to be regular users, and that the most common reasons for using kratom were being able to work harder for longer periods of time under direct sunlight, and to relieve fatigue. Women were nearly six times more likely to use kratom as a medication (as a cough suppressant, for treating diabetes, controlling weight, and relieving sleepiness).

In Myanmar, use of *M. speciosa* has been primarily documented in the south. As is done elsewhere, leaves are chewed or used to make a syrup or powder, which is eaten, smoked, or made into a tea (Fort 1965). More recently, Nyein (2013) also reported that use of *Mitragyna speciosa* (bein-sa-ywat) is prevalent in southern Burma's Tenasserim Division, where it is used for medicinal purposes and casual consumption. Although use of the plant is prohibited in Burma and the plant is categorized by the government as an illicit drug, it is widely used in southern Burma by laborers in the fisheries industry.

5.5 EXPANDED AVAILABILITY THROUGH THE INTERNET

The use of *Mitragyna speciosa* has expanded well beyond its natural distribution, and various formulations have become widely available through the Internet. The reasons given for using kratom are diverse (Prozialeck et al. 2012). An increase in online discussions regarding use of *M. speciosa* among individuals with chronic pain who were seeking to self-medicate for opioid withdrawal has been reported (Boyer et al. 2007, 2008; Ward et al. 2011). In some instances college students have found kratom tea parties to be an alternative to typical house parties (Osterhaus 2008). Numerous websites host discussion forums where topics range from the best vendors and highest potency "strains" of kratom, to preparation and dosing, reasons for use, and reports of experiences. *M. speciosa* is available in the form of whole and crushed leaves, leaf powder, encapsulated powder, concentrated extracts (5x to 100x; loose or encapsulated), solid resin and tinctures. Some vendors claim to standardize the alkaloid concentration in their preparations. Live plants and seeds are also available online, and the species is now being cultivated outside of Southeast Asia.

5.5.1 CATEGORIES

Although not described in the botanical literature, *M. speciosa* available through the Internet is often separated into categories based on vein color, "provenance" and potency. Suwanlert (1975) briefly mentioned that there were two kinds of kratom, red veined and white veined, used in Thailand and that the white-veined leaves had a stronger effect. Chittrakarn et al. (2010) reported that of the two main kinds of kratom used in Thailand, the red-veined variety is supposed to have a stronger effect. Many different categories or types of kratom are promoted by online vendors, including: premium kratom, commercial grade kratom, Bali kratom, enhanced Bali kratom, ultra enhanced Indo (U.E.I.) kratom, Indo red vein, Malaysian kratom, red vein Thai kratom, green- or white-vein Thai kratom, Maeng Da kratom, white-veined Borneo kratom, New Guinea kratom, Java kratom, Sumatra red, the Rifat strain, the bumblebee strain, Red Riau, and Green Riau. There is little clarity regarding the actual origin or potency of these many types of kratom, although there are extensive online discussions regarding this topic. Sukrong et al. (2007) noted the morphological and color variation in *M. speciosa* material from Thailand: "Specimens were grouped into 3 types, red-veining leaf (Kan daeng in Thai), white-veining leaf (Tang gua), and leaf with a pair of very small teeth exerted near apex (Yak yai)."

5.5.2 CONTENT VARIATION

Various researchers have found that alkaloid content in *Mitragyna* species can vary within individual plants depending on the provenance of the plant, plant part sampled, the age of the plant and/or the age of the plant tissue, as well as the time of year or season (Shellard and Houghton 1971, 1972; Shellard and Lala 1977, 1978; Shellard et al. 1978; Houghton et al. 1991). Although quantitative differences have been found between plant material of Thai and/or Malaysian origin, no extensive common garden studies have been conducted to assess the impact of genetic variation and environmental conditions on alkaloid production in *M. speciosa* (Takayama 2004; Temme 2011; Sabetghadam et al. 2013).

5.5.3 ADULTERANTS

In 2003 there was evidence that misidentified plant material was being sold as kratom in the United States (Hanna 2003). Kratom-containing products sold under the name "Krypton" were found to be adulterated with caffeine and synthetic *O*-desmethyltramadol (ODT), and these products were implicated in a number of reported fatalities (Arndt et al. 2011; Kronstrand et al. 2011). Recently, a number of studies have been published detailing the use of DNA analysis and chromatographic and spectroscopic techniques for the verification of *M. speciosa* (Chan et al. 2005; Sukrong et al. 2007; Kikura-Hanajiri et al. 2009; Maruyama et al. 2009; Kowalczuk et al. 2012).

5.6 CURRENT REGULATORY STATUS

In the United States *M. speciosa* and its alkaloids are not scheduled under the Controlled Substances Act (DEA 2013). Several states are trying to pass regulations restricting the plant or specific alkaloids, and three have already passed legislation. Indiana has banned the alkaloids mitragynine and 7-hydroxymitragynine, while Louisiana has made it "unlawful for any person to distribute any product containing *Mitragyna speciosa* to a minor." Tennessee made it illegal to possess the chemicals mitragynine and hydroxymitragynine as well (erowid.org 2013). *Mitragyna speciosa*, mitragynine, or 7-hydroxymitragynine are currently illegal or in some way regulated in a number of countries, including Australia, Bhutan, Denmark, Finland, Germany, Latvia, Lithuania, Malaysia, Myanmar, New Zealand, Poland, Romania, Russia, South Korea, Sweden, and Thailand (Chittrakarn 2010; EMCDDA 2011; Erowid 2013).

In Thailand, *M. speciosa* was first scheduled for control in 1943, and in 1979 it was included in the Thai Narcotics Act B.E. 2522, under Schedule 5. According to Chittrakarn et al. (2010), "This means that it is illegal to buy, sell, import, or grow and harvest. This law makes planting the tree illegal and requires existing trees to be cut down. However, it is not fully effective, since the tree is indigenous to the country and native people prefer to use them [leaves]. Hence, kratom remains a popular drug in Thailand, especially in southern regions." Tanguay (2011) and others have posited that the initial prohibition of *M. speciosa* was linked with declining revenues from the opium trade, and that kratom was made illegal in order to suppress competition in the opium market. In an attempt to curb a growing national problem with methamphetamines, the legal status of kratom is now under review in Thailand (Greenemeier 2013; Winn 2013).

REFERENCES

Adjanohoum, F.J., Ake Assi, L., Floret, J.J., Guinko, S., Koumare, M., Ahyi, A.M.R. and Raynal, J. 1985. *Médecine Traditionnelle et Pharmacopée. Contribution aux études ethnobotaniques et floristiques du Mali*. Paris: Agence de Cooperation Culturelle et Technique.

Ahmad, K. and Aziz Z. 2012. *Mitragyna speciosa* use in the northern states of Malaysia: A cross-sectional study. *J Ethnopharmacol* 141(1): 446–50.

Aji, B M., Onyeyili, P.A. and Osunkwo, U.A. 2001. The central nervous effects of *Mitragyna africanus* (Willd) stembark extract in rats. *J Ethnopharmacol* 77(2–3): 143–49.

Ajik, M. and Kimjus, K. 2010. Vegetative propagation of Sepat (*Mitragyna speciosa*). *Sepilok Bull* 12: 1–11

Arndt, T., Claussen, U., Güssregen, B., Schröfel, S., Stürzer, B., Werle, A. and Wolf, G. 2011. Kratom alkaloids and *O*-desmethyltramadol in urine of a 'Krypton' herbal mixture consumer. *Forensic Sci Int* 208: 47–52.

Asase, A., Oteng-Yeboah, A.A., Odamtten, G.T. and Simmonds, M.S.J. 2005. Ethnobotanical study of some Ghanaian anti-malarial plants. *J Ethnopharmacol* 99(2-3): 273–79.

Assanangkornchai, S., Muekthong, A., Sam-Angsri, N., and Pattanasattayawong, U. 2007. The use of *Mitragynine speciosa* ('Krathom'), an addictive plant, in Thailand. *Subst Use Misuse* 42(14): 2145–57.

Azmi, R. 1998. Natural Vegetation of the Kinabatangan Floodplain (Part 1). An introduction to its natural vegetation, including a preliminary plant checklist of the region. WWFM Project No. MYS 359/96. Kuala Lumpur: WWF Malaysia, and Sabah: Forest Research Centre (Sabah Forestry Department).

Badgujar, V.B. and Surana, S.J. 2010. In vitro investigation of anthelmintic activity of *Mitragyna parvifolia* (Roxb.) Korth. (Rubiaceae). *Vet. World* 3(7): 326–28.

Baffes, J., Lewin, B. and Varangis, P. 2004. Coffee: Market Setting and Policies. In *Global Agricultural Trade and Developing Countries*, ed. M.A. Aksoy and J.C. Beghin, 297–310. Washington, D.C.: World Bank Publications.

Balick, M.J. and Cox, P.A. 1996. *Plants, People, and Culture: The Science of Ethnobotany*. New York: Scientific American Library.

Barceloux, D.G. 2008. *Medical Toxicology of Natural Substances: Foods, Fungi, Medicinal Herbs, Plants, and Venomous Animals*. Hoboken: John Wiley & Sons, Inc.

Betti, J.L. 2002. Medicinal plants sold in Yaounde markets, Cameroon. *Afr Study Monogr* 23(2): 47–64.

Bidie, A.D., Koffi, E., N'Guessan, J.D., Djaman, A.J. and Guede-Guina, F. 2008. Influence of *Mitragyna ciliata* (MYTA) on the microsomal activity of ATPase Na⁺/K⁺ dependent extract on a rabbit heart. *Afr J Tradit Complement Altern Med* 5(3): 294–301.

Blanco, M. 1845. *Flora de Filipinas: según el sistema sexual de Linneo. Segunda impression*. Manila: D. Miguel Sanchez.

Boyer, E.W., Babu, K.M., Macalino, G.E. and Compton, W. 2007. Self-treatment of opioid withdrawal with a dietary supplement, kratom. *Am J Addict* 16(5): 352–56.

Boyer, E. W., Babu, K. M., Adkins, J. E., McCurdy, C.R., and Halpern, J.H. 2008. Self-treatment of opioid withdrawal using kratom (*Mitragynia speciosa* Korth). *Addiction* 103(6): 1048–50.

Brummitt, R.K. 1992. *Vascular Plant Families and Genera*. Royal Botanic Gardens, Kew.

Burkill, I.H. 1935. *A Dictionary of the Economic Products of the Malay Peninsula*. Vol. 2. London: Crown Agents for the Colonies.

Chan, K.B., Pakiam, C. and Rahim, R.A. 2005. Psychoactive plant abuse: The identification of mitragynine in ketum and in ketum preparations. *Bull Narc* 57(1-2): 249–56.

Chey, V.K. 2005. Happy caterpillars: A new record of *Cephonodes hylas* on *Mitragyna speciosa*. *Malays Nat* 59(1): 44–45.

Chittrakarn, S., Keawpradub, N., Sawangjaroen, K., Kansenalak, S., and Janchawee, B. 2010. The neuromuscular blockade produced by pure alkaloid, mitragynine and methanol extract of kratom leaves (*Mitragyna speciosa* Korth.). *J Ethnopharmacol* 129(3): 344–49.

Chua, L.S.L. and Schmelzer, G.H. 2001. *Mitragyna speciosa* (Korth.) Havil. [Internet] Record from Proseabase, van Valkenburg, J.L.C.H. and Bunyapraphatsara, N. (editors). PROSEA (Plant Resources of South-East Asia) Foundation, Bogor, Indonesia. http: //www.proseanet.org (accessed December 21, 2013).

Chuakul, W. Saralamp, P. and Boonpleng, A. 2002. Medicinal plants used in the Kutchum District, Yasothon Province, Thailand. *Thai J Phytopharmacy* 9(1): 22–49.

Chung, A.Y.C., Kimjus, K. and Ajik, M. 2012. Infestation of *Dysaethria quadricaudata* caterpillars (Lepidoptera: Uraniidae: Epipleminae) and its control measure, with reference to evaluation on different insecticide effectiveness. *Sepilok Bull* 15–16: 17–25.

Conn, B.J., Lee, L.L., and Kiapranis, R. 2004. PNG Plants database: Plant collections from Papua New Guinea (http: //www.pngplants.org/PNGdatabase) (accessed December 30, 2013).

DEA. 2013. Kratom (*Mitragyna speciosa* Korth.) Factsheet. January 2013. United States Drug Enforcement Administration. Office of Diversion Control, Drug & Chemical Evaluation Section. http: //www.deadiversion.usdoj.gov/drug_chem_info/kratom.pdf

Deng, Y. 2007. *Fleroya*, a substitute name for *Hallea* J.-F. Leroy (Rubiaceae). *Taxon* 56(1): 247–48.

Dongmo, A.B., Kamanyi, A., Dzikouk, G., Chungag-Anye Nkeh, B., Tan, P.V., Nguelefack, T., Nole, T., Bopelet, M. and Wagner, H. 2003. Antiinflammatory and analgesic properties of the stem bark extract of *Mitragyna ciliata* (Rubiaceae) Aubrév. & Pellegr. *J Ethnopharmacol* 84(1): 17–21.

eFloras. 2008. Published on the Internet http: //www.efloras.org Missouri Botanical Garden, St. Louis, & Harvard University Herbaria, Cambridge, Mass. (accessed December 24, 2013).

Erowid. 2013. Kratom Legal Status. Created by Erowid March 19, 2004, Last modified November 15, 2013. http: //www.erowid.org/plants/kratom/kratom_law.shtml (accessed December 20, 2013)

Evans, W.C. 2002. *Trease and Evans Pharmacognosy*. 15th edition. Edinburgh: WB Saunders.

Ferrars, M. and Ferrars, B. 1900. *Burma*. London: S. Low, Marston and Co.

Fort, J. 1965. Giver of delight or liberator of sin: Drug use and 'addiction' in Asia. *Bull Narc* 3: 1–11.

FRIM. 2013. Provisional Checklist of the Vascular Plants of Malaysia. Online Database. Forest Research Institute Malaysia. http: //www.chm.frim.gov.my/Data/Species/Flora/Provisional-Checklist-of-the-Vascular-Plants-of-Ma/Magnoliopsida/Gentianales/Rubiaceae/Mitragyna-speciosa.aspx (accessed December 22, 2013).

Govaerts, R., Frodin D.G., Ruhsam, M., Bridson, D.M. and Davis, A.P. 2007. *World Checklist & Bibliography of Rubiaceae*. The Trustees of the Royal Botanic Gardens, Kew.

Govaerts, R., Frodin D.G., Ruhsam, M., Bridson, D.M. and Davis, A.P. 2013a. *World Checklist of Rubiaceae*. Facilitated by the Royal Botanic Gardens, Kew. Published on the Internet; http: //apps.kew.org/wcsp/(accessed December 22, 2013)

Govaerts R. et al. (eds.). 2013b. WCSP: World Checklist of Selected Plant Families (version Oct 2011). In: Species 2000 & ITIS Catalogue of Life, 10 December 2013 (Roskov Y., Kunze T., Paglinawan L., Abucay L., Orrell T., Nicolson D., Culham A., et al., eds.). Digital resource at www.catalogueoflife .org/col. Species 2000: Naturalis, Leiden, the Netherlands.

Grewal, K.S. 1932a. The effect of mitragynine on man. *Br J Med Psychol* 12: 41–58.

Grewal, K.S. 1932b. Observations on the pharmacology of mitragynine. *J Pharmacol Exp Ther* 46(3): 251–71.

Hanna, J. 2003. Bogus kratom market exposed. *Vernal Equinox* 12(1): 26–28.

Haviland, G.D. 1897. A revision of the tribe Naucleeae. *Bot J Linn Soc* 33: 1–94.

Hooper, D. 1907. The anti-opium leaf. *Pharm J* 78: 453.

Houghton, P.J., Latiff, A. and Said, I.M. 1991. Alkaloids from *Mitragyna speciosa*. *Phytochemistry* 30(1): 347–50.

Huysmans, S., Robbrecht, E. and E. Smets. 1994. Are the genera *Hallea* and *Mitragyna* (Rubiaceae-Coptosapelteae) pollen morphologically distinct? *Blumea* 39: 321–40.

Igoli, J.O., Ogaji, O.G., Tor-Anyiin, T.A. and Igoli, N.P. 2005. Traditional medicine practice amongst the Igede people of Nigeria. Part II. *Afr J Tradit Complement Altern Med* 2(2): 134–52.

India Biodiversity Portal. 2013. http: //indiabiodiversity.org/(accessed December 21, 2013).

Jansen, K.L.R. and Prast, C.J. 1988. Ethnopharmacology of kratom and the *Mitragyna* alkaloids. *J Ethnopharmacol* 23: 115–19.

J.M.H. 1907. Anti-opium plants. *Kew Bulletin of Miscellaneous Information* n.5: 198-99.

Johns, R.J. 1995. Malesia: An introduction. *Curtis's Bot Mag* 12(2): 52–62.

Junsongduang, A., Balslev, H., Inta, A., Jampeetong, A. and Wangpakapattanawong, P. 2014. Karen and Lawa medicinal plant use: Uniformity or ethnic divergence? *J Ethnopharmacology* 151(1): 517–27.

Kikura-Hanajiri, R., Kawamura, M., Maruyama, T., Kitajima, M., Takayama, H. and Goda, Y. 2009. Simultaneous analysis of mitragynine, 7-hydroxymitragynine, and other alkaloids in the psychotropic plant 'kratom' (*Mitragyna speciosa*) by LC-ESI-MS. *Forensic Toxicol* 27(2): 67–74.

Korthals, P.W. 1839. *Observationes de Naucleis indicis*. Typis Caroli Georgi.

Korthals, P.W. 1842. Kruidkunde. In *Verhandelingen over de Natuurlijke Geschiedenis der Nederlandsche Overzeesche Bezittingen. Botanie*, ed. C. J. Temminck, 1–259. Leiden.

Kowalczuk, A.P., Lozak, A., Stewart, J., Kiljan, M., Baran, P. and Fijalek, Z. 2012. Identification of popular psychoactive plants in their powdered form. *Planta Med* 78(14): PL34.

Kronstrand, R., Roman, M., Thelander, G., and Eriksson, A. 2011. Unintentional fatal intoxications with mitragynine and O-desmethyltramadol from the herbal blend Krypton. *J Anal Toxicol* 35(4): 242–47.

Lace, J.H. and Rodger, A. 1922. *List of Trees, Shrubs, and Principal Climbers, etc. Recorded from Burma: With Vernacular Names*. Rangoon: Superintendent, Government Printing.

Lemmens, R.H.M.J., Louppe, D. and Oteng-Amoako, A.A. (eds). 2012. Plant Resources of Tropical Africa 7(2). Timbers 2. Wagenigen, Netherlands: Prota Foundation.

Leroy J.-F. 1975. Taxogénétique dans le genre *Hallea* sur la sous-tribu des Mitragyninae (Rubiaceae-Naucleae). *Adansonia série 2* 15: 65–88.

Low, J. 1836. *A Dissertation on the Soil & Agriculture of the British Settlement of Penang, or Prince of Wales Island, in the Straits of Malacca: Including Province Wellesley on the Malayan Peninsula. With Brief References to the Settlements of Singapore & Malacca.* Printed at the Singapore Free Press Office.

Maruyama, T., Kawamura, M., Kikura-Hanajiri, R., Takayama, H. and Goda, Y. 2009. The botanical origin of kratom (*Mitragyna speciosa*; Rubiaceae) available as abused drugs in the Japanese markets. *J Nat Med* 63(3): 340–44.

McNeill, J., Barrie, F.R., Buck, W.R., Demoulin, V., Greuter, W., Hawksworth, D.L., Herendeen, P.S., et al. (eds). 2012. *International Code of Nomenclature for Algae, Fungi, and Plants (Melbourne Code).* [Regnum vegetabile no. 154.] Königstein: Koeltz Scientific Books.

Miquel, F.A.W. 1856. Flora van Nederlandsch Indië. Vol. 2. Amsterdam: Van der Post.

NMCD. 2011. Natural Medicines Comprehensive Database Consumer Version [Internet]. Yohimbe. [reviewed 2011 Sept 12] Stockton, Cal: Therapeutic Research Faculty, 1995. http: //www.nlm.nih.gov/medlineplus/druginfo /natural/759.html (accessed December 12, 2013)

Ng, F. 1989. *Tree Flora of Malaya: A Manual for Foresters.* Vol. 4. Kuala Lumpur: Longman Malaysia.

Nilus, R., Fah, L-Y. and Hastie, A. 2011. Species selection trial in burnt peat swamp vegetation in southwest coast of Sabah, Malaysia. *Proceedings of the International Symposium on Rehabilitation of Tropical Rainforest Ecosystems*, October 24–25 2011, Kuala Lumpur.

Nyein, N. 2013. Opium keeps Burma in international spotlight on World Drug Day. *The Irrawaddy* Wednesday, June 26, 2013. http: //www.irrawaddy .org/z_education/opium-keeps-burma-in-international-spotlight-on-world -drug-day.html (accessed December 25, 2013).

Nyemb, N. 2011. *Mitragyna ledermannii* (K.Krause) Ridsdale. In: Prota 7(2): Timbers/Bois d'œuvre 2. [CD-Rom], Lemmens, R.H.M.J., Louppe, D. and Oteng-Amoako, A.A. (eds). PROTA, Wageningen, Netherlands. http: //database.prota.org/PROTAhtml/Mitragyna%20ledermannii_En.htm (accessed December 24, 2013)

Oliver-Bever, B. 1986. *Medicinal Plants in Tropical West Africa.* Cambridge: Cambridge University Press.

Ong, H.C. and Nordiana, M. 1999. Malay ethno-medico botany in Machang, Kelantan, Malaysia. *Fitoterapia* 70(5): 502–13.

Osterhaus, H. 2008. Natural herb flavors student's day. The University Daily Kansan. 2008 Jan 23 [online] http: //kansan.com/archives/2008/01/23 /osterhaus-natural-herb-flavors-students-day/(accessed December 23, 2013)

Pelser, P.B., Barcelona, J.F. and Nickrent, D.L. (eds.). 2011 onwards. Co's Digital Flora of the Philippines. www.philippineplants.org (accessed December 24, 2013)

Prozialeck, W.C., Jivan, J.K. and Andurkar, S.V. 2012. Pharmacology of kratom: An emerging botanical agent with stimulant, analgesic and opioid-like effects. *J Am Osteopath Assoc* 112(12): 792–99.

Rätsch, C. 2005. *The Encyclopedia of Psychoactive Plants: Ethnopharmacology and Its Applications*. Rochester, VT: Park Street Press.

Razafimandimbison, S. G. and Bremer, B. 2001. Tribal delimitation of Naucleeae (Cinchonoideae, Rubiaceae): Inference from molecular and morphological data. *Syst. Geogr. Plants* 71(2): 515–38.

Razafimandimbison, S.G. and Bremer, B. 2002. Phylogeny and classification of Naucleeae s.l. (Rubiaceae) inferred from molecular (ITS, *rBCL*, and *tRNT-F*) and morphological data. *Am J Bot* 89(7): 1027–41.

Reanmongkol, W., Keawpradub, N. and Sawangjaroen, K. 2007. Effects of the extracts from *Mitragyna speciosa* Korth. leaves on analgesic and behavioral activities in experimental animals. *Songklanakarin J Sci Technol* 29(S1): 39–48.

Ridley, H.N. 1923. *Flora of the Malay Peninsula*, Vol. 2: Gamopetalae. London: Reeve & Co., Ltd.

Ridsdale, C.E. 1978. A revision of *Mitragyna* and *Uncaria* (Rubiaceae). *Blumea* 24: 43–100.

Sabetghadam, A., Navaratnam, V. and Mansor, S.M. 2012. Dose–response relationship, acute toxicity, and therapeutic index between the alkaloid extract of *Mitragyna speciosa* and its main active compound mitragynine in mice. *Drug Dev Res* 74(1): 23–30.

Said, L.M. and Zakaria, R. 1992. *An Updated List of Wetland Plant Species of Peninsular Malaysia with Particular Reference to Those Having Socio-Economic Value*. Asian Wetland Bureau Publication No 79. Kuala Lumpur: Asian Wetland Bureau.

Sakaran, R., Othman, F., Jantan, I., Thent, Z.C. and Das, S. 2014. Effect of subacute dose of *Mitragyna speciosa* Korth crude extract in female Sprague Dawley rats. *J Med Bioengineering* 3(2): 98-101.

Schultes, R.E. and Hofmann, A. 1973. *The Botany and Chemistry of Hallucinogens*. Springfield, Ill.: Charles C. Thomas.

Shellard, E.J. and Houghton, P.J. 1972a. The *Mitragyna* species of Asia. Part XX. The alkaloidal pattern in *Mitragyna parvifolia* (Roxb.) Korth. from Burma. *Planta Med* 21(3): 263–66.

Shellard, E.J. and Houghton, P.J. 1972b. The *Mitragyna* species of Asia. Part XXII. The distribution of alkaloids in young plants of *Mitragyna parvifolia* grown from seeds obtained from Uttar Pradesh State of India. *Planta Med* 22(5): 97–102.

Shellard, E.J., and Lala, P.K. 1978. The Alkaloids of *Mitragyna rubrostipulata* (Schum.) Havil. *Planta Med* 33(1): 63–69.

Steenis-Kruseman, M.J. van. 1950. Alphabetical List of Collectors, *Flora Malesiana*, Series 1, volume 1, pp. 5–606.

Steudel, E.G. 1841. *Nomenclator Botanicus seu: Synonymia Plantarum Universalis, Enumerans Ordine Alphabetico Nomina Atque Synonyma, tum generica tum specifica, et a Linnaeo et a recentioribus de re botanica scriptiribus plantis phanerogamis imposita. Editio secunda ex nova elaborata et aucta*. 2nd. ed. Pars II. Stuttgartiae et Tubingae: typis et sumptibus J.G. Cottae.

Sukrong, S., Zhu, S., Ruangrungsi, N., Phadungcharoen, T., Palanuvej, C. and Komatsu, K. 2007. Molecular analysis of the genus *Mitragyna* existing in Thailand based on rDNA ITS sequences and its application to identify a narcotic species: *Mitragyna speciosa*. *Biol Pharm Bull* 30(7): 1284–88.

Suwanlert, S. 1975. A study of kratom eaters in Thailand. *Bull Narc* 27: 21–27.

Takhtajan, A. 1986. *Floristic Regions of the World*. Berkeley: University of California Press.

Temme, O. 2011. Alkaloide der Pflanze *Mitragyna speciosa* in Kratomzubereitungen. Doktor der Wissenschaften in der Medizin (Dr. rer. med.) der Universitätsmedizin der Ernst-Moritz-Arndt-Universität Greifswald.

Toure, H., Balansard, G., Pauli, A.M., and Scotto, A.M. 1996. Pharmacological investigation of alkaloids from leaves of *Mitragyna inermis* (Rubiaceae). *J Ethnopharmacol* 54(1): 59–62.

Tungtananuwat, W. and Lawanprasert, S. 2010. Fatal 4x100: Home-made kratom juice cocktail. *J Health Res* 24(1): 43–47.

Uddin, S.B., Mahabub-Uz-Zaman, M., Akter, R. and Ahmed, N.U. 2009. Antidiarrheal activity of ethanolic bark extract of *Mitragyna diversifolia*. *Bangladesh J Pharmacol* 4: 144–46.

Valeton, T. 1925. Rubiaceae. *Nova Guinea* 14: 258.

Verdcourt, B. 1985. A new combination in *Hallea* (Rubiaceae-Cinchoneae). *Kew Bull* 40: 508.

Ward, J., Rosenbaum, C., Hernon, C., McCurdy, C. R. and Boyer, E.W. 2011. Herbal medicines for the management of opioid addiction. *CNS Drugs* 25(12): 999–1007.

Wray, L. 1906. 'Biak': An opium substitute. *Journal of the Federated Malay States Museum* 2(2): 53–56.

6 Phytochemistry of *Mitragyna speciosa*

Vedanjali Gogineni, Francisco Leon,
Bonnie A. Avery, Christopher McCurdy,
and Stephen J. Cutler

CONTENTS

6.1 INTRODUCTION

Mitragyna (Rubiaceae) is a small Afro-Asian genus comprising nine species: four African and five Asian. The genus *Mitragyna* consists of trees growing exclusively in swampy humid territories in the tropical and subtropical regions (Razafimandimbison and Bremer 2002). Some of these plants have been used for centuries by the local people as medicine for a variety of diseases such as fever, malaria, diarrhea, cough, and muscular pains (Shellard and Phillipson 1964). *M. speciosa, M. hirsuta, M. tubulosa, M. diversifolia, M. parvifolia,* and *M. rotundifolia* are commonly found in the Malay Peninsula (Sukrong et al. 2007). *M. speciosa* is known by various names, including "kratom" in Thailand (Burkill et al. 1935) and "biak biak" and "ketum" in Malaysia (Chan et al. 2005). The leaves of *M. speciosa* (Figure 6.1) have been used for centuries in Thailand for their known opioid-like effects (Vicknasingam et al. 2010), being chewed, smoked, or drunk as a tea. It has been used for the treatment of cough, hypertension, diarrhea, depression, analgesia, and fever reduction, and in relieving muscle pain associated with opiate withdrawal (Jansen and Prast 1988). The plant is known to possess unusual dual properties resulting in analgesia

FIGURE 6.1   Leaves of *Mitragyna speciosa*.

and stimulation and reports suggest that *M. speciosa* possesses antioxidant and antibacterial activities (Parthasarathy et al. 2009). However, due to its narcotic effects, *M. speciosa* was federally prohibited in Thailand in the 1940s and in Malaysia and Australia in 2003. Despite its listing as a drug of concern by the US Drug Enforcement Administration in 2005, kratom's popularity persists on the Internet and is now considered one of the top ten "legal highs" in the United States (Microgram Bulletin 2006).

Several natural products exist in various *Mitragyna* species, and extensive phytochemical investigations have been done regarding the natural occurrence of terpenoids, flavonoids, and alkaloids. The predominant compounds are indole alkaloids, mainly of the corynanthe type, including oxindole derivatives, which can occur in tetra- or penta-cyclic rings (Takayama et al. 2004a, 2005). Here, we summarize the scientific progress in determining all the known phytochemical compounds isolated from *M. speciosa*, including a special review on the constituents of all *Mitragyna* species.

6.2 CHEMICAL CONSTITUENTS

6.2.1 ALKALOIDS

Various methods have been developed for the extraction, isolation, and purification of alkaloids from *Mitragyna*. The most popular one is the extraction from the plant material by alcoholic (MeOH, EtOH, iProp, n-But) or alcoholic–water mixture by maceration, sonication, or the Soxhlet technique. The crude extract is further subjected to an acidic or basic extraction or acid-base partition to yield the alkaloid fraction. There are other techniques, such as ultrasound-assisted extraction (UAE), microwave-assisted extraction (MAE), and supercritical carbon-dioxide extraction (SFE-CO_2), which show an increase in the alkaloidal concentrations (Orio et al. 2012).

Extensive phytochemical investigations have revealed the natural occurrence of indole alkaloids in the genus *Mitragyna*. The major constituent of *M. speciosa* is mitragynine [1], an indole alkaloid that constitutes about 66% of the total alkaloidal content (Ponglux et al. 1994). The compound was first isolated and named by Field in 1921 (Field 1921), but not until 1965 was the structure elucidated (Beckett et al. 1965a) and its absolute stereochemistry confirmed by X-ray crystallographic analysis (Zacharias et al. 1965). The percentage of mitragynine [1] varies between younger and older plants, being much more abundant in older plants than the younger ones. The difference in the presence of the alkaloidal content in various species can be accounted for through environmental factors (León et al. 2009); moreover, with some alkaloids, the contents varied month to month (Shellard et al. 1978a). The content of mitragynine [1] present in a Thai specimen is around 66% higher than in a Malaysian specimen (Takayama et al. 1998). Similarly, the alkaloidal contents of *M. speciosa* in other locations such as London, England, are different, containing alkaloids previously not reported in the Southeast Asian specimens (Houghton et al. 1991). The predominant alkaloid in *M. speciosa* grown in the United States at the University of Mississippi is the oxindole-type mitraphylline [2] and not mitragynine [1] (Leon et al. 2009). Mitragynine [1] and 7-hydroxymitragynine [20] have shown high affinity towards opioid receptors (Takayama et al. 2004a). Both exhibit antinociceptive activity, but it was found that 7-hydroxymitragynine [20] showed 46-fold more and 13-fold greater potency than mitragynine [1] and morphine [56], respectively (Matsumoto et al. 2004; Adkins et al. 2011).

Several corynanthe-type indole alkaloids (Figure 6.2) have been isolated from the leaves of *M. speciosa*, including diastereoisomers of mitragynine [1]: speciogynine [3] (Beckett et al. 1965a), speciociliatine [4] (Shellard et al. 1978b), and mitraciliatine [5] (Shellard et al. 1978b). Indole alkaloids without the methoxyl group at C-9 are also common, which include corynantheidine [6] (Beckett et al. 1966) and 3-isocorynantheidine [7] (Shellard et al. 1978b). Some examples of the alkaloids with unsaturated carbons at C-18 are paynantheine [8] (Shellard et al. 1978b) and isopaynantheine [9] (Shellard et al. 1978b); at C-3 is 3-dehydromitragynine [10] (Houghton and Said 1986), and at C-3 and C-5 is 3,4,5,6-tetrahydromitragynine [11] (Takayama et al. 1998). The sulfonate derivative mitrasulgynine [12] (Takayama et al. 1998) is the first example of an indole alkaloid with a sulfonate group at the C-14 position. The oxidized derivatives in ring D generate an unusual skeleton, as seen in mitragynaline [13] (Takayama et al. 2001), corynantheidaline [14] (Takayama et al. 2001), corynantheidalinic acid [15] (Houghton et al. 1991), and mitragynalinic acid [16] (Houghton et al. 1991). The lactone derivatives include mitralactonal [17] (Takayama et al. 1998), mitralactonine [18] (Takayama et al. 1999), and 9-methoxymitralactonine [19] (Takayama et al.

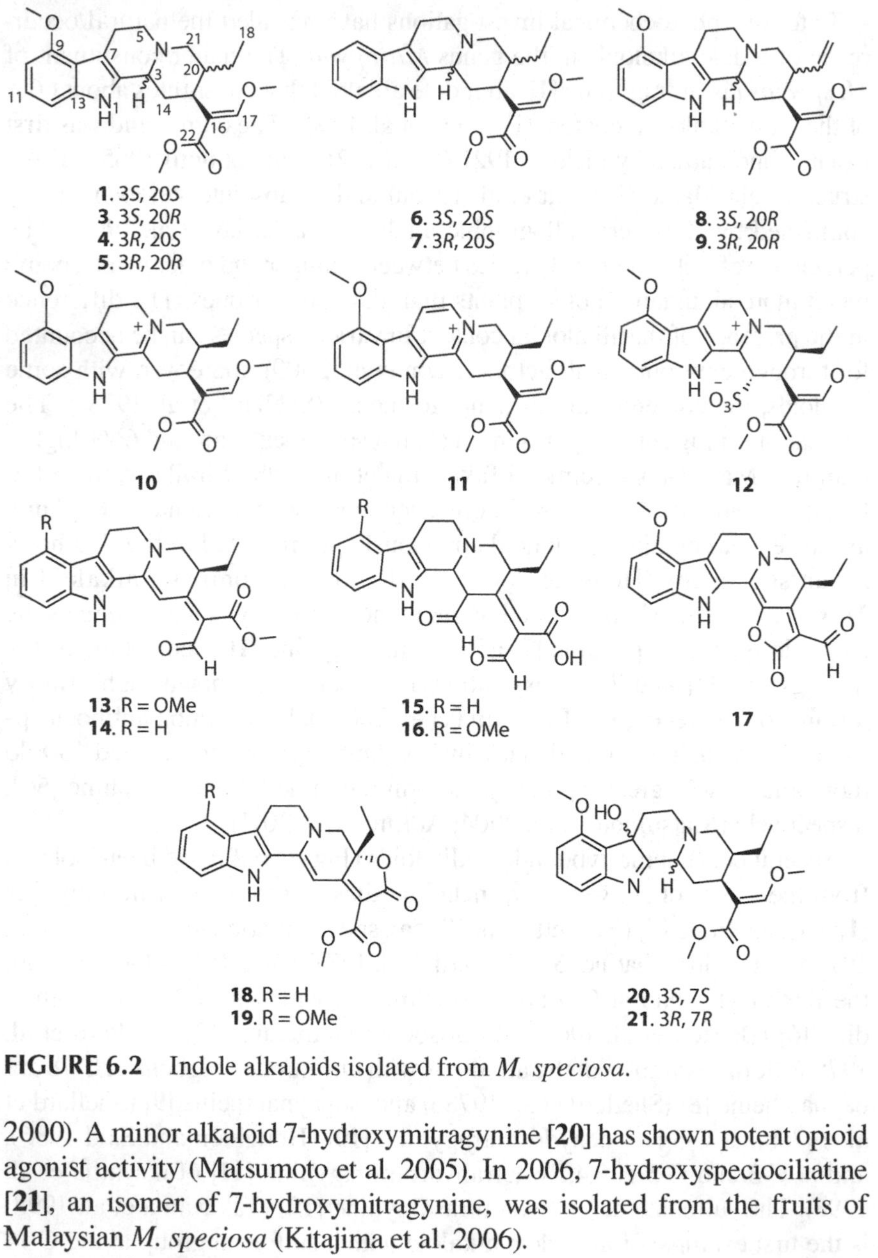

1. 3S, 20S
3. 3S, 20R
4. 3R, 20S
5. 3R, 20R

6. 3S, 20S
7. 3R, 20S

8. 3S, 20R
9. 3R, 20R

10

11

12

13. R = OMe
14. R = H

15. R = H
16. R = OMe

17

18. R = H
19. R = OMe

20. 3S, 7S
21. 3R, 7R

FIGURE 6.2 Indole alkaloids isolated from *M. speciosa*.

2000). A minor alkaloid 7-hydroxymitragynine [**20**] has shown potent opioid agonist activity (Matsumoto et al. 2005). In 2006, 7-hydroxyspeciociliatine [**21**], an isomer of 7-hydroxymitragynine, was isolated from the fruits of Malaysian *M. speciosa* (Kitajima et al. 2006).

In our investigations of the secondary metabolites from the leaves of *M. speciosa* growing in the gardens of the University of Mississippi, we showed that the predominant type of alkaloids are oxindole alkaloids, with mitraphylline [**2**] as the predominant one at 45% of the total alkaloids isolated. We suggested that our US-grown *M. speciosa* displayed a chemotype different from that of the Asian-African plants (León et al. 2009). Along with mitraphylline,

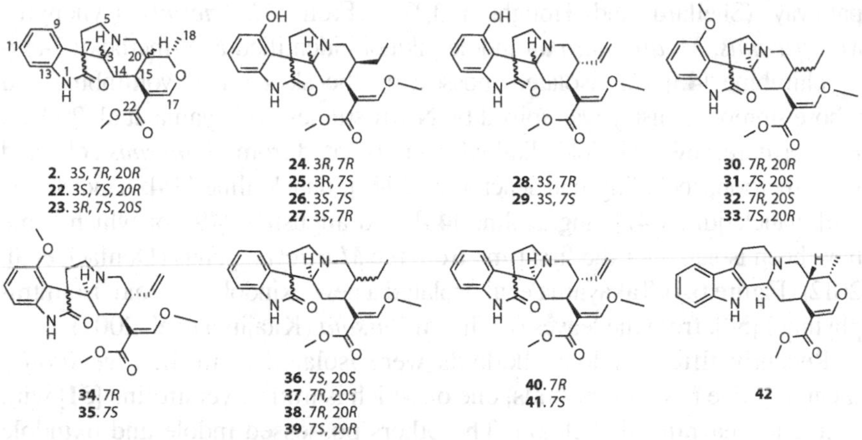

FIGURE 6.3 Oxindole and indole alkaloids isolated from *M. speciosa*.

several oxindole alkaloids were described from *M. speciosa* (Figure 6.3), including isomitraphylline [22] (Beckett et al. 1966) and speciophylline [23] (Beckett et al. 1966). The C-9 hydroxy isomers are speciofoline [24] (Beckett et al. 1965b), isospeciofoline [25] (Hemingway et al. 1975), mitrafoline [26] (Hemingway et al. 1975), and isomitrafoline [27] (Hemingway et al. 1975), as well as rotundifoleine [28] (Hemingway et al. 1975) and isorotundifoleine [29] (Hemingway et al. 1975). The C-9 methoxy derivatives include ciliaphylline [30] (Shellard et al. 1978b), mitragynine oxindole A [31] (Shellard et al. 1978b), mitragynine oxindole B [32] (Shellard et al. 1978b), rhynchociline [33] (Shellard et al. 1978b), specionoxeine [34] (Trager et al. 1968), and isospecionoxeine [35] (Trager et al. 1968).

Corynoxine [36] (Shellard et al. 1978b), isocorynoxine [37] (Shellard et al. 1978b), rhynchophylline [38] (Shellard et al. 1978b), isorhynchophylline [39] (Shellard et al. 1978b), corynoxeine [40] (Kitajima et al. 2006), and isocorynoxeine [41] (Kitajima et al. 2006) are some of the other alkaloids found in *M. speciosa*.

The heteroyohimbine-type alkaloid is not common in *M. speciosa*; however, this heterocyclic system is common in other *Mitragyna* species. The compound ajmalicine [42] has been isolated from *M. speciosa* (Beckett et al. 1965b) and is known to be used as an α-adrenergic receptor antagonist; it is marketed under such names as circolene and lamuran (Schmeller and Wink 1998).

In the 1960s and 70s, the Shellard group reported extensive investigations on the presence of alkaloids, mainly of the indole and oxindole types, from the African-Asian *Mitragyna* species (Shellard and Houghton 1974; Shellard et al. 1977, 1983) as well as biogenetic studies supporting the idea of the conversion of indole to oxindole alkaloids in *M. parvifolia* as part of their biosynthesis pathway (Shellard and Houghton 1973). From *M. inermis* (synonyms *M. africanus*, *M. africana*) an indole alkaloid identified as 9-methoxy-3-*epi*-α-yohimbine [43] was isolated, possessing the skeleton of yohimbine and whose stereochemistry was solved by NMR studies (Takayama et al. 2004b). Furthermore, several indole alkaloids were isolated from *M. inermis* collected in Cameroon, including naucleactonin D [44], nauclefiline [45], nauclefidine [46], nauclefidine [47], angustoline [48], and angustine [49], of which some have been isolated for the first time from the *Mitragyna* genus (Donfack et al. 2012) (Figure 6.4). Takayama et al. isolated a new oxindole alkaloid, isomitraphyllinol [50], from the leaves of Thai *M. hirsuta* (Kitajima et al. 2007).

Recently fifteen indole alkaloids were isolated from *M. diversifolia*, including five new compounds, one of which is mitradiversifoline [51] with a unique, rearranged skeleton. The others possessed indole and oxindole skeletons, including specionoxeine-*N*(4)-oxide [52], 7-hydroxyisopaynan-theine [53], 3-dehydropaynantheine [54], and 3-isopaynantheine-*N*(4)-oxide [55] (Cao et al. 2013) (Figure 6.5).

The biosynthesis of mitragynine [1] (Figure 6.6) proceeds in the typical indole alkaloid pathway where tryptamine and the monoterpenoid gluco-side secologanin are the building blocks. Tryptamine is produced by the shikimate pathway with the amino acid tryptophan as an intermediate, while secologanin is derived from loganin via the methyl-erythritol phos-phate pathway (MEP) (Ndagijimana et al. 2013; O'Connor et al. 2006). The condensation of tryptamine and secologanin is regulated by the enzyme strictosidine synthase (STR1) to generate strictosidine. This enzyme plays a primary role in generating the entire monoterpenoid indole alkaloidal family, which includes several skeletons such as corynanthe, quinoline,

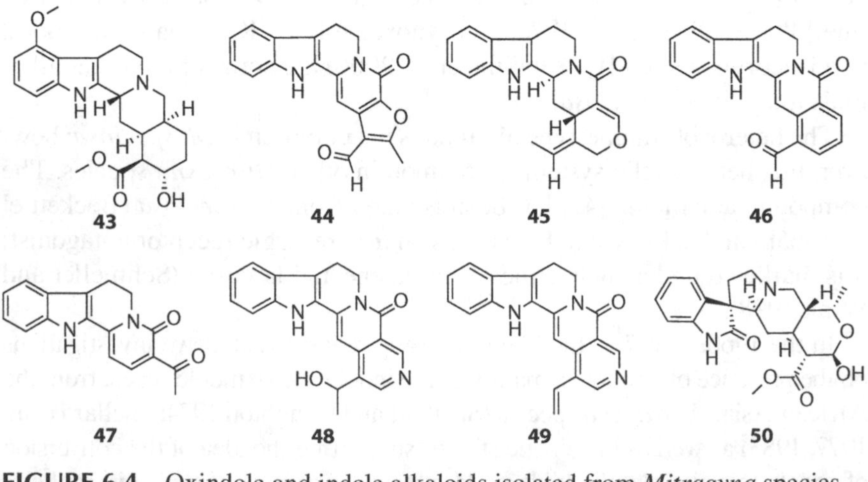

FIGURE 6.4 Oxindole and indole alkaloids isolated from *Mitragyna* species.

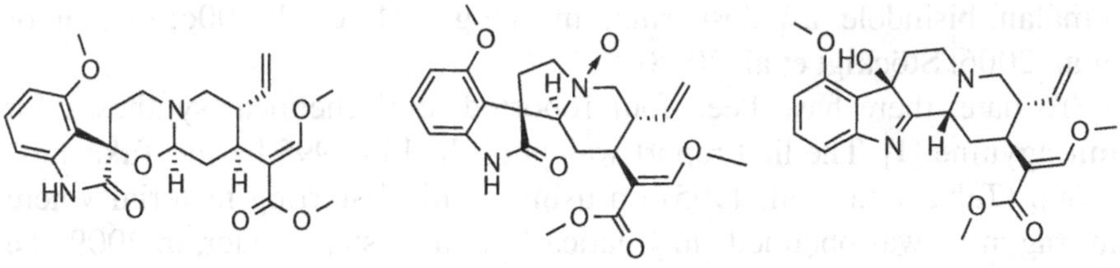

51 52 53

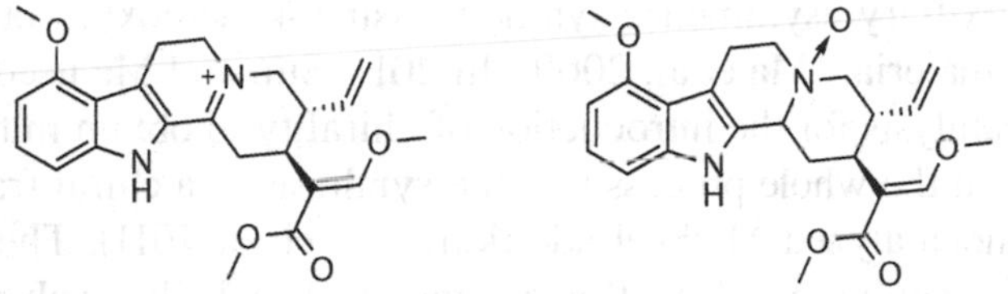

54 55

FIGURE 6.5 Oxindole and indole alkaloids isolated from *M. diversifolia*.

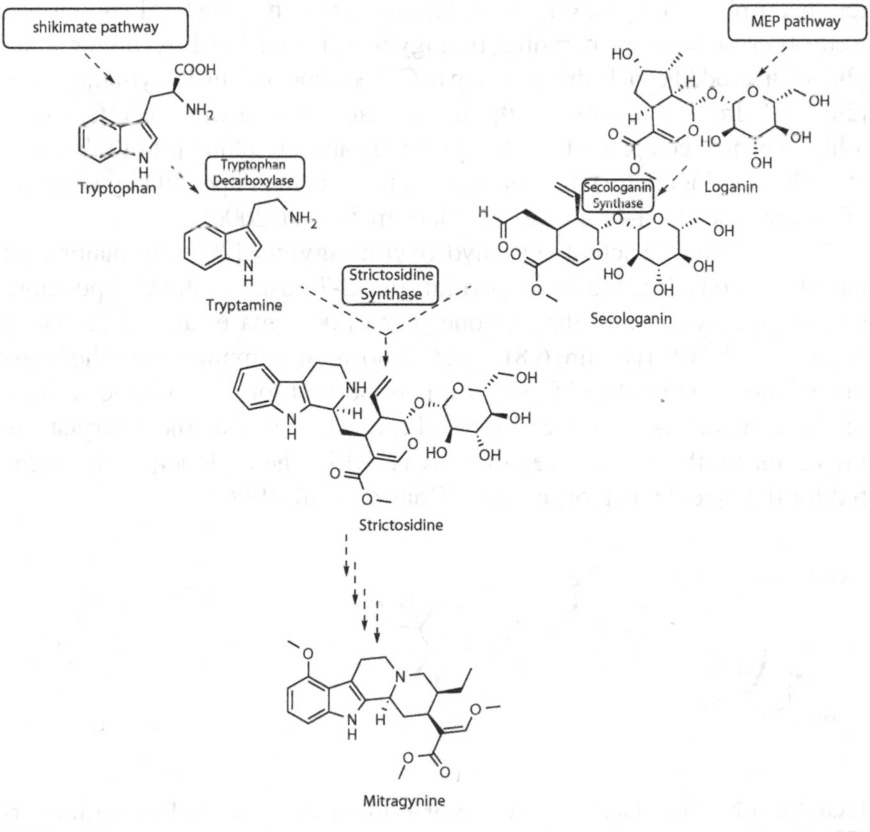

FIGURE 6.6 Biosynthesis of mitragynine.

ajmalan, bisindole, aspidosperma, and iboga (Ma et al. 2006; O'Connor et al. 2006; Stöckigt et al. 2008).

To date, there have been four reported total chemical syntheses for mitragynine [1]. The first report was published in 1995 by the Takayama group (Takayama et al. 1995) on using a chiral starting material where mitragynine was obtained enzymatically in nine steps. Later, in 2009, the Cook group reported the total synthesis of mitragynine in 22 steps via chiral-auxiliary asymmetric synthesis using 4-methoxytryptophan as the starting material (Ma et al. 2009). In 2011, Sun and Ma used asymmetric organocatalysis for the introduction of chirality to obtain mitragynine [1]. The key in the whole process was the synthesis of a chiral fragment using an organocatalyzed Michael addition (Sun et al. 2011). The most recent synthesis was reported by Kerschgens et al., which involved nine steps using an enantioselective Pictet–Spengler reaction catalyzed by chiral thiourea with 4-methoxytryptamine as a starter (Kerschgens et al. 2012).

With respect to the structure–activity relationship (SAR), the three major functional sites considered important for opioid receptor binding are a benzene residue, a phenolic hydroxyl, and a tertiary nitrogen. Structural similarities were observed between morphine, mitragynine [1], and 7-hydroxymitragynine [20], with an additional hydroxyl group at C-7 position on 7-hydroxymitragynine [20]. 7-Hydroxymitragynine [20] showed rapid antinociceptive effect with a high potency compared to mitragynine [1] and morphine [56] (Takayama et al. 2004a) (Figure 6.7) that could result from its high lipophilicity and ease of crossing the blood–brain barrier (Matsumoto et al. 2006).

To determine the activity of 7-hydroxymitragynine [20], four major sites have been considered to be important: the C-7 position, the C-9 position, β-methoxyacrylate, and the N_b lone pair (Takayama et al. 2002; Taufik Hidayat et al. 2010) (Figure 6.8). The C-7 position is important for the binding affinities, while the C-9 position is important for the intrinsic activity of the compounds. Modification of the residue of β-methoxyacrylate is important for the opioid receptor activity, while the N_b lone pair is essential for the opioid receptor binding (Dhawan et al. 1996).

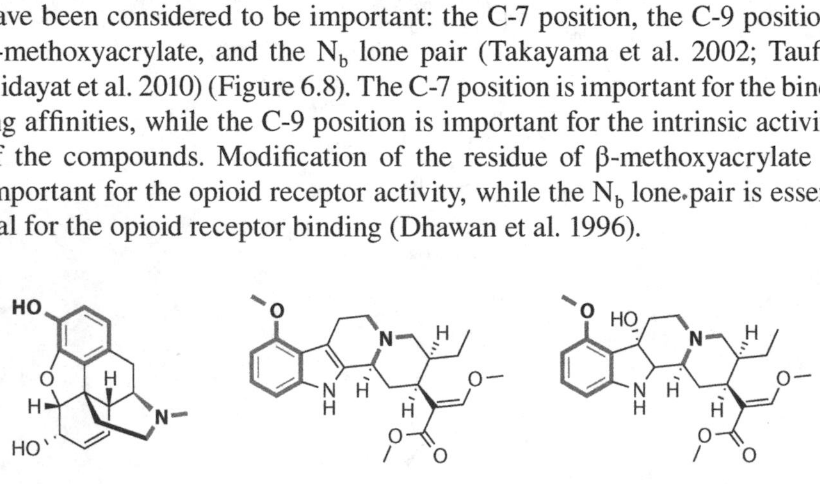

56   1   20

FIGURE 6.7   Functional sites responsible for opioid receptor binding in morphine, mitragynine, and 7-hydroxymitragynine.

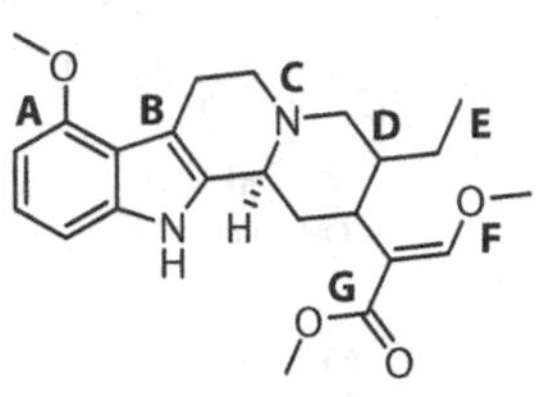

A. Presence of long ethers, absence of $-CH_3$ and $-OH$ or acetylation could lead to decrease in activity.
B. Presence of $-OH$ improves activity.
C. Oxidation of nitrogen eliminates activity (N-oxidation).
D. Presence of double bond reduces activity.
E. Changing stereochemistry eliminates activity.
F. Replacement with a $-NH$ group eliminates activity.
G. Activity is decreased when ester is hydrolyzed or alcohol is reduced.

FIGURE 6.8   SAR in mitragynine.

6.2.2 FLAVONOIDS AND POLYPHENOLIC COMPOUNDS

Although some studies showed the presence of phenolic compounds, including flavonoids, in the leaves of *M. speciosa* (Figure 6.9) (Parthasarathy et al. 2009), there are only a few reports that mention the isolation or structural elucidation of these compounds. Flavonoids such as apigenin [57] and its 7-glycosides [58, 59] have been identified from *M. speciosa* leaves (Hinou and Harvala 1988) along with flavonol derivatives such as quercetin [60] and its glycosides: quercitrin [61], rutin [62], isoquercitrin [63], hyperoside [64] and quercetin-3-galactoside-7-rhamnoside [65], kaempferol [66], and its 3-glucoside derivative [67] (Harvala and Hinou 1988), and epicatechin [68]. In relation to the phenolic compounds, caffeic acid [69] and chlorogenic acid [70] were isolated (Hinou and Harvala 1988) along with 1-*O*-feruloyl-β-D-glucopyranoside [71] and benzyl-β-D-glucopyranoside [72].

With respect to other *Mitragyna* species (Figure 6.10), such as *M. rotundifolia*, several phenolic compounds have been isolated and their antioxidant activity studied. The compounds isolated include 3,4-dihydroxybenzoic acid [73], catechin [74], caffeic acid [69], epicatechin [68], kaempferol [66], 4'-*O*-methyl-gallocatechin [75], 4-hydroxy-3-methyloxybenzoic acid [76], and 3-hydroxy-4-methyloxybenzoic acid [77] (Kang et al. 2006, 2010; Kang and Li 2009). Quercetin [60] has been isolated from *M. inermis* (Asase et al. 2008), while dihydrodehydrodiconiferyl alcohol [78], isolariciresinol [79], and isolariciresinol-3α-*O*-β-D-glucopyranoside [80] were isolated from *M. africanus* synonym for *M. inermis* (Takayama et al. 2004b).

6.2.3 TRITERPENOIDS AND TRITERPENOID SAPONINS

Pentacyclic triterpenoids of ursane, oleane, and taraxane skeletons are usually isolated from plants of *Mitragyna* species. Ursolic acid [81] and oleanoic acid [82] were isolated from *M. speciosa* root cultures infected with *Agrobacterium rhizogenes*, and the plants generated threefold more mitragynine than normal plants (Phongprueksapattana et al. 2008). The

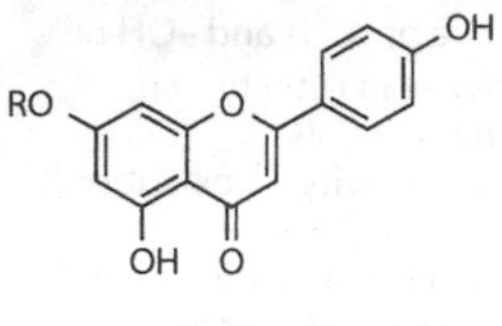

57. R = H
58. R = β-D-Glcp
59. R = β-D-Rhap

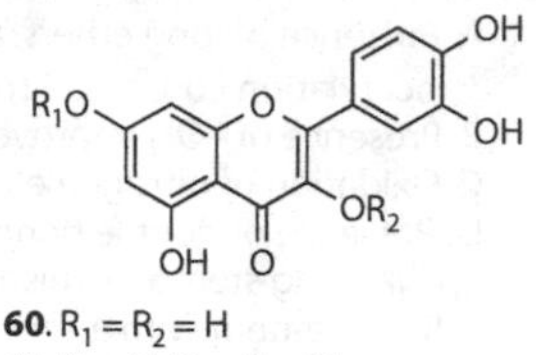

60. $R_1 = R_2 = H$
61. $R_1 = H, R_2 = $ β-D-Rhap
62. $R_1 = H, R_2 = $ α-L-Rhap`
    (1→6)β-D-Glcp
63. $R_1 = H, R_2 \equiv $ β-D-Glcp
64. $R_1 = H, R_2 = $ β-D-Galp
65. $R_1 = $ β-D-Galp, $R_2 = $ β-D-Rhap

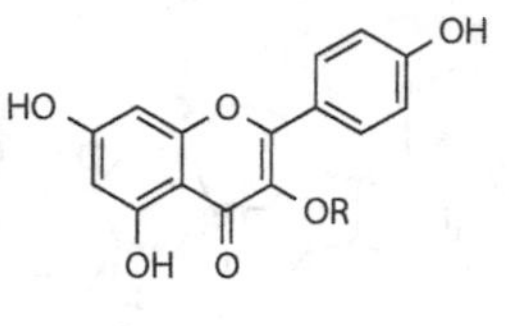

66. R = H
67. R = β-D-Glcp

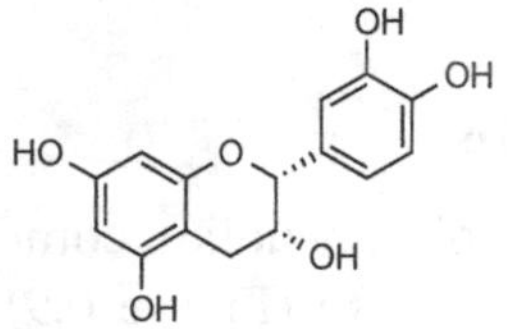

68 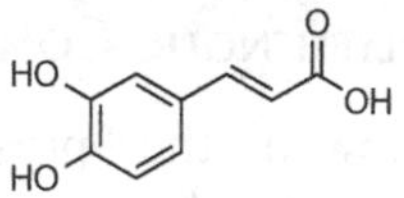   **69** 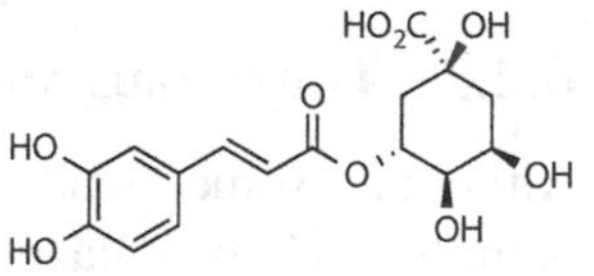   **70**

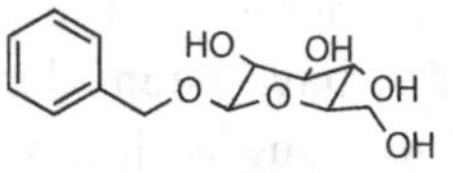

71 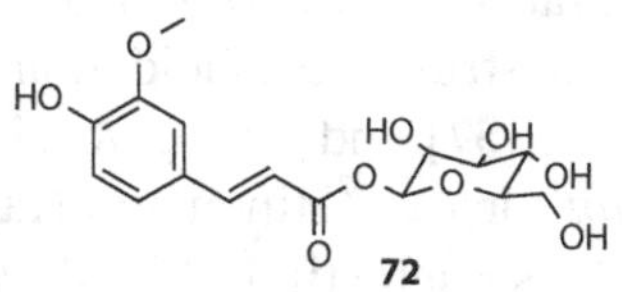   **72**

FIGURE 6.9   Polyphenolic compounds isolated from *M. speciosa*.

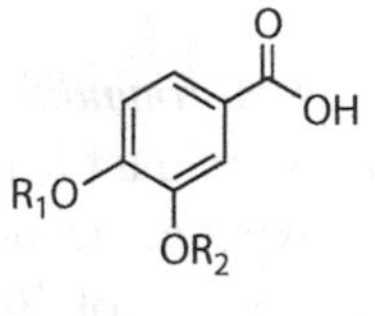

73. $R_1 = R_2 = H$
76. $R_1 = H; R_2 = $ Me
77. $R_1 = $ Me; $R_2 = H$

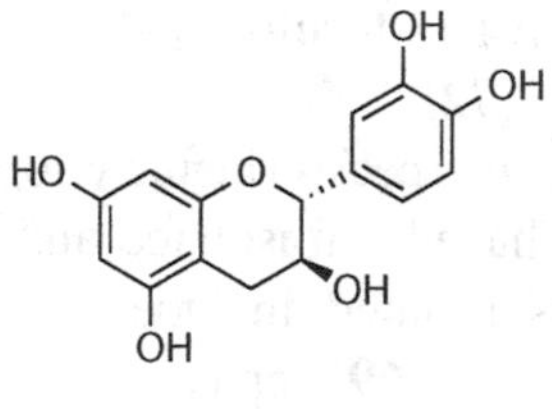

74 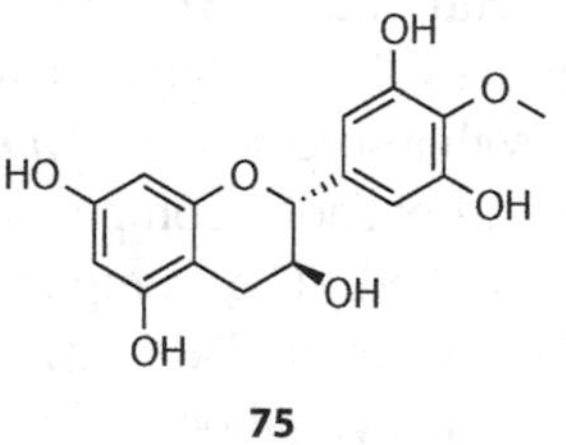   **75**

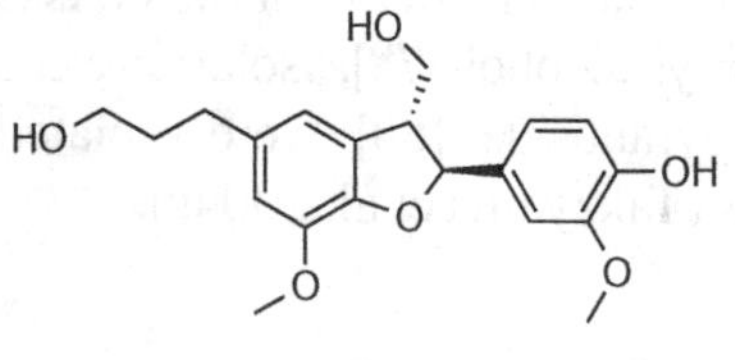

78

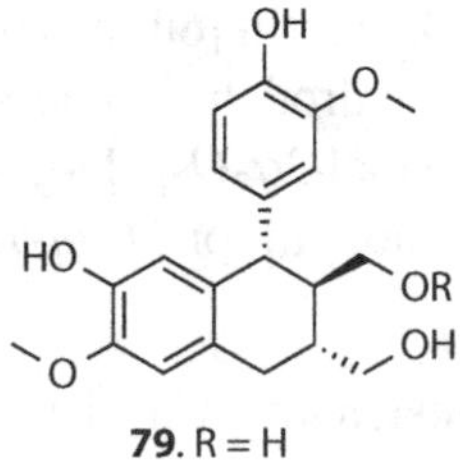

79. R = H
80. R = β-D-Glcp

FIGURE 6.10   Polyphenolic compounds isolated from *Mitragyna* species.

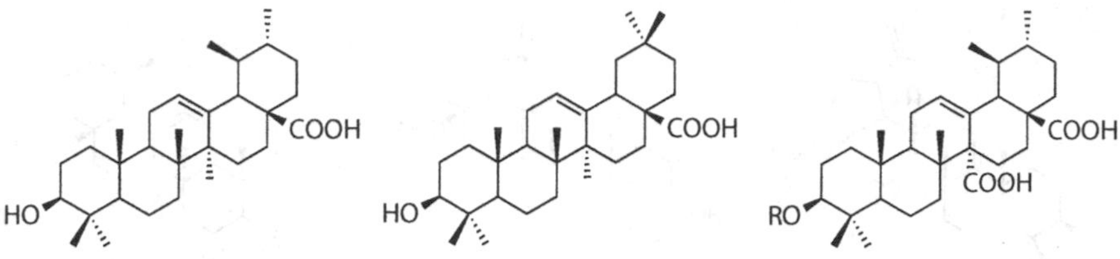

81   **82**   **83.** R = 6-deoxy-β-ᴅ-Glup
                                84. R = β-ᴅ-Glup

FIGURE 6.11   Triterpenoids and triterpenoid saponins isolated from *Mitragyna* species.

triterpenoid saponins quinovic acid 3-*O*-β-D-quinovopyranoside [**83**] and quinovic acid 3-*O*-β-D-glucopyranoside [**84**] were isolated from the leaves of *M. speciosa* grown in the United States (Figure 6.11).

In other *Mitragyna* species, the triterpenoids and triterpenoid saponins are the predominant compounds compared to the indole alkaloids; for example the triterpenes ursolic acid [**81**], oleanoic acid [**82**], betulinic acid [**85**], barbinervic acid [**86**], and the triterpene saponin quinovic acid 3-*O*-α-L-rhamnopyranoside [**87**] were isolated from the roots and fruits of *M. inermis* (Donfack et al. 2012). The triterpenoid saponins quinovic acid 3β-β-D-glucopyranosyl-(1→4)-α-L-rhamnopyranosyl-28-*O*-β-D-glucopyranoside [**88**], quinovic acid 3β-*O*-β-D-quinovopyranoside [**83**], and quinovic acid 3β-*O*-β-D-glucopyranoside [**84**] were obtained from the bark of *M. inermis* (Cheng et al. 2002a). Two new nor-triterpenoid glycosides, inermiside I [**89**] and inermiside II [**90**], along with quinovic acid [**91**], 3-oxoquinovic acid [**92**], 3-*O*-[β-D-glucopyranosyl-(1→4)-α-L-rhamnopyranosyl]-quinovic acid [**93**], quinovic acid 3-*O*-β-D- glucopyranosyl-28-*O*-β-D-glucopyranoside [**94**], quinovic acid 3-*O*-α-L-rhamnopyranoside [**87**], quinovic acid 3-*O*-β-D-quinovopyranosyl-28-*O*-β-D-glucopyranoside [**95**], and quinovic acid 3-*O*-β-D-quinovopyranoside [**83**], were obtained from the bark and stem of *M. inermis* (Cheng et al. 2002b; Bishay et al. 1988; Takayama et al. 2004b) (Figure 6.12).

M. rotundifolia has been reported as a good source of triterpenoid saponins with the aglycone quinovic and cincholic acids, thus quinovic acid 3-*O*-β-D-6-deoxyglucopyranoside 28-*O*-β-D-glucopyranosyl [**96**], quinovic acid 27-*O*-α-L-rhamnopyranosyl [**97**], quinovic acid 3-*O*-α-L-rhamnopyranoside [**87**], quinovic acid 27-*O*-β-D-glucopyranosyl [**98**], quinovic acid 3-*O*-β-D-quinovopyranoside [**83**], quinovic acid 27-*O*-β-D-6-deoxyglucopyranosyl [**99**], cincholic acid 3-*O*-β-D-6-deoxy-glucopyranoside [**100**], cincholic acid 3-*O*-β-D-6-deoxyglucopyranoside [**101**], and cincholic acid 28-*O*-β-D-glucopyranosyl [**102**] were isolated (Kang et al. 2006) (Figure 6.13).

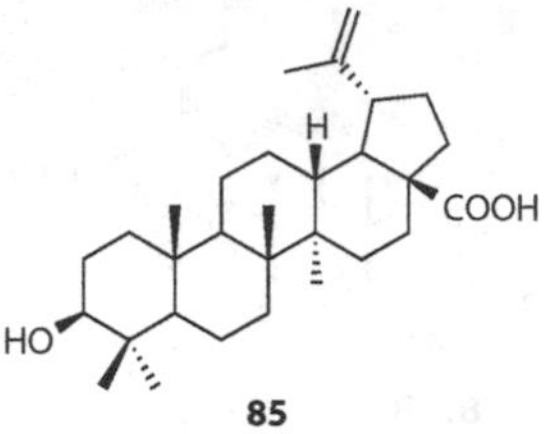

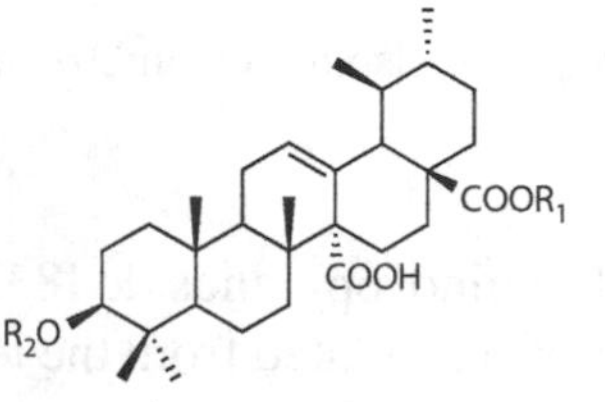

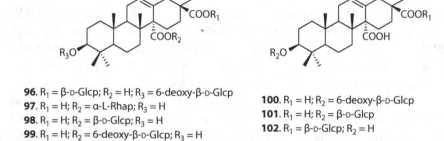

87. R$_1$ = H; R$_2$ = α-L-Rhap
88. R$_1$ = β-D-Glcp; R$_2$ = β-D-Glcp–(1→4)-α-L-Rhap
93. R$_1$ = H; R$_2$ = β-D-Glcp–(1→4)–α-L-Rhap
94. R$_1$ = R$_2$ = β-D-Glcp
95. R$_1$ = β-D-Glcp; R$_2$ = 6-deoxy-β-D-Glcp

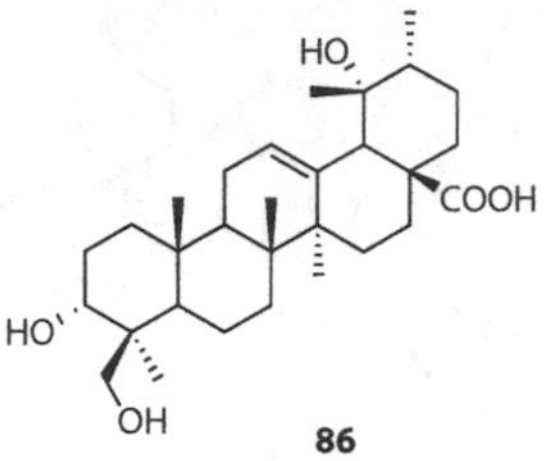

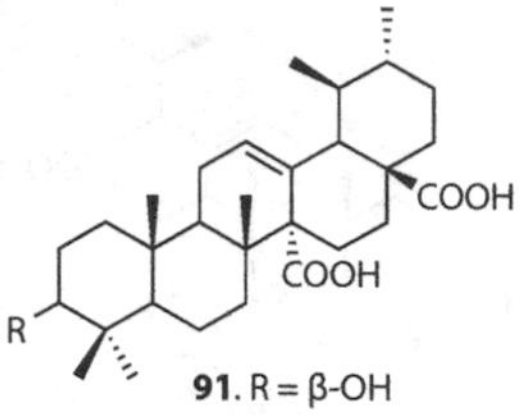

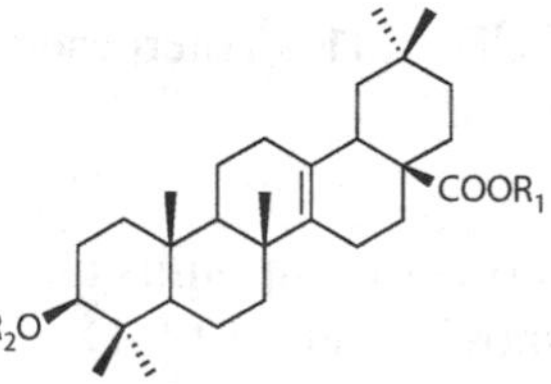

89. R$_1$ = 6-deoxy-β-D-Glcp; R$_2$ = β-D-Glcp–(1→6)
                                             –β-D-Glcp
90. R$_1$ = H; R$_2$ = 6-deoxy-β-D-Glcp

FIGURE 6.12   Triterpenoids and triterpenoid saponins isolated from *M. inermis*.

96. R$_1$ = β-D-Glcp; R$_2$ = H; R$_3$ = 6-deoxy-β-D-Glcp
97. R$_1$ = H; R$_2$ = α-L-Rhap; R$_3$ = H
98. R$_1$ = H; R$_2$ = β-D-Glcp; R$_3$ = H
99. R$_1$ = H; R$_2$ = 6-deoxy-β-D-Glcp; R$_3$ = H

100. R$_1$ = H; R$_2$ = 6-deoxy-β-D-Glcp
101. R$_1$ = H; R$_2$ = β-D-Glcp
102. R$_1$ = β-D-Glcp; R$_2$ = H

FIGURE 6.13   Triterpenoid saponins isolated from *M. rotundifolia*.

From the stem and bark of *M. stipulosa*, ursolic acid [81], quinovic acid [91], quinovic acid 3-*O*-β-D-glucopyranoside [84], quinovic acid 3-*O*-[(2-*O*-sulfo)-β-D-quinovopyranoside [103], and quinovic acid 3-*O*-β-D-6-deoxyglucopyranoside 27-*O*-β-D-glucopyranoside [104] were isolated (Tapondjou et al. 2002) (Figure 6.14). Compounds [91] and [104] were known to be the first examples of natural snake venom phosphodiesterase (PDE) I inhibitors (Fatima et al. 2002). PDEs are located

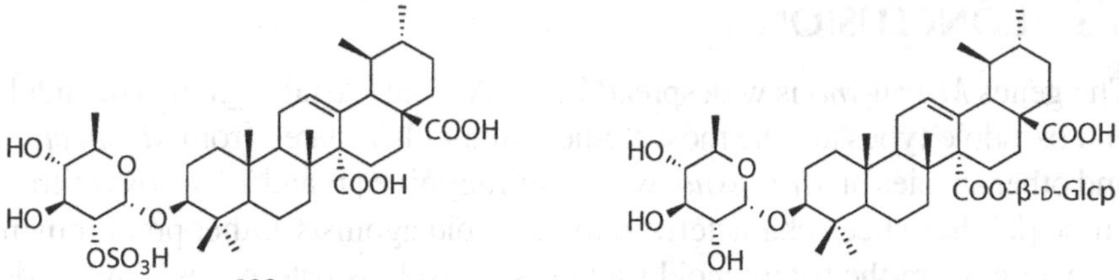

103 104

FIGURE 6.14 Triterpenoids saponins isolated from *M. stipulosa*.

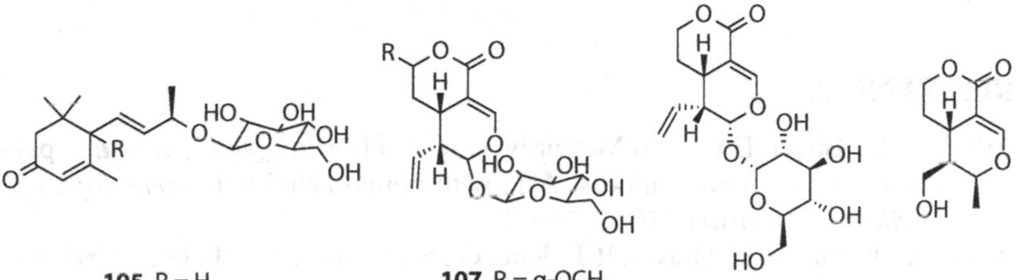

105. R = H
106. R = α-OH

107. R = α-OCH$_3$
108. R = β-OCH$_3$

109

110

FIGURE 6.15 Monoterpenes and secoiridoids isolated from *M. speciosa* and *M. inermis*.

principally in the cytosol or in the subcellular compartments, and they are responsible for the degradation of the phosphodiester bond in cyclic adenosine monophosphate (cAMP) and cyclic guanosine monophosphate (cGMP), leading to hydrolysis resulting in the formation of 5'-mononucleotides. PDEs have been used as tools for sequence studies in nucleic acids (Cheepala et al. 2013).

6.2.4 Miscellaneous Compounds

Minor compounds such as the monoterpenes 3-oxo-α-ionyl-*O*-β-D-glucopyranoside [**105**] and roseoside [**106**] and secoiridoid glycosides such as vogeloside [**107**] and epivogeloside [**108**] are present in the leaves of *M. speciosa*, while two secoiridoids, sweroside [**109**] and dihydroepinaucledal [**110**] are present in *M. inermis* (Figure 6.15)

Sitosterol, stigmasterol, and daucosterol are some of the common phytosterols that are reported in many species of *Mitragyna*, such as *M. speciosa* (Phongprueksapattana et al. 2008), *M. rotundifolia* (Kang et al. 2006b), *M. inermis* (Donfack et al. 2012), and *M. stipulosa* (Fatima et al. 2002).

6.3 CONCLUSION

The genus *Mitragyna* is widespread in the African-Asian region. The indole and oxindole types are the most studied alkaloids isolated from *M. speciosa* and other species of *Mitragyna*, where mitragynine [1] and 7-hydroxymitragynine [20] have been characterized to be opioid agonists. Other predominant compounds are the triterpenoid saponins, as well as triterpenes, flavonoids, steroids, monoterpenes, and secoiridoids. The present chapter shows the majority of the compounds isolated from *Mitragyna*. But the structures of very common compounds such as steroids have not been included.

REFERENCES

Adkins, J.E., Boyer, E.W., and McCurdy, C.R. 2011. *Mitragyna speciosa*, a psychoactive tree from Southeast Asia with opioid activity. *Current Topics in Medicinal Chemistry* 11(9): 1165–75.

Asase, A., Kokubun, T., Grayer, R.J., Kite, G., Simmonds, M.S.J., Oteng-Yeboah, A.A., and Odamtten, G.T. 2008. Chemical constituents and antimicrobial activity of medicinal plants from Ghana: *Cassia sieberiana, Haematostaphis barteri, Mitragyna inermis* and *Pseudocedrela kotschyi*. *Phytotherapy Research* 22(8): 1013–16.

Beckett, A.H., Shellard, E.J., and Tackie, A.N. 1965a. The *Mitragyna* species of Asia. Part IV: The alkaloids of the leaves of *M. speciosa* Korth.: Isolation of mitragynine and speciofoline. *Planta Medica* 13(2): 241–45.

Beckett, A.H., Shellard, E.J., Phillipson, J.D., and Lee, C.M. 1965b. Alkaloids from *Mitragyna speciosa* (Korth.). *Journal of Pharmacy and Pharmacology* 17(11): 753–55.

Beckett, A.H., Shellard, E.J., Phillipson, J.D., and Lee, C.M. 1966. The *Mitragyna* species of Asia. VI. Oxindole alkaloids from the leaves of *Mitragyna speciosa* Korth. *Planta Medica* 14(3): 266–76.

Bishay, D.W., Che, C.T., Gonzalez, A., Pezzuto, J.M., Kinghorn, Douglas A., and Farnsworth, N.R. 1988. Further chemical constituents of *Mitragyna inermis* stem bark. *Fitoterapia* 59(5): 397–98.

Burkill, J.H., Birtwistle, W., Foxworthy, F.W., Scrivenor, J.B., and Watson, J.G. 1935. *A Dictionary of the Economic Products of the Malay Peninsula*. Vol 2: 1480–83. London, England: Crown Agents for the Colonies.

Cao, X.F., Wang, J.S., Wang, X.B., Luo, J., Wang, H.Y., and Kong, L.Y. 2013. Monoterpene indole alkaloids from the stem bark of *Mitragyna diversifolia* and their acetylcholine esterase inhibitory effects. *Phytochemistry* 96: 389–96.

Chan, K.B., Pakiam, C., and Rahim, R.A. 2005. Psychoactive plant abuse: The identification of mitragynine in ketum and in ketum preparations. *Bulletin on Narcotics* 57(1–2): 249–56.

Cheepala, S., Hulot, J.S., Morgan, J.A., Sassi, Y., Zhang, W., Naren, A.P., and Schuetz, J.D. 2013. Cyclic nucleotide compartmentalization: Contributions of phosphodiesterases and ATP-binding cassette transporters. *The Annual Review of Pharmacology and Toxicology* 53: 231–53

Cheng, Z., Yu, B., and Yang, X. 2002a. 27-nor-triterpenoid glycosides from *Mitragyna inermis*. *Phytochemistry* 61(4): 379–82.

Cheng, Z., Yu, B., Yang, X., and Zhang, J. 2002b. Triterpenoid saponins from the bark of *Mitragyna inermis*. *Zhongguo Zhongyao Zazhi* 27(4): 274–77.

Dhawan, B. N., Cesselin, F., Raghubir, R., Reisine, T., Bradley, P.B., Portoghese, P.S., and Hamon, M. 1996. International Union of Pharmacology. XII. Classification of opioid receptors. *Pharmacological Reviews* 48(4): 567–92.

Donfack, E.V., Lenta, B.N., Kongue, M.D.T., Fongang, Y.F., Ngouela, S., Tsamo, E., Dittrich, B., and Laatsch, H. 2012. Naucleactonin D, an indole alkaloid and other chemical constituents from roots and fruits of *Mitragyna inermis*. *Zeitschrift für Naturforschung B-Journal of Chemical Sciences* 67(11): 1159–65.

Fatima, N., Tapondjou, L.A., Lontsi, D., Sondengam, B.L., Atta-Ur-Rahman, and Choudhary, I.M. 2002. Quinovic acid glycosides from *Mitragyna stipulosa*: First examples of natural inhibitors of snake venom phosphodiesterase I. *Natural Product Letters* 16(6): 389–93.

Field, E. 1921. Mitragynine and mitraversine, two new alkaloids from species of mitragyne. *Journal of the Chemical Society, Transactions* 119: 887–91.

Harvala, C., and Hinou, J. 1988. Flavonol derivatives from the leaves of *Mitragyna speciosa*. *Pharmazie* 43(5): 372.

Hemingway, S.R., Houghton, P.J., Phillipson, D.J., and Shellard, E.J. 1975. 9-Hydroxyrhynchophylline-type oxindole alkaloids. *Phytochemistry* 14(2): 557–63.

Hidayat, T., Apryani, E.M., Nabishah, B.M., Moklas, M.A.A., Sharida, F., and Farhan, M.A. 2010. Determination of mitragynine bound opioid receptors. *Advances in Medical and Dental Sciences* 3(3): 65–70.

Hinou, J., and Harvala, C. 1988. Polyphenolic compounds from the leaves of *Mitragyna speciosa*. *Fitoterapia* 59(2): 156.

Houghton, P.J., and Said, I.M. 1986. 3-Dehydromitragynine: An alkaloid from *Mitragyna speciosa*. *Phytochemistry* 25(12): 2910–12.

Houghton, P.J., Latiff, A., and Said, I.M. 1991. Alkaloids from *Mitragyna speciosa*. *Phytochemistry* 30(1): 347–50.

Jansen, K.L.R., and Prast, C.J. 1988. Psychoactive properties of mitragynine (kratom). *Journal of Psychoactive Drugs* 20(4): 455–57.

Kang, W., and Hao, X. 2006a. Triterpenoid saponins from *Mitragyna rotundifolia*. *Biochemical Systematics and Ecology* 34(7): 585–87.

Kang, W., and Li, C. 2009. Study on the chemical constituents of *Mitragyna rotundifolia* and their bioactivities. *Zhong Cheng Yao* 31(7): 1104–106.

Kang, W., Li, C., and Liu, Y. 2010. Antioxidant phenolic compounds and flavonoids of *Mitragyna rotundifolia* (Roxb.) Kuntze in vitro. *Medicinal Chemistry Research* 19(9): 1222–32.

Kang, W., Zhang, B., Xu, Q., and Hao, X. 2006b. Study on the chemical constituents of *Mitragyna rotundifolia*. *Zhong Yao Cai* 29(6): 557–60.

Kerschgens, I.P., Claveau, E., Wanner, M.J., Ingemann, S., van Maarseveen, J.H., and Hiemstra, H. 2012. Total syntheses of mitragynine, paynantheine and speciogynine via an enantioselective thiourea-catalysed Pictet–Spengler reaction. *Chemical Communications* 48(100): 12243–45.

Kitajima, M., Misawa, K., Kogure, N., Said, I. M., Horie, S., Hatori, Y., Murayama, T., and Takayama, H. 2006. A new indole alkaloid, 7-hydroxyspeciociliatine, from the fruits of Malaysian *Mitragyna speciosa* and its opioid agonistic activity. *Journal of Natural Medicines* 60(1): 28–35.

Kitajima, M., Nakayama T., Kogure, N., Wongserioioatana, S., and Takayama, H. 2007. New heteroyohimbine-type oxindole alkaloid from the leaves of Thai *Mitragyna hirsuta*. *Journal of Natural Medicines* 61(2): 192–95.

León, F., Habib, E., Adkins, J.E., Furr, E.B., McCurdy, C.R., and Cutler, S.J. 2009. Phytochemical characterization of the leaves of *Mitragyna speciosa* grown in USA. *Natural Product Communications* 4(7): 907–10.

Ma, J., Yin, W., Zhou, H., Liao, X., and Cook, J. M. 2009. General approach to the total synthesis of 9-methoxy-substituted indole alkaloids: Synthesis of mitragynine, as well as 9-methoxygeissoschizol and 9-methoxy-methylgeissoschizol. *Journal of Organic Chemistry* 74(1): 264–73.

Ma, X., Panjikar, S., Koepke, J., Loris, E., and Stöckigt, J. 2006. The structure of *Rauvolfia serpentina* strictosidine synthase is a novel six-bladed β-propeller fold in plant proteins. *The Plant Cell Online* 18(4): 907–20.

Matsumoto, K., Hatori, Y., Murayama, T., Tashima, K., Wongseripipatana, S., Misawa, K., Kitajima, M., Takayama, H., and Horie, S. 2006. Involvement of μ-opioid receptors in antinociception and inhibition of gastrointestinal transit induced by 7-hydroxymitragynine, isolated from Thai herbal medicine *Mitragyna speciosa*. *European Journal of Pharmacology* 549(1): 63–70.

Matsumoto, K., Horie, S., Takayama, H., Ishikawa, H., Aimi, N., Ponglux, D., Murayama, T., and Watanabe, K. 2005. Antinociception, tolerance and withdrawal symptoms induced by 7-hydroxymitragynine, an alkaloid from the Thai medicinal herb *Mitragyna speciosa*. *Life Sciences* 78(1): 2–7.

Matsumoto, K., Syunji H., Hayato I., Hiromitsu T., Norio A., Dhavadee P., and Kazuo W. 2004. Antinociceptive effect of 7-hydroxymitragynine in mice: Discovery of an orally active opioid analgesic from the Thai medicinal herb *Mitragyna speciosa*. *Life Sciences* 74(17): 2143–55.

Microgram Bulletin. 2006. Edited by U.S. Department of Justice, 25–40. Washington, D.C.: The Drug Enforcement Administration.

Ndagijimana, A., Wang, X., Pan, G., Zhang, F., Feng, H., and Olaleye, O. 2013. A review on indole alkaloids isolated from *Uncaria rhynchophylla* and their pharmacological studies. *Fitoterapia* 86: 35–47.

O'Connor, S.E., and Maresh, J.J. 2006. Chemistry and biology of monoterpene indole alkaloid biosynthesis. *Natural Product Reports* 23(4): 532–47.

Orio, L., Alexandru, L., Cravotto, G., Mantegna, S., and Barge, A. 2012. UAE, MAE, SFE-CO$_2$ and classical methods for the extraction of *Mitragyna speciosa* leaves. *Ultrasonics Sonochemistry* 19(3): 591–95.

Parthasarathy, S., Azizi, J. B., Ramanathan, S., Ismail, S., Sasidharan, S., Said, M.I.M., and Mansor, S.M. 2009. Evaluation of antioxidant and antibacterial activities of aqueous, methanolic and alkaloid extracts from *Mitragyna speciosa* (Rubiaceae family) leaves. *Molecules* 14(10): 3964–74.

Phongprueksapattana, S., Putalun, W., Keawpradub, N., and Wungsintaweekul, J. 2008. *Mitragyna speciosa*: Hairy root culture for triterpenoid production and high yield of mitragynine by regenerated plants. *Zeitschrift für Naturforschung C. A Journal of Biosciences* 63(9/10): 691–98.

Ponglux, D., Wongseripipatana, S., Takayama, H., Kikuchi, M., Kurihara, M., Kitajima, M., Aimi, N., and Sakai, S. 1994. A new indole alkaloid, 7 alpha-hydroxy-7h-mitragynine, from *Mitragyna speciosa* in Thailand. *Planta Medica* 60: 580–81.

Razafimandimbison, S.G., and Bremer, B. 2002. Phylogeny and classification of Naucleeae S.L. (Rubiaceae) inferred from molecular (ITS, *r*BCL, and *t*RNT-F) and morphological data. *American Journal of Botany* 89(7): 1027–41.

Schmeller, T., and Wink, M. 1998. Utilization of alkaloids in modern medicine. In: *Alkaloids: Biochemistry, Ecology, and Medicinal Applications*, Roberts, M.F., and Wink, M, eds., 450–51. Springer.

Shellard, E.J. 1983. *Mitragyna*: A note on the alkaloids of African species. *Journal of Ethnopharmacology* 8(3): 345–47.

Shellard, E.J., and Houghton, P.J. 1973. *Mitragyna* species of Asia. XXV. In vivo studies, using carbon-14-labeled alkaloids, in the alkaloidal pattern in young plants of *Mitragyna parvifolia* grown from seed obtained from Ceylon. *Planta Medica* 24(4): 341–52.

Shellard, E.J., and Houghton, P.J. 1974. *Mitragyna* species of Asia. XXVII. Alkaloidal N-oxides in the leaves of *Mitragyna parvifolia* from Sri Lanka. *Planta Medica* 25(2): 172–74.

Shellard, E.J., Houghton, P.J., and Resha, M. 1978a. The *Mitragyna* species of Asia. Part XXXI. The alkaloids of *Mitragyna speciosa* Korth from Thailand. *Planta Medica* 34(1): 26–36.

Shellard, E.J., Houghton, P.J., and Resha, M. 1978b. The *Mitragyna* species of Asia. Part XXXII. The distribution of alkaloids in young plants of *Mitragyna speciosa* Korth grown from seed obtained from Thailand. *Planta Medica* 34(3): 253–63.

Shellard, E.J., and Lala, P.K. 1977. The *Mitragyna* species of Asia. Part XXIX. The alkaloidal pattern in the leaves, stem bark and root bark of *Mitragyna parvifolia* from the Kerala State, India. *Planta Medica* 31(4): 395–99.

Shellard, E.J., and Phillipson, J.D. 1964. The *Mitragyna* species of Asia. I. The alkaloids of the leaves of *Mitragyna rotundifolia*. *Planta Medica* 12 (1): 27–32.

Stöckigt, J., Barleben, L., Panjikar, S., and Loris E.A. 2008. 3D-structure and function of strictosidine synthase: The key enzyme of monoterpenoid indole alkaloid biosynthesis *Plant Physiology and Biochemistry* 46: 340–55.

Sukrong, S., Zhu, S., Ruangrungsi, N., Phadungcharoen, T., Palanuvej, C., and Komatsu, K. 2007. Molecular analysis of the genus *Mitragyna* existing in Thailand based on rDNA: Its sequences and its application to identify a narcotic species: *Mitragyna speciosa*. *Biological and Pharmaceutical Bulletin* 30(7): 1284–88.

Sun, X., and Ma, D. 2011. Organocatalytic approach for the syntheses of corynan-theidol, dihydrocorynantheol, protoemetinol, protoemetine, and mitragy-nine. *Chemistry: An Asian Journal* 6(8): 2158–65.

Takayama, H. 2004a. Chemistry and pharmacology of analgesic indole alka-loids from the rubiaceous plant, *Mitragyna speciosa. Chemical and Pharmaceutical Bulletin* 52(8): 916–28.

Takayama, H., Ishikawa, H., Kitajima, M., Aimi N., and Baba, M. 2004b. A new 9-methoxyyohimbine-type indole alkaloid from *Mitragyna africanus. Chemical & Pharmaceutical Bulletin* 52(3): 359–61.

Takayama, H., Ishikawa, H., Kurihara, M., Kitajima, M., Aimi, N., Ponglux, D., Koyama, F., Matsumoto, K., Moriyama, T., Yamamoto, L.T., Watanabe, K., Murayama T., and Horie, S. 2002. Studies on the synthesis and opioid agonistic activities of mitragynine-related indole alkaloids: Discovery of opioid agonists structurally different from other opioid ligands. *Journal of Medicinal Chemistry* 45(9): 1949–56.

Takayama, H., Ishikawa, H., Kurihara, M., Kitajima, M., Sakai, S., Aimi, N., Seki, H., Yamaguchi, K., Said, I.M., and Houghton, P. J. 2001. Structure revision of mitragynaline, an indole alkaloid in *Mitragyna speciosa. Tetrahedron Letters* 42(9): 1741–43.

Takayama, H., Kurihara, M., Kitajima, M., Said, I.M., and Aimi, N. 1998. New indole alkaloids from the leaves of Malaysian *Mitragyna speciosa. Tetrahedron* 54(29): 8433–40.

Takayama, H., Kurihara, M., Kitajima, M., Said, I.M., and Aimi, N. 1999. Isolation and asymmetric total synthesis of a new mitragyna indole alkaloid, (−)-mitralactonine. *Journal of Organic Chemistry* 64(6): 1772–73.

Takayama, H., Kurihara, M., Kitajima, M., Said, I.M., and Aimi, N. 2000. Structure elucidation and chiral-total synthesis of a new indole alka-loid, (−)-9-methoxymitralactonine, isolated from *Mitragyna speciosa* in Malaysia. *Tetrahedron* 56(20): 3145–51.

Takayama, H., Maeda, M., Ohbayashi, S., Kitajima, M., Sakai, S., and Aimi, N. 1995. The first total synthesis of (−)-mitragynine, an analgesic indole alka-loid in *Mitragyna speciosa. Tetrahedron Letters* 36(51): 9337–40.

Tapondjou, L.A., Lontsi, D., Sondengam, B.L., Choudhary, M.I., Park, H., Choi, J., and Lee, K. 2002. Structure-activity relationship of triterpenoids isolated from *Mitragyna stipulosa* on cytotoxicity. *Archives of Pharmacal Research* 25(3): 270–74.

Trager, W.F., Lee, C.M., Phillipson, J.D., Haddock, R.E., Dwuma-Badu, D., and Beckett, A.H. 1968. Configurational analysis of rhynchophylline-type oxindole alkaloids: The absolute configuration of ciliaphylline, rhyncho-ciline, specionoxeine, isospecionoxeine, rotundifoline and isorotundifoline. *Tetrahedron* 24(2): 523–43.

Vicknasingam, B., Narayanan, S., Beng, G.T., and Mansor, S.M. 2010. The infor-mal use of ketum *Mitragyna speciosa* for opioid withdrawal in the northern states of Peninsular Malaysia and implications for drug substitution therapy. *International Journal of Drug Policy* 21(4): 283–88.

7 Chemistry of Mitragynines
A Brief Introduction

Robert B. Raffa

CONTENTS

7.1 INTRODUCTION

Among the many chemical compounds that are derived from plants, the indole alkaloids provide a particularly large category of biologically active materials. An indole alkaloid is an alkaloid—a naturally occurring chemical substance that usually contains mostly basic nitrogen atoms—that also contains an indole, or closely related, structural feature (Figure 7.1). Indole alkaloids can be further described based on source as either isoprenoids (derivatives of indole, or β-carboline, or pyrroloindole alkaloids) or non-isoprenoids (monoterpenoids or hemiterpenoids—better known in medicine as ergot alkaloids). The variety and diversity of sources and chemical structures of indole alkaloids gives rise to a variety and diversity of biological activities both of endogenous (e.g., neurotransmitter) and exogenous (e.g., drug) substances.

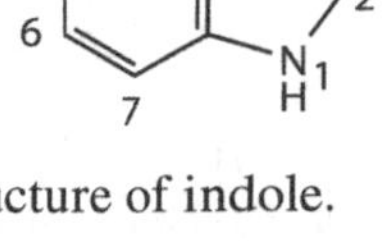

FIGURE 7.1 The bicyclic structure of indole.

7.2 BIOSYNTHESIS

In nature, the amino acid tryptophan (Trp, W) is the precursor of indole alkaloids. For example, the synthesis of serotonin (5-hydroxytryptamine, 5-HT) is shown in Figure 7.2. Similarly, Figure 7.3 shows the biosynthesis of psilocybin, the psychedelic compound found in >200 species of mushrooms. Examples of the biosynthesis of β-carbolines, ergot alkaloids, and other indole alkaloids are shown in Figures 7.4–7.6.

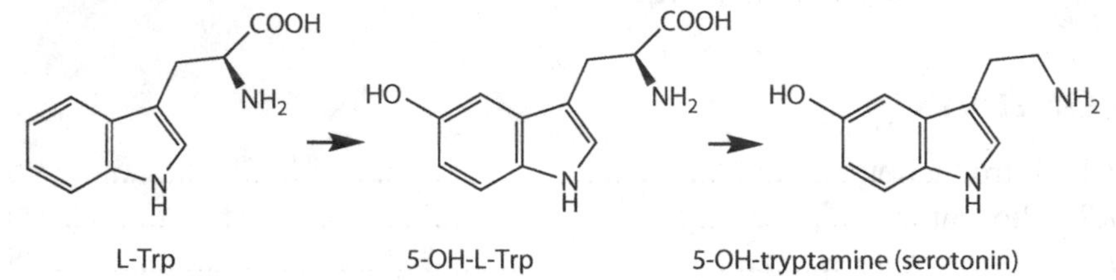

FIGURE 7.2   The synthesis of serotonin (5-hydroxytryptamine).

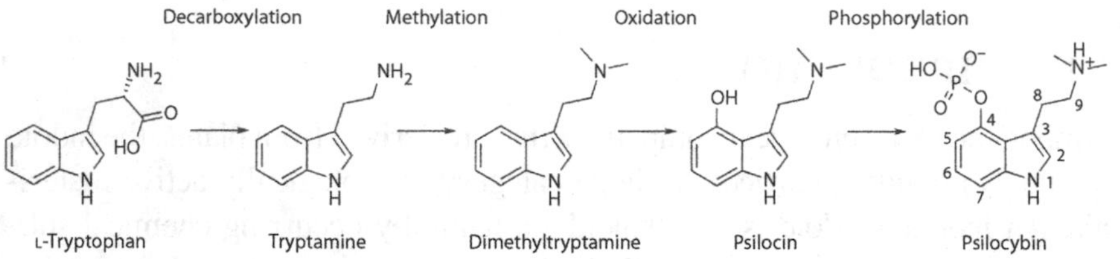

FIGURE 7.3   The biosynthesis of psilocybin.

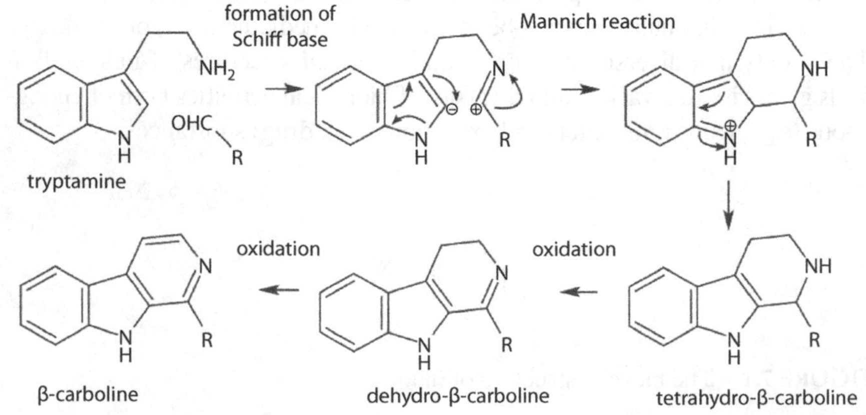

FIGURE 7.4   The biosynthesis of β-carboline.

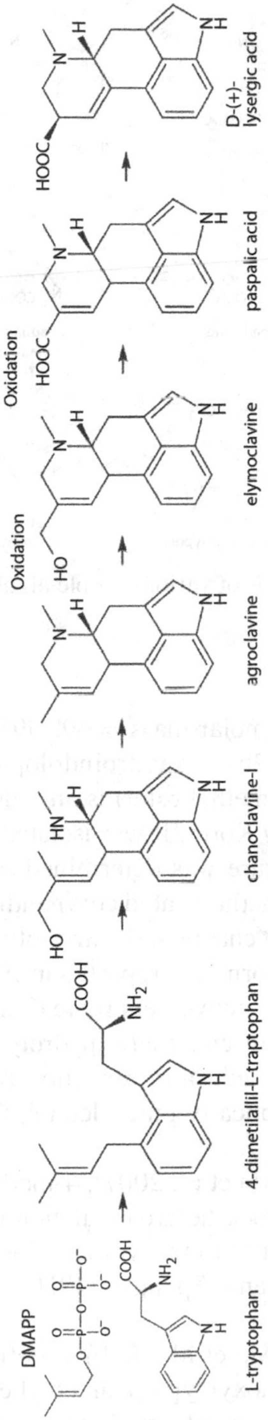

FIGURE 7.5 The biosynthesis of an ergot alkaloid (lysergic acid).

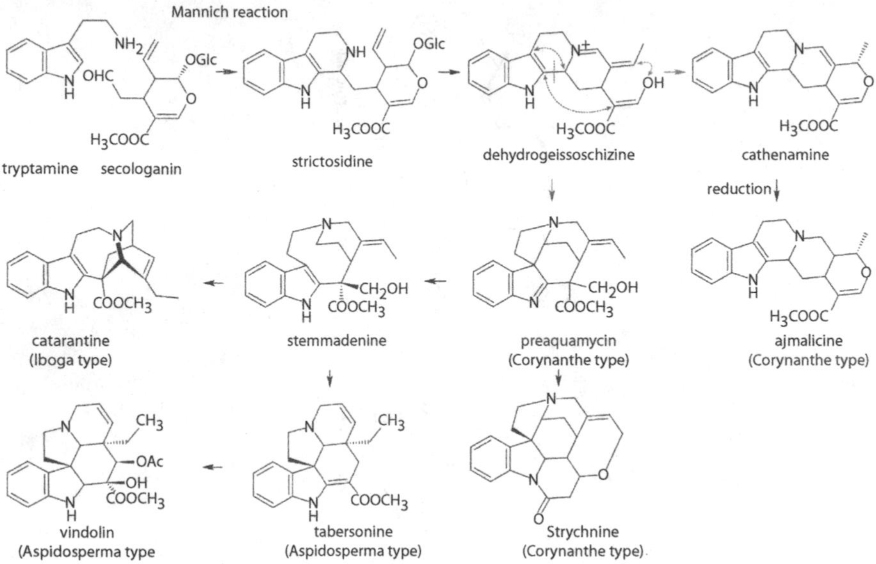

FIGURE 7.6   The biosynthesis of various indole alkaloids.

7.3   MITRAGYNINE

Mitragynine ($C_{23}H_{30}N_2O_4$; molar mass = 398.495; (*E*)-2-[(2*S*,3*S*)-3-ethyl-8-methoxy-1,2,3,4,6,7,12,12b-octahydroindolo[3,2-h]quinolizin-2-yl]-3-methoxyprop-2-enoic acid methyl ester) is an indole alkaloid (Figure 7.7) that is found in *M. speciosa* Korth. It was isolated in 1907 (Hooper 1907), and its X-ray crystal structure was determined in 1965 (Zacharias et al. 1965), which revealed a 9-methoxylated corynanthe skeleton. As described in more detail in subsequent chapters, the first total synthesis of (–)-mitragynine in the optically pure form was reported in 1995 using an enantiometrically and stereochemically convergent route (Takayama et al. 1995). The synthesis was initiated from enzymatic hydrolysis of the commercially available 6-chloronicotinic acid and reduction of the ketone derivative. Additional steps yielded optically pure alcohol, from which mitragynine was synthesized.

In an alternative route (Ma et al. 2007), 4-methoxy-D-tryptophan ethyl ester was prepared via Larock heteroannulation (Larock and Yum 1991) and was used as a precursor to mitragynine, using the asymmetric Pictet–Spengler reaction (Pictet and Spengler 1911) and Ni(COD)$_2$-mediated cyclization as key steps.

In a subsequent study (Ma et al. 2009), an efficient synthesis of optically active D- or L-4-methoxytryptophan ethyl ester was developed, and this was transformed into an intermediate tetracycle. The stereocenter

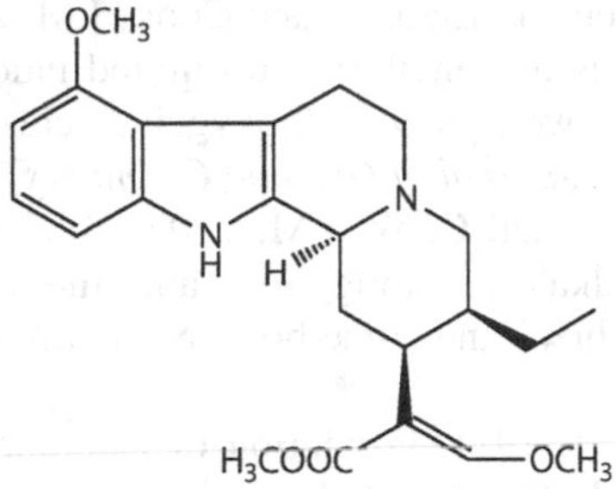

Mitragynine

FIGURE 7.7 The indole alkaloid mitragynine.

at C-3 was installed using the Pictet–Spengler reaction, and the *cis* configurations at C-3 and at C-5 were achieved using $Ni(COD)_2$-mediated cyclization.

Further details of these synthetic routes and additional routes are described and illustrated in a comprehensive review (Edwankar et al. 2009) and in other chapters in this book, including chapters written by the originators of the syntheses.

7.4 CONCLUSION

An extensive collection of mitragynine analogs (mitragynines) have been synthesized, and their biological activities have been described in several models of therapeutic (e.g., antinociception) and some side-effect (e.g., constipation) endpoints. These ligands can now serve as valuable pharmacologic tools for further elucidation of the mechanism(s) of action of these substances and as chemical templates for possible drug discovery and development efforts.

ACKNOWLEDGEMENT

Figures 7.3–7.6 are from Wikipedia Commons (http://en.wikipedia.org /wiki/Indole_alkaloid).

REFERENCES

Edwankar, C.R., Edwankar, R.V., Namjoshi, O.A., Rallapalli, S.K., Yang, J. and Cook, J.M. 2009. Recent progress in the total synthesis of indole alkaloids. *Curr Opin Drug Discov Devel* 12: 752–71.

Hooper, D. 1907. The anti-opium leaf. *Pharmaceutical Journal* 78: 453.

Larock, R.C. and Yum, E.K. 1991. Synthesis of indoles via palladium-catalyzed heteroannulation of internal alkynes. *Journal of the American Chemical Society* 113: 6689–90.

Ma, J., Wenyuan, Y., Zhou, H., Liao, X. and Cook, J.M. 2009. General approach to the total synthesis of 9-methoxy-substituted indole alkaloids: Synthesis of mitragynine, as well as 9-methoxygeissoschizol and 9-methoxy-N$_b$-methygeissoschizol. *Journal of Organic Chemistry* 74: 264–73.

Ma, J., Yin, W., Zhou, H. and Cook, J.M. 2007. Total synthesis of the opioid agonistic indole alkaloid mitragynine and the first total syntheses of 9-methoxygeissoschizol and 9-methoxy-N$_b$-methylgeissoschizol. *Org Lett* 9: 3491–94.

Pictet, A. and Spengler, T. 1911. Formation of isoquinoline derivatives by the action of methylal on phenylethylamine, phenylalanine and tyrosine. *Berichte der Deutschen Chemischen Gesellschaft* 44.

Takayama, H., Maeda, M., Ohbayashi, S., Kitajima, M., Sakai, S.-I. and Aimi, N. 1995. The first total synthesis of (–)-mitragynine, an analgesic indole alkaloid in *Mitragyna speciosa*. *Current Opinion in Drug Discovery & Development* 12: 752–71.

Zacharias, D.E., Rosenstein, R.D. and Jeffrey, G.A. 1965. The structure of mitragynine hydroiodide. *Acta Crystallography* 18: 1039–43.

8 Chemistry of *Mitragyna* Alkaloids

Mariko Kitajima and Hiromitsu Takayama

CONTENTS

8.1 INTRODUCTION

Mitragyna speciosa Korth., endemic to tropical Southeast Asia, is a species of particular medicinal importance (Jansen and Prast 1988; Raffa et al. 2013). Known as "kratom" in Thailand and "biak-biak" in Malaysia, the leaves have been traditionally used by natives for their opium-like effect and coca-like stimulant ability. A number of Corynanthe-type indole alkaloids have been identified from this plant. Among them, mitragynine (**1**) and a minor constituent, 7-hydroxymitragynine (**2**) (Ponglux et al. 1994), are novel opioid agonists having potent analgesic activity and structures that differ from that of morphine (Figure 8.1) (Horie et al. 1995, 2005; Matsumoto et al. 1996a,b, 1997, 2004, 2005a,b, 2006; Takayama 2004; Takayama et al. 2000a, 2002a; 2005; Thongpraditchote et al. 1998; Tohda et al. 1997; Tsuchiya et al. 2002; Watanabe et al. 1997, 1999). For this reason, a myriad of chemical studies of *Mitragyna* alkaloids have been carried out over the years. This chapter covers the syntheses and chemical reactions of *Mitragyna* alkaloids.

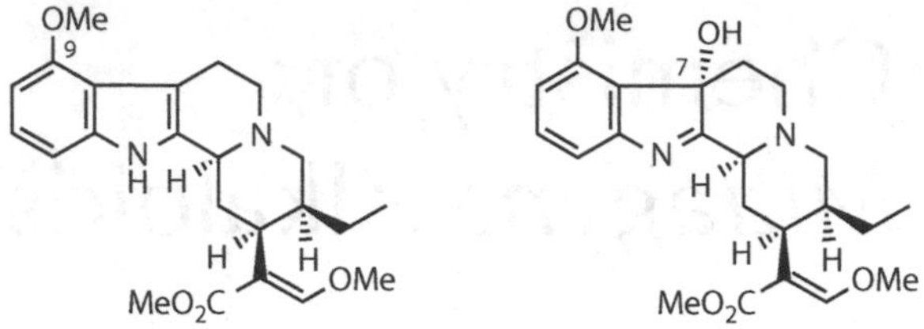

mitragynine (1)   7-hydroxymitragynine (2)

FIGURE 8.1 Structures of mitragynine (1) and 7-hydoxymitragynine (2).

8.2 TOTAL SYNTHESES OF MITRAGYNINE AND RELATED ALKALOIDS

8.2.1 FIRST TOTAL SYNTHESIS OF MITRAGYNINE; TAKAYAMA 1995

The first total synthesis of mitragynine (1) was accomplished by using the lipase-catalyzed enzymatic resolution and the Johnson–Claisen rearrangement as key steps (Figure 8.2) (Takayama et al. 1995). The synthesis started from the preparation of optically pure alcohol (R)-(+)-3. Racemic acetate 4 was subjected to enzymatic hydrolysis with lipase SAM II under phosphate-buffered (pH 7.0) conditions to produce secondary alcohol (+)-3 (32% chemical yield, 100% ee) and acetate (–)-4 (38% chemical yield, 100% ee). Alternatively, by reducing ketone 5 in the presence of a chiral oxazaborolidine catalyst, optically active alcohol (+)-3 (93% ee) was obtained in 80% yield and its optical purity was raised to 100% ee via separation of the diastereomeric ester derivatives with (R)-O-methylmandelic acid. On the other hand, 4-methoxytryptophylbromide (6) was prepared from 4-hydroxyindole via a five-step operation (i. Me_2SO_4, nBu_4NHSO_4, benzene, NaOH aq.; ii. $(COCl)_2$, Et_2O; iii. EtOH, Et_3N; iv. $LiAlH_4$, THF; and v. PBr_3, Et_2O). The thus obtained bromide 6 and optically pure pyridine derivative (R)-(+)-3 were condensed in heated benzene in the presence of a catalytic amount of NaI to produce pyridinium salt 7, which was then reduced with $NaBH_4$ to generate two diastereomers 8 and 9 in 33% and 27% yields, respectively. In order to install an acetic acid residue at the C15 position, each allylic alcohol (8 and 9) was subjected to the Johnson–Claisen rearrangement. By heating with trimethyl orthoacetate in the presence of a catalytic amount of benzoic acid in o-xylene, 8 and 9 produced acetates 10 and 11, respectively, as the sole product. The absolute configurations at C3 and the stereochemistry at C15 and C19 in 10 and 11 were determined by analysis of CD spectra and NMR spectra, respectively. Compound 10 had the appropriate absolute configurations at C3 and C15 for further transformation into mitragynine (1), whereas isomer 11 had the opposite configuration at C3 that nevertheless could be inverted into 10 by an oxidation–reduction sequence via a

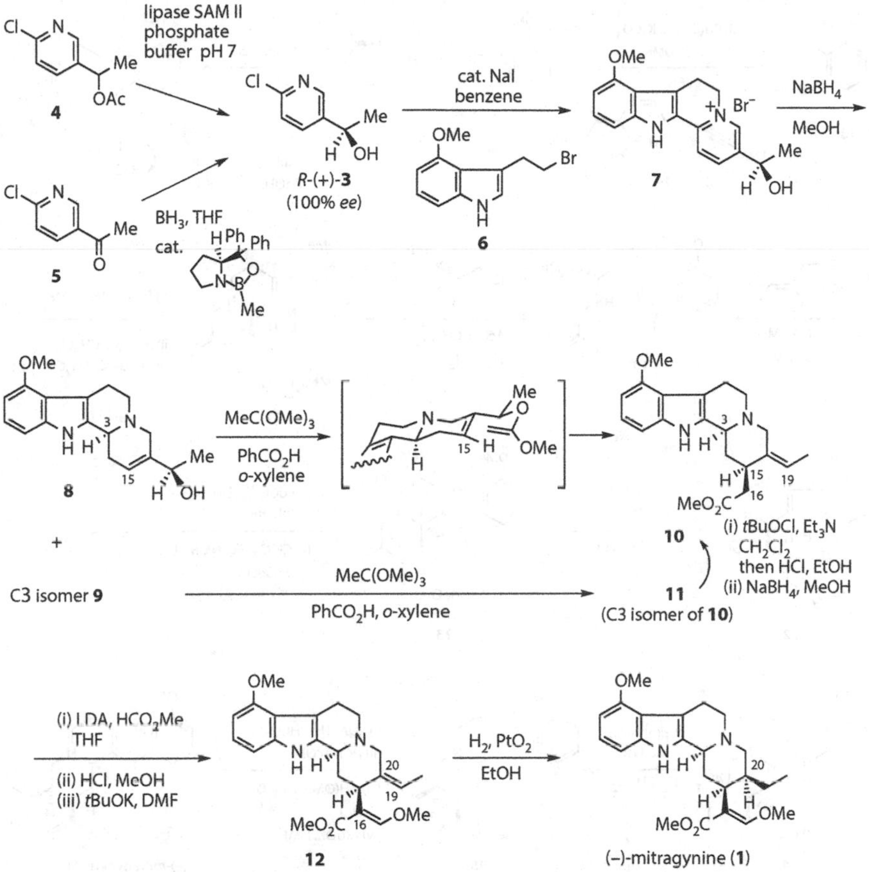

FIGURE 8.2 First total synthesis of mitragynine (**1**) by Takayama.

3,4-dehydroimmonium salt. In this manner, optically pure Corynanthe-type compound **10** could be convergently prepared. Next, according to the conventional method (LDA, HCO$_2$Me, THF), a formyl group was introduced to C16 in **10**, which was then converted into the dimethyl acetal derivative. Treatment of the acetal with *t*BuOK in DMF gave methyl enol ether **12**. Finally, by stereoselective reduction of the double bond at the C19—C20 positions over PtO$_2$ under H$_2$ atmosphere, mitragynine (**1**) having the natural absolute configuration was obtained.

8.2.2 SYNTHESIS OF MITRAGYNINE; COOK 2007

Cook and coworkers reported a new total synthesis of mitragynine (**1**) involving two key steps: the regiospecific Larock heteroannulation and the Ni(COD)$_2$-mediated Heck-type cyclization (Figure 8.3) (Ma et al. 2007, 2009). 2-Iodo-3-methoxyaniline **13** and TMS-propargyl-substituted

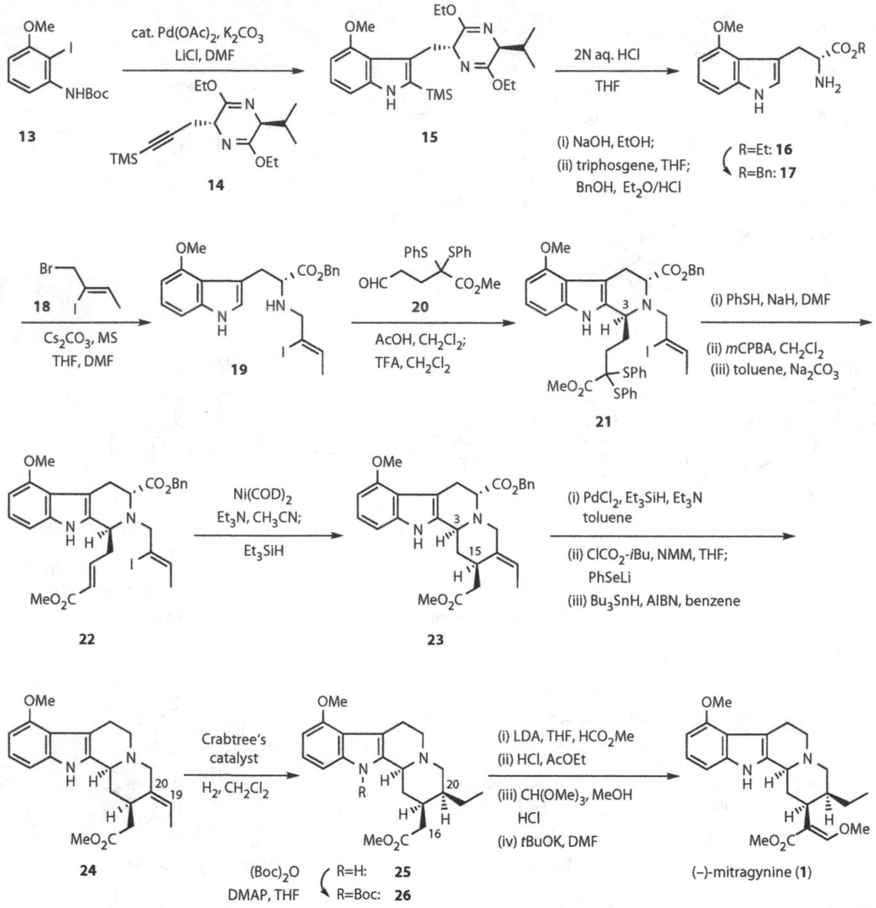

FIGURE 8.3   Cook's synthesis of mitragynine (**1**).

Schöllkopf chiral auxiliary **14** were subjected to the Larock indole synthesis by treatment with Pd(OAc)$_2$, K$_2$CO$_3$, and LiCl in DMF to provide indole **15** in a regiospecific manner. Hydrolysis of the chiral auxiliary in **15** and concomitant loss of the sily group gave 4-methoxy-D-tryptophan ethyl ester (**16**). The ethyl ester group in **16** was converted into benzyl ester to give **17**. Monoalkylation of the primary amine in **17** with allyl bromide **18** gave **19**, which was then subjected to the Pictet–Spengler reaction with aldehyde **20** to afford tetrahydro-β-carboline derivative **21** having the desired configuration at C3 in a highly diastereoselective manner. Then, **21** was converted into α,β-unsaturated ester **22** via a three-step operation (i. removal of one thiophenyl group; ii. *m*CPBA oxidation; and iii. elimination of resultant sulfoxide). Cyclization of α,β-unsaturated ester **22** was performed by treatment with Ni(COD)$_2$ and Et$_3$N in CH$_3$CN followed by the addition of Et₃SiH to give desired Corynanthe skeleton **23** having a *cis* configuration at C3 and C15. Removal of the benzyloxycarbonyl group in **23** by applying the Martin modification of the Barton–Crich decarboxylation provided compound **24**. The C19—C20 double bond in **25** was stereoselectively hydrogenated by using Crabtree's catalyst to give compound **25**. Construction of the β-methoxyacrylate residue in **25** via a five-step conventional operation afforded mitragynine (**1**).

8.2.3 FORMAL SYNTHESIS OF MITRAGYNINE; SUN AND MA 2011

Sun and Ma reported a formal synthesis of mitragynine (**1**) that used optically active aldehyde **27** generated from the Michael addition reaction of alkylidene malonate **28** and *n*-butanal in the presence of an organocatalyst (Figure 8.4) (Sun and Ma 2011). The Michael addition reaction of **28** and *n*-butanal in the presence of *O*-TMS-protected diphenylprolinol gave aldehyde **27** in >95% conversion yield with diastereomeric ratio 6.2:1 and 91.0% *ee*. Thus obtained aldehyde **27** with 81.6% *ee* was reacted with 4-methoxytryptamine (**29**) in CH₂Cl₂ at room temperature to give enamine **30** as a single diastereomer. Diastereoselective hydrogenation of the

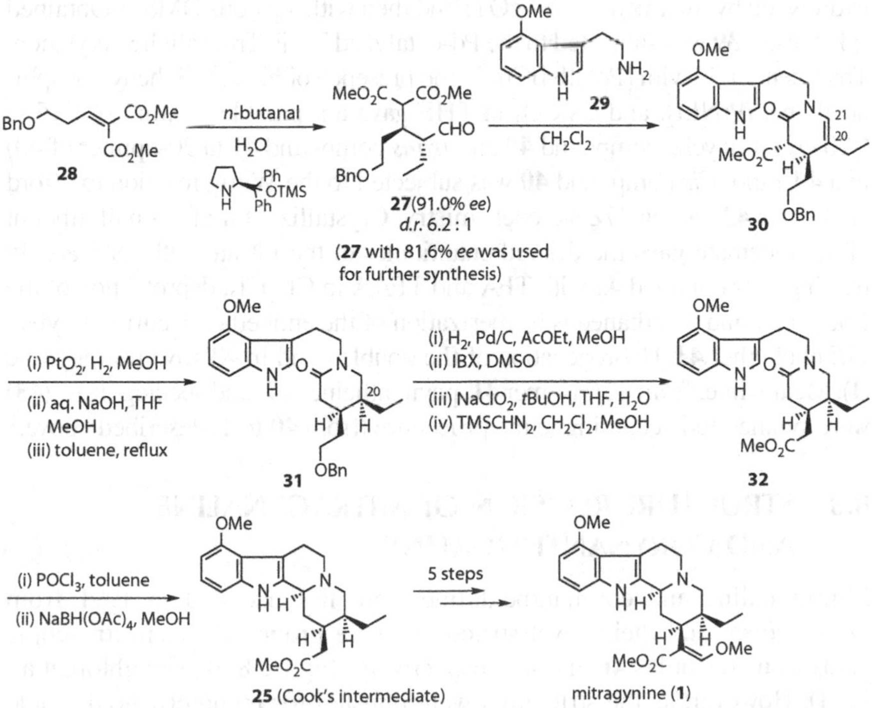

FIGURE 8.4 Ma's formal synthesis of mitragynine (**1**).

C20—C21 double bond in **30** by using PtO_2 followed by alkaline hydrolysis and decarboxylation yielded lactam **31**. Ester **32** was obtained via a four-step operation (i. debenzylation; ii. oxidation of a primary alcohol; iii. Pinnick oxidation; and iv. methyl esterification). The Bischler–Napieralski cyclization of ester **32** followed by reduction gave tetracyclic compound **25**, an intermediate in the total synthesis of mitragynine (**1**) by Cook's group (Ma et al. 2007, 2009), as described above.

8.2.4 SYNTHESES OF MITRAGYNINE, PAYNANTHEINE AND SPECIOGYNINE; HIEMSTRA 2012

Hiemstra and coworkers reported the synthesis of mitragynine (**1**) that involved two key steps: the enantioselective thiourea-catalyzed Pictet–Spengler reaction, and the Pd-catalyzed Tsuji–Trost allylic alkylation (Figure 8.5) (Kerschgens et al. 2012). By applying nosyl chemistry (Kan and Fukuyama 2004), the primary amine in 4-methoxytryptamine (**29**) was cleanly alkylated with allyl bromide **33** to give secondary amine **34**. The enantioselective Pictet–Spengler reaction of secondary amine **34** with aldehyde **35** utilizing quinine-derived thiourea **36** as catalyst afforded tetrahydro-β-carboline **37** in 90% yield with 89% *ee*. After protection of the N_a position with Boc, dithioacetal in **38** was hydrolyzed by treatment with AgOTf and then with aqueous DMSO. Obtained α-ketoester **39** was subjected to the Pd-catalyzed Tsuji–Trost allylic alkylation. Treatment of **39** with $[Pd(allyl)Cl]_2$ in the presence of bis-1,2-diphenylphosphinoethane, *i*PrNEt₂, and Cs_2CO_3 in THF gave a separable mixture of C15—C20 *cis* tetracyclic compound **40** and *trans* compound **41** (a 20-epimer of **40**) in a 4:1 ratio. *Cis* compound **40** was subjected to the Wittig reaction to afford enol ether **42** having 17Z stereochemistry. Crystallization of a small amount of the racemate gave the desired enantiomer in the filtrate with 98% *ee*. By treating thus obtained **42** with TFA and TFAA in CH_2Cl_2, deprotection of the Boc group and simultaneous isomerization of the enol ether occurred to yield 17*E* enol ether **43**. Hydrogenation of the double bond in **43** gave mitragynine (**1**). Meanwhile, from C20 isomer **41**, paynantheine (**44**) and speciogynine (**45**) were synthesized according to the procedure (from **40** to **1**) described above.

8.3 STRUCTURE REVISION OF MITRAGYNALINE AND CORYNANTHEIDALINE

Mitragynaline and corynantheidaline were first isolated in 1991 from *M. speciosa*, and their novel structures were proposed by spectroscopic analysis as formulas **46** and **47**, respectively (Figure 8.6) (Houghton et al. 1991). However, as the structures were unusual for monoterpenoid indole alkaloids, re-investigation through spectroscopic and chemical studies was

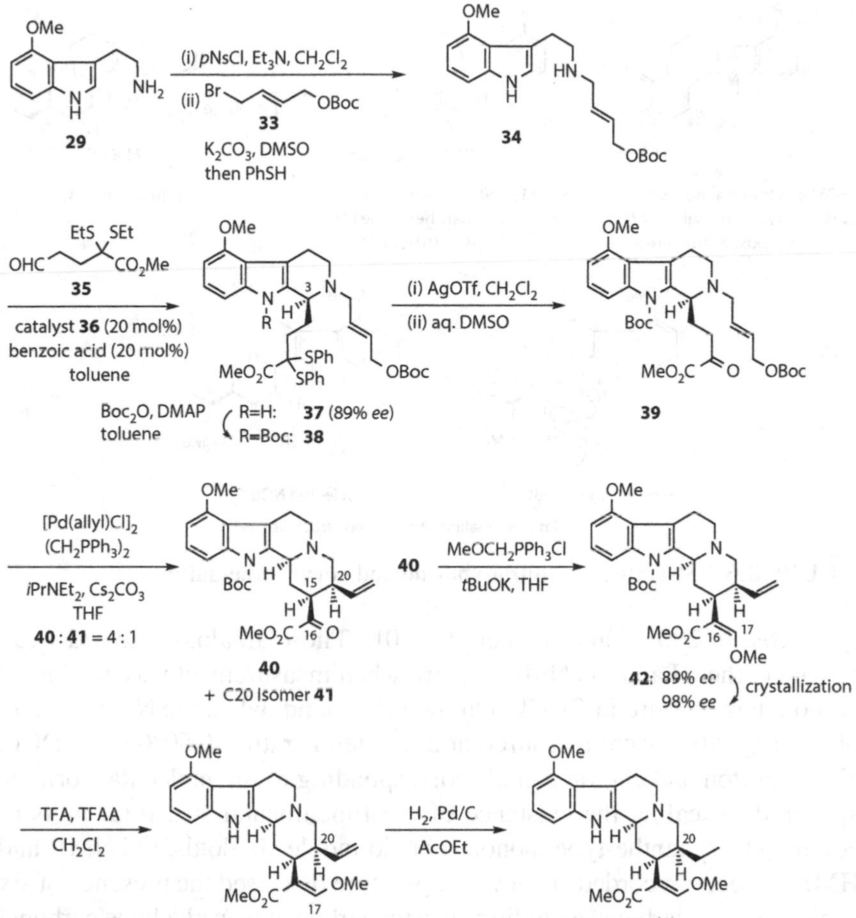

FIGURE 8.5 Hiemstra's synthesis of mitragynine (**1**), paynantheine (**44**), and speciogynine (**45**).

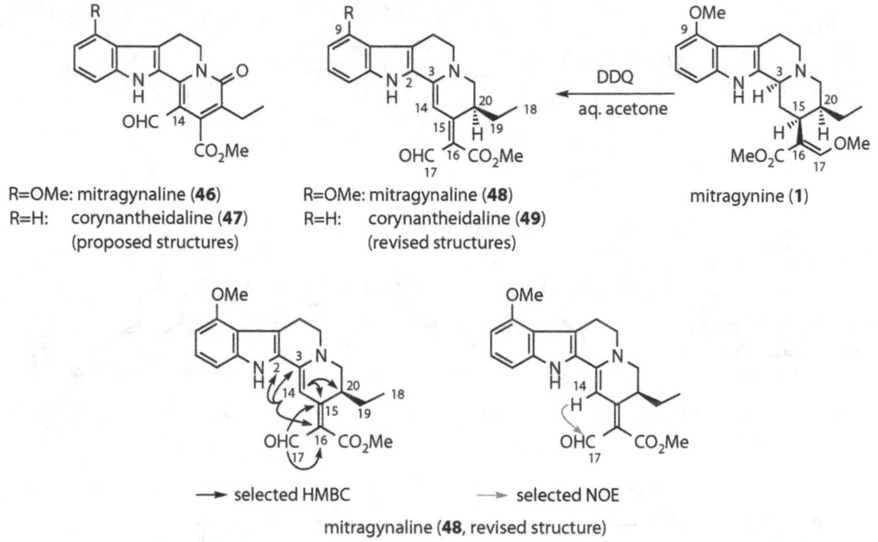

R=OMe: mitragynaline (**46**)
R=H: corynantheidaline (**47**)
(proposed structures)

R=OMe: mitragynaline (**48**)
R=H: corynantheidaline (**49**)
(revised structures)

mitragynine (**1**)

DDQ
aq. acetone

⟶ selected HMBC

⟶ selected NOE

mitragynaline (**48**, revised structure)

FIGURE 8.6 Structures of mitragynaline and corynantheidaline.

conducted in 2001 (Takayama et al. 2001). Those alkaloids showed weak signals in the ^{1}H and ^{13}C NMR spectra when measurement was conducted at room temperature in CDCl$_3$. On the other hand, when the NMR spectra of mitragynaline were measured at a low temperature (–50°C) in CDCl$_3$, all the proton and carbon signals corresponding to the molecular formula appeared, revealing the existence of the fundamental structural units of common Corynanthe-type monoterpenoid indole alkaloids. ^{13}C NMR and HMBC spectra recorded at a low temperature disclosed the presence of six conjugated sp^2 carbons, including an ester carbon and an aldehyde carbonyl carbon, besides the aromatic carbons assignable to the indole nucleus. The long-wavelength absorption at 485 nm in the UV spectrum also indicated a high degree of unsaturation in the molecule. The characteristic proton signal observed at δ 8.28 ppm (1H, singlet), which was finally detected at a low temperature, was unambiguously assigned to the proton at C14 by HMQC measurement. In addition, this proton showed HMBC correlations with C2, C3, C15, C16, and C20 carbons. The HMBC correlations between the aldehyde proton at δ 10.00 ppm and the carbons at C15 and C16 indicated that the aldehyde group was located at C16. Based on the above findings as well as biogenetic consideration, the molecular structure of mitragynaline was proposed to be that shown in formula **48** (Takayama et al. 2001). The geometry at C16 was confirmed from the NOE correlation between H14 and the aldehyde proton. Finally, the structure was determined by x-ray crystallographic analysis (Takayama 2004). When mitragynine (**1**) was treated with DDQ in aqueous acetone, mitragynaline (**48**) was obtained in 13% yield as one of the oxidation products. The semi-synthetic compound was completely identical in all respects to natural **48**. The optical rotation of the semi-synthetic compound was levorotatory, establishing the absolute stereochemistry at the C20 position. Based on the spectroscopic comparison of mitragynaline (**48**) and corynantheidaline, new structure **49** was proposed for corynantheidaline.

8.4 TOTAL SYNTHESES AND OPTICAL PURITY OF 9-METHOXYMITRALACTONINE AND MITRALACTONINE

9-Methoxymitralactonine (**50**), a new alkaloid isolated from the young leaves of *M. speciosa*, has a highly conjugated pentacyclic 9-methoxy-Corynanthe skeleton (Figure 8.7) (Takayama et al. 1999). In the ^{13}C NMR spectrum, signals corresponding to the sp^2 carbons at C14 and C16 resonated at unusually high-field positions, δ 87.08 and 94.91 ppm, respectively.

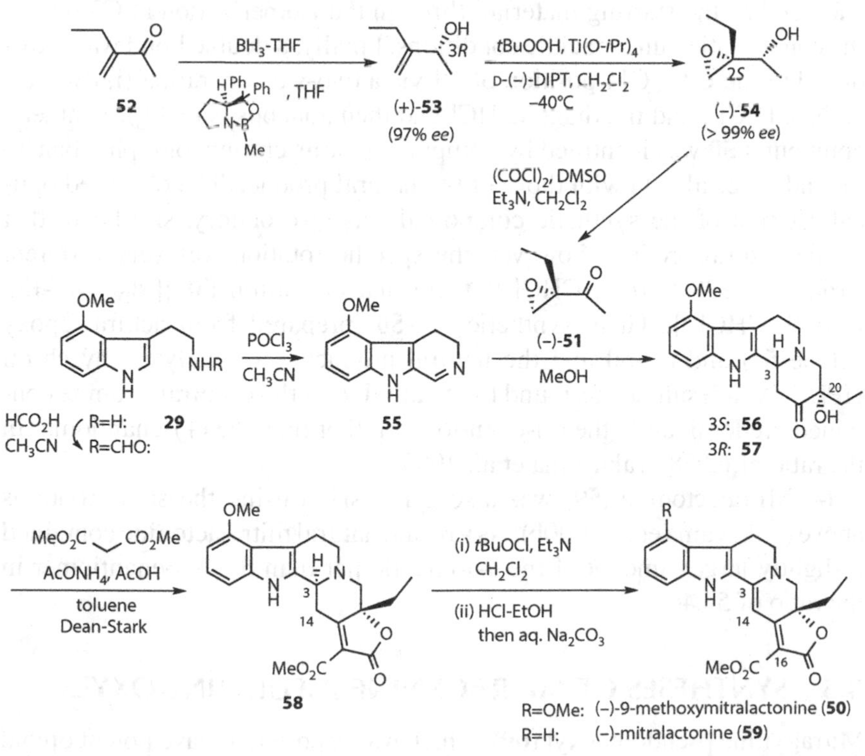

FIGURE 8.7 Syntheses of (−)-9-methoxymitralactonine (**50**) and (−)-mitralactonine (**59**).

In order to confirm the novel chemical structure proposed by spectroscopic analysis and the absolute configuration of **50**, the asymmetric total synthesis of 9-methoxymitralactonine (**50**) was conducted. First, chiral epoxy ketone **51** was prepared as follows. By reducing enone **52** in the presence of a chiral oxazaborolidine catalyst, optically active alcohol (*R*)-(+)-**53** was obtained with 97% *ee*. Next, allylic alcohol **53** (97% *ee*) was subjected to the Sharpless asymmetric epoxidation under kinetic resolution conditions to give (–)-epoxide **54** with >99% *ee*. The Swern oxidation of the secondary alcohol in **54** gave (–)-epoxy ketone **51**. Thus obtained epoxide (–)-**51** and 5-methoxy-3,4-dihydro-β-carboline (**55**), which was prepared from 4-methoxytryptamine (**29**) through *N*-formylation and the Bischler–Napieralski reaction, were condensed in heated MeOH to afford two diastereomeric tetracyclic compounds **56** and **57** in 33% and 17% yields, respectively. Their C3 configurations were deduced by comparison of CD spectra. Major isomer **56** was subjected to the Knoevenagel condensation with dimethyl malonate in refluxing toluene in the presence of AcONH$_4$ and AcOH to give directly pentacyclic product **58** having a lactone residue. Interestingly, the same product **58** was obtained when minor isomer **57** was used as the starting material, through the isomerization at C3 during the condensation under acidic conditions. Finally, a double bond was introduced to the C3—C14 position of **58** via a two-step operation (i. *t*BuOCl, Et$_3$N, CH$_2$Cl$_2$; and ii. ethanolic HCl and then aqueous Na$_2$CO$_3$). Synthetic compound **50** was identified by comparison of its chromatographic behavior and spectral data with those of the natural product. The observed optical rotation of the synthetic compound was levorotatory, similar to that of the natural product; however, the specific rotation was very different $\{[\alpha]_D^{24} = -838 \ (c\ 0.10, \text{CHCl}_3)\}$ from that of natural **50** $\{[\alpha]_D^{18} = -123 \ (c\ 0.19, \text{CHCl}_3)\}$. Then, synthetic (±)-**50** (prepared from achiral epoxy ketone **51**) and (–)-**50** and the natural product were analyzed by chiral HPLC. As a result, it was found that natural 9-methoxymitralactonine contained predominantly the (–)-enantiomer rather than the (+)-enantiomer in the ratio of 62:38 (Takayama et al. 1999).

(–)-Mitralactonine (**59**) was also synthesized using the same route as above (Takayama et al. 2000b). As a result, natural mitralactonine contained a slightly larger amount of the (–)-enantiomer than the (+)-enantiomer in the ratio of 54:46.

8.5 SYNTHESES OF MITRAGYNINE PSEUDOINDOXYLS

Mitragynine pseudoindoxyl (**60**), which was reported to have potent opioid agonistic activity (Takayama et al. 2002a; Yamamoto et al. 1999), was synthesized by the chemical transformation of mitragynine (**1**) (Figure 8.8)

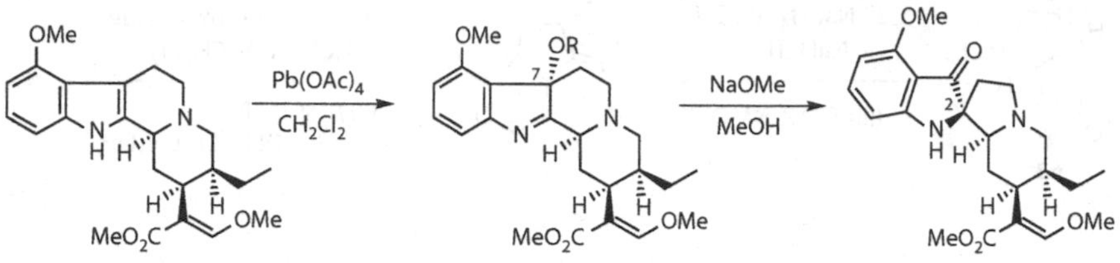

FIGURE 8.8 Transformation of mitragynine (1) into mitragynine pseudoindoxyl (60).

(Takayama et al. 1996). Mitragynine (1) was oxidized by $Pb(OAc)_4$ to give 7-acetoxymitragynine (61). By treatment with NaOMe, 61 yielded mitragynine pseudoindoxyl (60) via the semi-pinacol rearrangement of 7-hydroxymitragynine (2). The stereochemistry at the C2 spirocenter was determined by NOE and CD measurements.

Sorensen and coworkers reported the synthesis of unnatural 11-methoxy mitragynine pseudoindoxyl (62) starting from a derivative of the Geissmann–Waiss lactone (Figure 8.9) (Kim et al. 2012). Cbz-protected Geissmann–Waiss lactone 63 was converted into ketone 64 via a three-step operation (i. reductive cleavage of the lactone ring; ii. formation of the cyclic carbamate; and iii. oxidation of the secondary alcohol). Tetracyclic indoxyl 65 was diastereoselectively prepared from ketone 64 via the "interrupted Ugi reaction," which combines the mechanistic elements of the Ugi reaction and the Houben–Hoesch reaction. Thus, condensation of ketone 64 with 3,5-dimethoxyaniline followed by treatment of the resultant ketimine with $HBF_4 \bullet OEt_2$ and *p*-tosylmethyl isocyanide afforded spiro-fused tosylmethyl imine via a face-selective attack of an isocyanide on an iminium ion and an intramolecular addition of an electron-rich arene to a nitrilium ion in an intermediate 65. The resultant imine was hydrolyzed to give tetracyclic indoxyl 66. Protection of N_a in 66 with a Boc group followed by hydrolytic cleavage of the cyclic carbamate and N_b-alkylation with 1-bromo-2-butyne yielded alkyne 67. Ene–yne compound 68 was prepared by oxidation of the primary alcohol and successive homologation with the Wittig reaction. The nickel-induced reductive cyclization of ene–yne 68 was achieved by treatment with $Ni(COD)_2$ and DBU in THF to produce tetracyclic compound 69 diastereoselectively. The conversion of 69 into methyl enol ether 70 was accomplished via the same three-step sequence that was performed in the synthesis of mitragynine (1) (Takayama et al. 1995; Ma et al. 2007, 2009; Sun and Ma 2011). The C19—C20 double bond was stereoselectively hydrogenated by PtO_2 to give 11-methoxy mitragynine pseudoindoxyl (62).

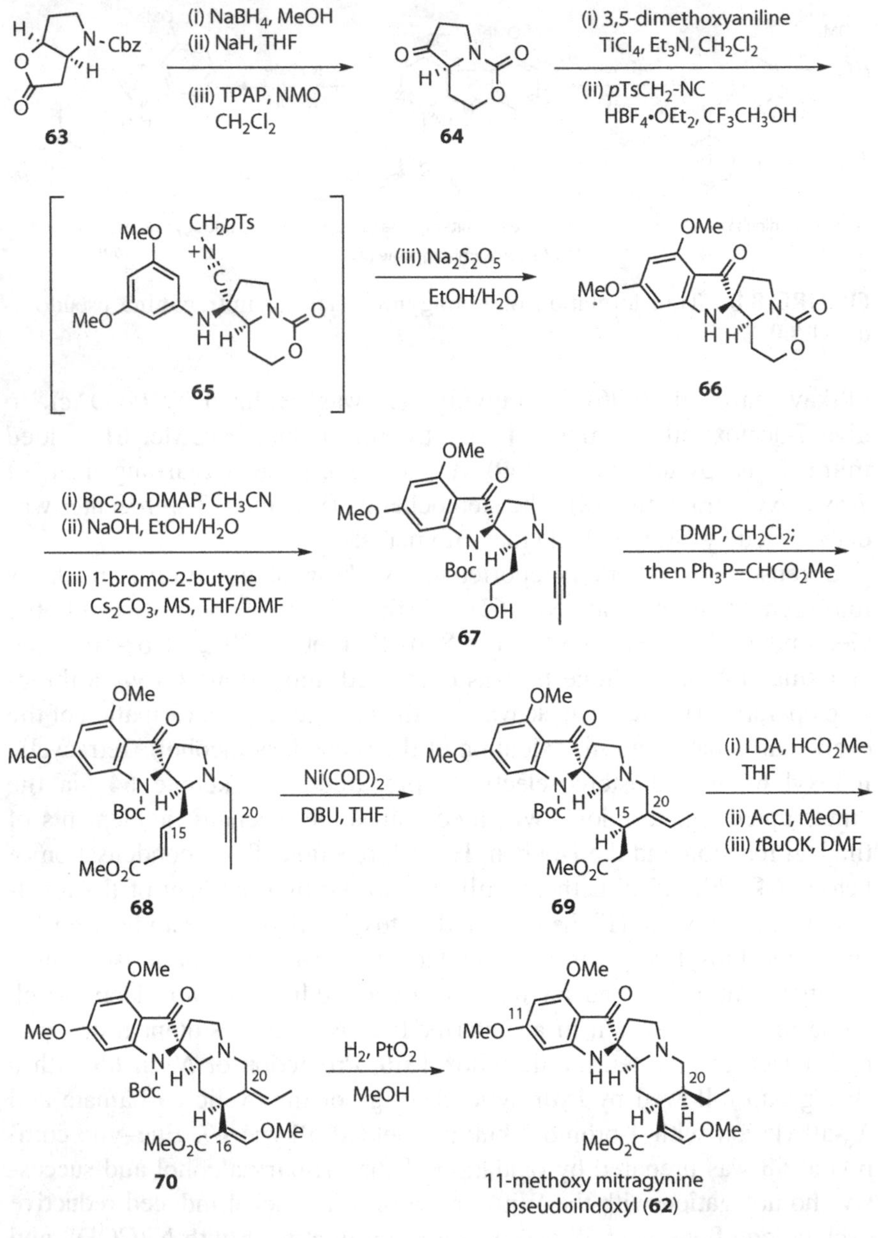

FIGURE 8.9 Synthesis of 11-methoxy mitragynine pseudoindoxyl (**62**).

8.6 OXIDATIVE DIMERIZATION OF MITRAGYNINE

In general, the oxidation of indoles by Pb(OAc)$_4$ produces 7-acetoxyindolenine derivatives (Finch et al. 1963, 1965). In the case of mitragynine (**1**), the yield of desired indolenine **61** was 50% at most, and

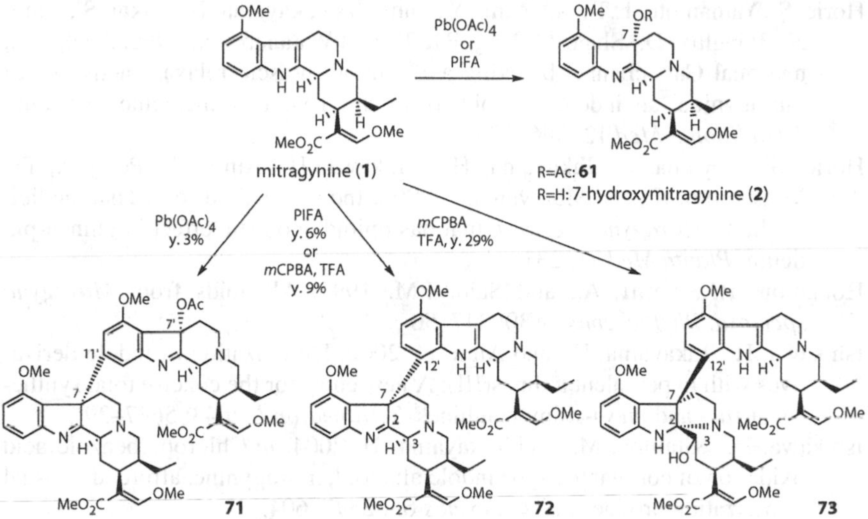

FIGURE 8.10 Oxidative dimerization reaction of mitragynine (**1**).

an unusual dimeric compound **71** was obtained as the minor product (3% yield) (Figure 8.10) (Takayama et al. 2002b). When mitragynine (**1**) was exposed to a hypervalent iodinc(III) reagent (PIFA) in aqueous CH_3CN, with the intent to introduce a hydroxyl group at C7, unusual dimeric compound **72** and the desired product, 7-hydroxymitragynine (**2**), were formed in 6% and 50% yields, respectively (Ishikawa et al. 2002). Gueller and Borschberg reported an efficient procedure for transforming Aristotelia-type indole alkaloids into 7-hydroxyindolenine derivatives using *m*CPBA in the presence of TFA (Gueller and Borschberg 1992). When this method was applied to mitragynine (**1**), unexpected dimerization products (**72** and **73**) were obtained in 9% and 29% yields, respectively (Ishikawa et al. 2004).

REFERENCES

Finch, N., Gemenden, C.W., Hsu, I. H.-C., and Taylor, W.I. 1963. Oxidative transformations of indole alkaloids. II. The preparation of oxindoles from *cis*-DE-yohimbinoid alkaloids: The partial synthesis of carapanaubine. *J Am Chem Soc* 85: 1520–23.

Finch, N., Gemenden, C.W., Hsu, I.H.-C., Kerr, A., Sim, G.A., and Taylor, W.I. 1965. Oxidative transformations of indole alkaloids. III. Pseudoindoxyls from yohimbinoid alkaloids and their conversion to 'invert' alkaloids. *J Am Chem Soc* 87: 2229–35.

Gueller, G., and Borschberg, H.J. 1992. Synthesis of Aristotelia-type alkaloids. Part X. Biomimetic transformation of synthetic (+)-aristotelline into (−)-alloaristoteline. *Tetrahedron: Asymmetry* 3: 1197–204.

Horie, S., Yamamoto, L.T., Futagami, Y., Yano, S., Takayama, H., Sakai, S., Aimi, N., Ponglux, D., Shan, J., Pang, P.K.T., and Watanabe, K. 1995. Analgesic, neuronal Ca^{2+} channel-blocking and smooth muscle relaxant activities of mitragynine, an indole alkaloid, from the Thai folk medicine 'Kratom.' *J Traditional Med* 12: 366–67.

Horie, S., Koyama, F., Takayama, H., Ishikawa, H., Aimi, N., Ponglux, D., Matsumoto, K., and Murayama, T. 2005. Indole alkaloids of a Thai medicinal herb, *Mitragyna speciosa*, that has opioid agonistic effect in guinea-pig ileum. *Planta Med* 71: 231–36.

Houghton, P.J., Latiff, A., and Said, I.M. 1991. Alkaloids from *Mitragyna speciosa*. *Phytochemistry* 30: 347–50.

Ishikawa, H., Takayama, H., and Aimi, N. 2002. Dimerization of indole derivatives with hypervalent iodines(III): A new entry for the concise total synthesis of *rac*- and *meso*-chimonanthines. *Tetrahedron Lett* 43: 5637–39.

Ishikawa, H., Kitajima, M., and Takayama, H. 2004. *m*-Chloroperbenzoic acid oxidation of corynanthe-type indole alkaloid, mitragynine, afforded unusual dimerization products. *Heterocycles* 63: 2597–604.

Jansen, K.L.R., and Prast, C.J. 1988. Ethnopharmacology of kratom and the *Mitragyna* alkaloids. *J Ethnopharmacol* 23: 115–19 and references cited therein.

Kan, T., and Fukuyama, T. 2004. Ns strategies: A highly versatile synthetic method for amines. *Chem. Commun.* 353–59.

Kerschgens, I.P., Claveau, E., Wanner, M.J., Ingemann, S., van Maarseveen, J.H., and Hiemstra, H. 2012. Total syntheses of mitragynine, paynantheine and speciogynine *via* an enantioselective thiourea–catalysed Pictet–Spengler reaction. *Chem Commun* 48: 12243-45.

Kim, J., Schneekloth, J.S., Jr., and Sorensen, E.J. 2012. A chemical synthesis of 11-methoxy mitragynine pseudoindoxyl featuring the interrupted Ugi reaction. *Chem Sci* 3: 2849–52.

Ma, J., Yin, W., Zhou, H., and Cook, J.M. 2007. Total synthesis of the opioid agonistic indole alkaloid mitragynine and the first total syntheses of 9-methoxygeissoschizol and 9-methoxy-N_b-methylgeissoschizol. *Org Lett* 9: 3491–94.

Ma, J., Yin, W., Zhou, H., Liao, X., and Cook, J.M. 2009. General approach to the total synthesis of 9-methoxy-substituted indole alkaloids: Synthesis of mitragynine, as well as 9-methoxygeissoschizol and 9-methoxy-N_b-methylgeissoschizol. *J Org Chem* 74: 264–73.

Matsumoto, K., Mizowaki, M., Thongpradichote, S., Murakami, Y., Takayama, H., Sakai, S., Aimi, N., and Watanabe, H. 1996a. Central antinociceptive effects of mitragynine in mice: Contribution of descending noradrenergic and serotonergic systems. *Eur J Pharmacol* 317: 75–81.

Matsumoto, K., Mizowaki, M., Thongpradichote, S., Takayama, H., Sakai, S., Aimi, N., and Watanabe, H. 1996b. Antinociceptive action of mitragynine in mice: Evidence for the involvement of supraspinal opioid receptors. *Life Sci* 59: 1149–55.

Matsumoto, K., Mizowaki, M., Takayama, H., Sakai, S., Aimi, N., and Watanabe, H. 1997. Suppressive effect of mitragynine on the 5-methoxy-N,N-dimethyltryptamine-induced head-twitch response in mice. *Pharmacol Biochem Behav* 57: 319–23.

Matsumoto, K., Horie, S., Ishikawa, H., Takayama, H., Aimi, N., Ponglux, D., and Watanabe, K. 2004. Antinociceptive effect of 7-hydroxymitragynine in mice: Discovery of an orally active opioid analgesic from the Thai medicinal herb *Mitragyna speciosa*. *Life Sci* 74: 2143–55.

Matsumoto, K., Horie, S., Takayama, H., Ishikawa, H., Aimi, N., Ponglux, D., Murayama, T., and Watanabe, K. 2005a. Antinociception, tolerance and withdrawal symptoms induced by 7-hydroxymitragynine, an alkaloid from the Thai medicinal herb *Mitragyna speciosa*. *Life Sci* 78: 2–7.

Matsumoto, K., Yamamoto, L. T., Watanabe, K., Yano, S., Shan, J., Pang, P. K. T., Ponglux, D., Takayama, H., and Horie, S. 2005b. Inhibitory effect of mitragynine, an analgesic alkaloid from Thai herbal medicine, on neurogenic contraction of the vas deferens. *Life Sci* 78: 187–94.

Matsumoto, K., Hatori, Y., Murayama, T., Tashima, K., Wongseripipatana, S., Misawa, K., Kitajima, M., Takayama, H., and Horie, S. 2006. Involvement of μ-opioid receptors in antinociception and inhibition of gastrointestinal transit induced by 7-hydroxymitragynine, isolated from Thai herbal medicine *Mitragyna speciosa*. *Eur J Pharmacol* 549: 63–70.

Ponglux, D., Wongseripipatana, S., Takayama, H., Kikuchi, M., Kurihara, M., Kitajima, M., Aimi, N., and Sakai, S. 1994. A new indole alkaloid, 7α-hydroxy-7*H*-mitragynine, from *Mitragyna speciosa* in Thailand. *Planta Medica* 60: 580–81.

Raffa, R.B., Beckett, J.R., Brahmbhatt, V.N., Ebinger, T.M., Fabian, C.A., Nixon, J.R., Orlando, S.T., Rana, C.A., Tejani, A.H., and Tomazic, R.J. 2013. Orally active opioid compounds from a non-poppy source. *J Med Chem* 56: 4840–48 and references cited therein.

Sun, X., and Ma, D. 2011. Organocatalytic approach for the syntheses of corynantheidol, dihydrocorynantheol, protoemetinol, protoemetine, and mitragynine. *Chem Asian J* 6: 2158–65.

Takayama, H. 2004. Chemistry and pharmacology of analgesic indole alkaloids from the rubiaceous plant, *Mitragyna speciosa*. *Chem Pharm Bull* 52: 916–28.

Takayama, H., Maeda, M., Ohbayashi, S., Kitajima, M., Sakai, S., and Aimi, N. 1995. The first total synthesis of (–)-mitragynine, an analgesic indole alkaloid in *Mitragyna speciosa*. *Tetrahedron Lett* 36: 9337–40.

Takayama, H., Kurihara, M., Subhadhirasakul, S., Kitajima, M., Aimi, N., and Sakai, S. 1996. Stereochemical assignment of pseudoindoxyl alkaloids. *Heterocycles* 42: 87–92.

Takayama, H., Kurihara, M., Kitajima, M., Said, I.M., and Aimi, N. 1999. Isolation and asymmetric total synthesis of a new *Mitragyna* indole alkaloid, (–)-mitralactonine. *J Org Chem* 64: 1772–73.

Takayama, H., Aimi, N., and Sakai, S. 2000a. Chemical studies on the analgesic indole alkaloids from the traditional medicine (*Mitragyna speciosa*) used for opium substitute. *Yakugaku Zasshi* 120: 959–67.

Takayama, H., Kurihara, M., Kitajima, M., Said, I.M., and Aimi, N. 2000b. Structure elucidation and chiral-total synthesis of a new indole alkaloid, (–)-9-methoxymitralactonine, isolated from *Mitragyna speciosa* in Malaysia. *Tetrahedron* 56: 3145–51.

Takayama, H., Ishikawa, H., Kurihara, M., Kitajima, M., Sakai, S., Aimi, N., Seki, H., Yamaguchi, K., Said, I.M., and Houghton, P. J. 2001. Structure revision of mitragynaline, an indole alkaloid in *Mitragyna speciosa*. *Tetrahedron Lett* 42: 1741–43.

Takayama, H., Ishikawa, H., Kurihara, M., Kitajima, M., Aimi, N., Ponglux, D., Koyama, F., Matsumoto, K., Moriyama, T., Yamamoto, L.T., Watanabe, K., Murayama, T., and Horie, S. 2002a. Studies on the synthesis and opioid agonistic activities of mitragynine-related indole alkaloids: Discovery of opioid agonists structurally different from other opioid ligands. *J Med Chem* 45: 1949–56.

Takayama, H., Ishikawa, H., Kitajima, M., and Aimi, N. 2002b. Formation of an unusual dimeric compound by lead tetraacetate oxidation of a Corynanthe-type indole alkaloid, mitragynine. *Chem Pharm Bull* 50: 960–63.

Takayama, H., Kitajima, M., and Kogure, N. 2005. Chemistry of Corynanthe-type related indole alkaloids from the *Uncaria, Nauclea,* and *Mitragyna* plants. *Curr Org Chem* 9: 1445–64.

Thongpraditchote, S., Matsumoto, K., Tohda, M., Takayama, H., Aimi, N., Sakai, S., and Watanabe, H. 1998. Identification of opioid receptor subtypes in antinociceptive actions of supraspinally-administered mitragynine in mice. *Life Sci* 62: 1371–78.

Tohda, M., Thongpraditchote, S., Matsumoto, K., Murakami, Y., Sakai, S., Aimi, N., Takayama, H., Tongroach, P., and Watanabe, H. 1997. Effects of mitragynine on cAMP formation mediated by δ-opiate receptors in NG108-15 cells. *Biol Pharm Bull* 20: 338–40.

Tsuchiya, S., Miyashita, S., Yamamoto, M., Horie, S., Sakai, S., Aimi, N., Takayama, H., and Watanabe, K. 2002. Effect of mitragynine, derived from Thai folk medicine, on gastric acid secretion through opioid receptor in anesthetized rats. *Eur J Pharmacol* 443: 185–88.

Watanabe, K., Yano, S., Horie, S., and Yamamoto, L.T. 1997. Inhibitory effect of mitragynine, an alkaloid with analgesic effect from Thai medicinal plant, *Mitragyna speciosa*, on electrically stimulated contraction of isolated guinea-pig ileum through the opioid receptor. *Life Sci* 60: 933–42.

Watanabe, K., Yano, S., Horie, S., Yamamoto, L.T., Takayama, H., Aimi, N., Sakai, S., Ponglux, D., Tongroach, P., Shan, J., and Pang, P.K.T. 1999. Pharmacological properties of some structurally related indole alkaloids contained in the Asian herbal medicines, hirsutine and mitragynine, with special reference to their Ca^{2+} antagonistic and opioid-like effects. In *Pharmacological Research on Traditional Herbal Medicines*, Watanabe H., and Shibuya T. (eds.), 163–77. Amsterdam: Harwood Academic Publishers.

Yamamoto, L.T., Horie, S., Takayama, H., Aimi, N., Sakai, S., Yano, S., Shan, J., Pang, P.K.T., Ponglux, D., and Watanabe, K. 1999. Opioid receptor agonistic characteristics of mitragynine pseudoindoxyl in comparison with mitragynine derived from Thai medicinal plant *Mitragyna speciosa*. *Gen Pharmacol* 33: 73–81.

9 Medicinal Chemistry Based on Mitragynine

Hiromitsu Takayama and Mariko Kitajima

CONTENTS

9.1 INTRODUCTION

Mitragynine (**1**) is the major alkaloid in the traditional medicinal herb *Mitragyna speciosa* Korth. (Adkins et al. 2011; Raffa et al. 2013; Takayama 2004; Takayama et al. 2005; Watanabe et al. 1999), and has been proven to exhibit analgesic activity (Horie et al. 1995, 2005; Matsumoto et al. 1996a,b, 1997, 2005; Thongpraditchote et al. 1998; Tohda et al. 1997; Watanabe et al. 1997) mediated by opioid receptors. This natural product has been used as a seed molecule for the development of new analgesic in medicinal chemistry. This chapter covers published work on the syntheses of mitragynine derivatives, the evaluation of their analgesic activities, and the structure–activity relationship studies of mitragynine analogs.

9.2 STRUCTURAL REQUIREMENT OF MITRAGYNINE TO EXHIBIT THE ACTIVITY

Mitragynine (**1**) is categorized into the Corynanthe-type monoterpenoid indole alkaloids, more than 3,000 of which have been found mainly in the plant kingdom (Saxton 1994). The Corynanthe-type alkaloids having A, B, C, and D rings are further classified into four groups, *normal*, *pseudo*, *allo*, and *epi-allo*, according to the stereochemistry at the C3 and C20 positions of the molecule. Mitragynine (**1**), the major alkaloid in *Mitragyna speciosa*, has the 3*S*, 20*R* configuration and belongs to the *allo* group (Figure 9.1). Minor constituents of this plant, speciogynine (**2**), having the

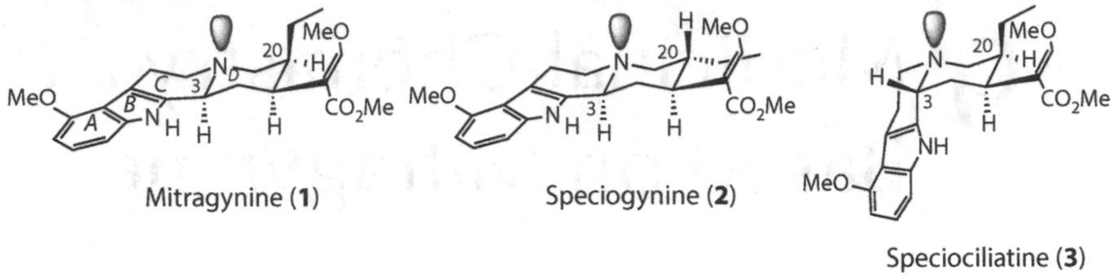

FIGURE 9.1 Stereostructures of mitragynine (**1**), speciogynine (**2**), and specio-ciliatine (**3**).

3*S*, 20*S* configuration (*normal* group), and speciociliatine (**3**), having the 3*R*, 20*R* configuration (*epi-allo* group), show lower activity toward opioid receptors than mitragynine (**1**) (Takayama et al. 2002). The results indicate that both the stereochemistry of the ethyl group at C20 and the flat *trans*-quinolizidine conformation in the Corynanthe skeleton are important for exhibiting the opioid agonistic activity.

Beckett and Casy (1954) proposed that the spatial arrangement of the three functional groups—nitrogen atom, benzene residue, and phenol function—in the morphine structure is a critical factor for the analgesic activity. Portoghese (1965) suggested the new concept "different binding modes for different opioid skeletons." Those concepts have since been refined by researchers, and today receptor-based models are used to interpret the interactions between the opioid receptors and the agonists, including the non-morphinan skeletons thus far developed (Dhawan et al. 1996). Inspection of the structural similarity between mitragynine (**1**) and morphine (**4**) using molecular modeling techniques revealed that all the three functions (nitrogen atom, benzene ring, and oxygen function on the benzene ring) of the two molecules could not be superimposed (Figure 9.2). Thus it was ascertained that *Mitragyna* alkaloids are structurally different from the morphine skeleton.

The N_b oxide derivative (**5**) prepared by the *m*-CPBA oxidation of **1** (Scheme 9.1) showed no opioid agonistic activity in guinea pig ileum. The N_b lone electron pair was essential for exhibiting the opioid agonistic activity.

To investigate the effect of the β-methoxyacrylate residue on the opioid agonistic activity, the three derivatives **6**, **7**, and **8** were prepared, as shown in Scheme 9.1. These compounds exhibited very weak or no opioid agonistic activity, indicating that the β-methoxyacrylate function is essential for binding to the opioid receptors.

The influence of the methoxy group at the C9 position of the indole ring in **1** was examined because the presence of the function was a structural characteristic of *Mitragyna* alkaloids compared with other common Corynanthe-type indole alkaloids. As shown in Scheme 9.2, various derivatives concerning

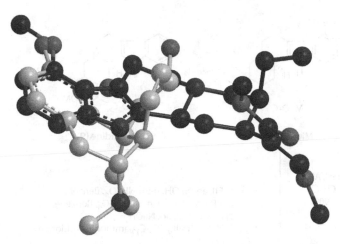

FIGURE 9.2 Overlays of the low-energy conformations of morphine (**4**) (gray) and mitragynine (**1**) (black). Hydrogen atoms are omitted.

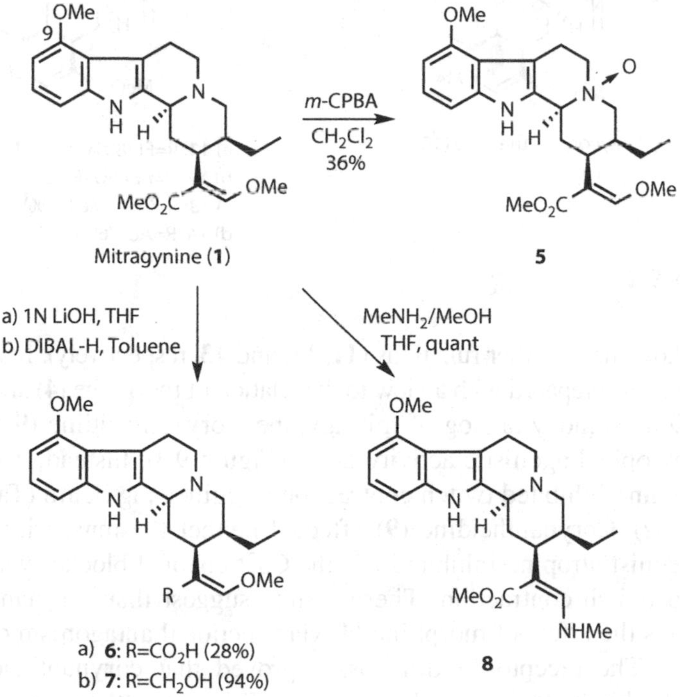

SCHEME 9.1

the C9-methoxy group were prepared. The methyl ether at C9 was selectively cleaved with EtSH and AlCl₃ in CH₂Cl₂ to give 9-hydroxycorynantheidine (**10**). The thus obtained phenolic function was converted into ethyl, *i*-propyl,

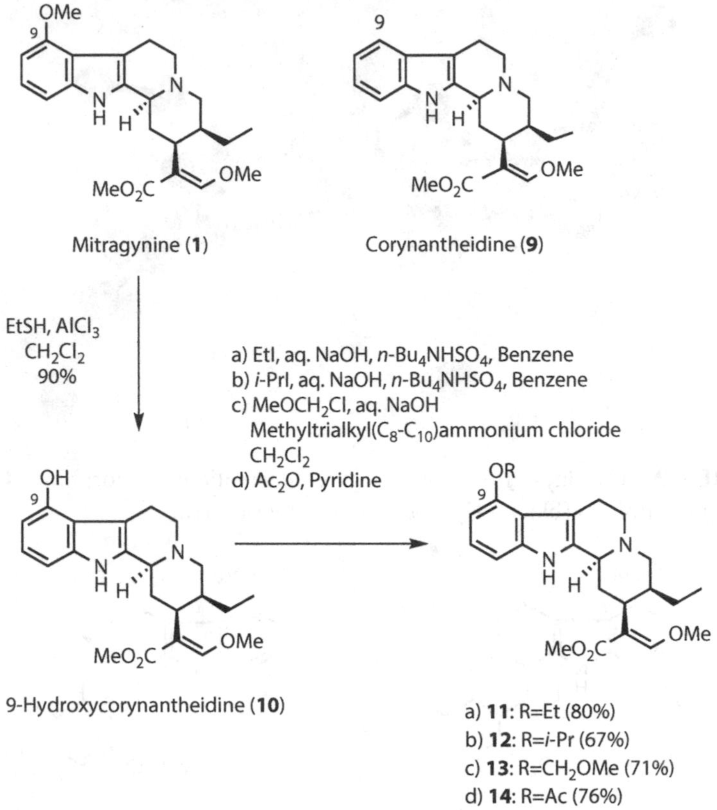

SCHEME 9.2

and methoxymethyl ether functions (**11**, **12**, and **13**, respectively). In addition, acetate **14** was prepared with a view to the relation of morphine (**4**) and heroin.

The 9-demethoxy analog of mitragynine, corynantheidine (**9**), did not show any opioid agonistic activity at all (Figure 9.3). Instead, it reversed the morphine-inhibited twitch contraction in guinea pig ileum (Takayama et al. 2002). Corynantheidine (**9**) affected neither the muscarinic receptor antagonist atropine-inhibited nor the Ca^{2+} channel blocker verapamil-inhibited twitch contraction. These results suggest that corynantheidine (**9**) inhibits the effect of morphine (**4**) via functional antagonism of opioid receptors. The receptor-binding assay proved that corynantheidine (**9**) selectively binds to μ-opioid receptors. Taken together, corynantheidine (**9**) was found to have an antagonistic effect on μ-opioid receptors. The 9-demethyl analog of mitragynine, 9-hydroxycorynantheidine (**10**), also inhibited the electrically stimulated twitch contraction in guinea pig ileum, but its maximum percentage inhibition was lower than that of mitragynine (Figure 9.3). The receptor-binding assay clarified that

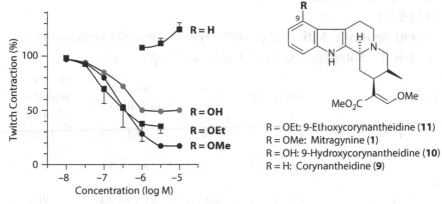

FIGURE 9.3   Effect of substituents at C9 in mitragynine on twitch contraction induced by electrical stimulation in guinea pig ileum.

9-hydroxycorynantheidine (**10**) binds to μ-opioid receptors. Taken together, the results suggest that 9-hydroxycorynantheidine (**10**) is a partial agonist of opioid receptors (Matsumoto et al. 2006). It is interesting that a transformation of the substituent at C9, i.e., from OMe to OH or to H, led to a shift of activity from a full agonist via a partial agonist to an antagonist of opioid receptors. Thus it was found that the functional group at C9 of mitragynine-related compounds is responsible for the relative inhibitory activity, which means the intrinsic activity toward opioid receptors. The introduction of an acetoxy group at C9 of the indole ring (compound **14**) led to marked reductions of both intrinsic activity and potency compared with those of mitragynine (**1**). Compounds **11** and **12**, having ethyl and *i*-propyl ether groups, induced naloxone-insensitive inhibition of the twitch contraction (Table 9.1), suggesting an inhibitory effect via mechanisms distinct from those of the stimulation of opioid receptors. Compound **13** did not show any opioid agonistic activity. The results demonstrate that the intrinsic activities of the compounds toward opioid receptors are determined by the functional groups at the C9 position, and that the methoxy group at the C9 position is the most suitable functional group for pharmacophore binding to opioid receptors.

9.3   OXIDATIVE DERIVATIVES OF INDOLE NUCLEUS IN MITRAGYNINE

In 1974, Zarembo et al. (1974) reported that mitragynine pseudoindoxyl (**15**), which was obtained by the microbial transformation of **1** by the fungus *Helminthosporium* sp., exhibited an almost ten-fold higher analgesic activity than mitragynine in the D'Amour–Smith test (Scheme 9.3). Compound

TABLE 9.1

Opioid Agonistic Activities of Mitragynine-Related Compounds in Electrically Stimulated Guinea Pig Ileum Preparations

Compound	n	pD$_2$ Value	Relative Potency	Maximum Inhibition (%)	Relative Inhibitory Activity
4 (Morphine)	5	7.17 ± 0.05	100%	87.2 ± 1.8	100%
1 (Mitragynine)	5	6.59 ± 0.13	26%	83.1 ± 3.7	95%
15	6	8.71 ± 0.07	3467%	83.5 ± 3.3	96%
9	5	—	—	−18.1 ± 8.6	−21%
10	5	6.78 ± 0.23	41%	49.4 ± 3.1	57%
11[a]	5	NS	NS	NS	NS
12[a]	5	NS	NS	NS	NS
13[b]	5	NE	NE	NE	NE
14	5	5.39 ± 0.12	2%	33.2 ± 8.8	38%
5	5	—	—	−123.5 ± 39.2	−142%
16	6	6.50 ± 0.16	21%	13.4 ± 12.7	15%
17	5	8.20 ± 0.14	1071%	86.3 ± 4.8	99%
18	5	6.45 ± 0.04	19%	60.9 ± 7.2	70%
19	5	5.29 ± 0.12	1%	22.9 ± 1.1	26%
22	5	6.70 ± 0.04	34%	74.1 ± 5.6	85%
3	5	5.40 ± 0.07	2%	85.9 ± 2.7	101%

Relative potency is expressed as percentage of the pD$_2$ value of each compound relative to that of morphine. Maximum inhibition (%), which is elicited by a compound when the response reaches a plateau, is calculated by regarding the electrically stimulated contraction as 100%. The concentration range of the tested compounds is from 100 pM to 30 μM. Relative inhibitory activity, which means intrinsic activity toward opioid receptors, is expressed as percentage of the maximum inhibition by each compound relative to that by morphine.

[a] In the case of naloxone-insensitive inhibition, the effect is regarded as "non-specific" (NS).

[b] In the case that significant inhibition is not obtained at the compound concentration of 30 μM, the effect is regarded as "no effect" (NE).

15 was also prepared by chemical synthesis as follows. Oxidation of **1** with lead tetraacetate (Pb(OAc)$_4$) gave 7-acetoxyindolenine derivative **16**, which was subjected to alkaline hydrolysis to yield 7-hydroxymitragynine (**17**), a minor constituent in the leaves of *M. speciosa*. Treatment of **17** with sodium methoxide in methanol gave pseudoindoxyl derivative **15** (Takayama et al. 1996). Two oxindole derivatives (**20** and **21**) were also

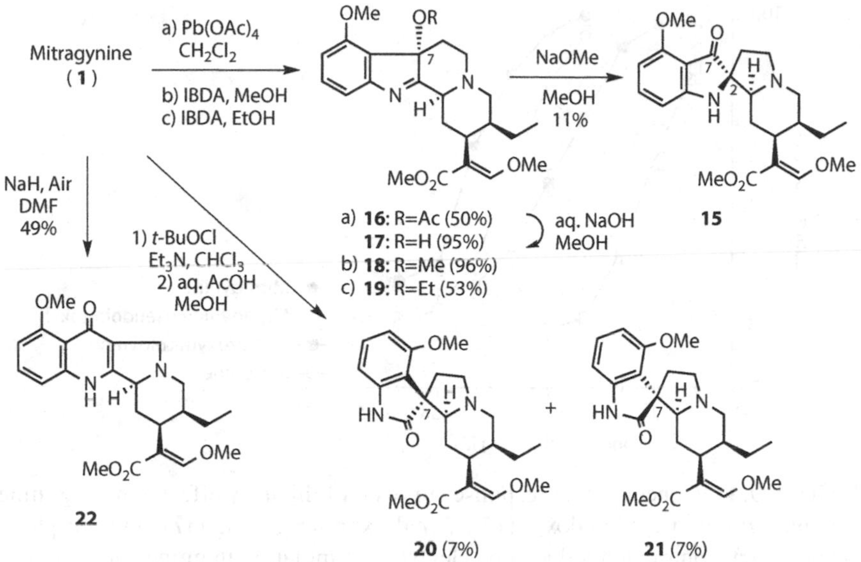

SCHEME 9.3

prepared from mitragynine (**1**) via oxidation with *t*-BuOCl and subsequent treatment of the resultant 7-chloroindolenine derivative with aqueous acetic acid in methanol. For comparison with 7-hydroxymitragynine (**17**), 7-methoxy and 7-ethoxy indolenine derivatives (**18** and **19**) were prepared from **1** by treatment with iodobenzene diacetate in an appropriate solvent. Treatment of mitragynine (**1**) with NaH in DMF in air yielded 4-quinolone (Winterfeld 1971) derivative **22**.

Among the derivatives prepared above, mitragynine pseudoindoxyl (**15**) inhibited the electrically stimulated ileum contraction in a concentration-dependent manner (from 100 pM–30 nM) (Yamamoto et al. 1999) and its pD_2 value was 8.71 ± 0.07 (Table 9.1). This potency was approximately 100- and 20-fold higher than those of mitragynine (**1**) and morphine (**4**), respectively. On the other hand, mitragynine oxindoles (**20** and **21**) showed quite weak opioid agonistic activities. 7-Hydroxymitragynine (**17**), a minor constituent of *M. speciosa* (Ponglux et al. 1994), was found to exhibit high potency toward opioid receptors: approximately 13- and 46-fold higher than morphine (**4**) and mitragynine (**1**), respectively (Figure 9.4). The intrinsic activity of 7-hydroxymitragynine (**17**) suggests its full agonistic effect on opioid receptors. The introduction of a hydroxyl group at the C7 position increased the potency compared with mitragynine (**1**). In contrast, 7-acetoxy derivative (**16**) had lower intrinsic activity than mitragynine (**1**), although its potency was nearly equal to that of mitragynine (**1**)

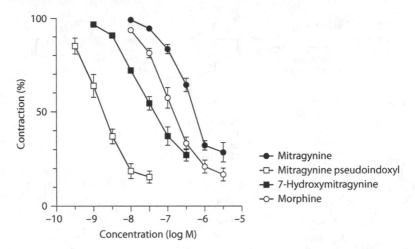

FIGURE 9.4 Concentration response curves for inhibitory effect of mitragynine (**1**), mitragynine pseudoindoxyl (**15**), 7-hydroxymitragynine (**17**), and morphine (**4**) on twitch contraction induced by electrical stimulation in guinea pig ileum.

(Table 9.1). The introduction of a methoxy or an ethoxy group at the C7 position (compounds **18** and **19**) led to a marked reduction in both intrinsic activity and potency toward opioid receptors (Table 9.1). It is conceivable that the hydroxyl group at the C7 position in the mitragynine skeleton is necessary for the increased potency toward opioid receptors (Matsumoto et al. 2004, 2005, 2006).

As described above, the hydroxyl group at the C7 position in mitragynine (**1**) was a suitable functional group for pharmacophore binding to the opioid receptors. However, 7-hydroxyspeciociliatine (**23**), which also has a hydroxyl group at the C7 position in the Corynanthe skeleton, exhibited approximately 28-fold lower potency than 7-hydroxymitragynine (**17**). As shown in Figure 9.5, 7-hydroxymitragynine (**17**), with the 3*S*, 7*S* configuration, and 7-hydroxyspeciociliatine (**23**), with the 3*R*, 7*R* configuration, have antipodal structures with respect to the 7-hydroxyl indolenine moiety. Therefore, the appropriate spatial disposition for pharmacophore binding to the opioid receptors is thought to be the 3*S*, 7*S* form in the Corynanthe skeleton (Kitajima et al. 2006).

Among the mitragynine derivatives described above, mitragynine pseudoindoxyl (**15**) and 7-hydroxymitragynine (**17**) emerged as the compounds with potential analgesic activity in the in vitro assay. In this regard, the antinociceptive activities of these compounds were evaluated in mice. Mitragynine pseudoindoxyl (**15**) administered by i.c.v. injection showed antinociceptive activity in the tail-flick test in mice (Takayama et al. 2002), its activity reaching a maximum between about 15 to 45 minutes after

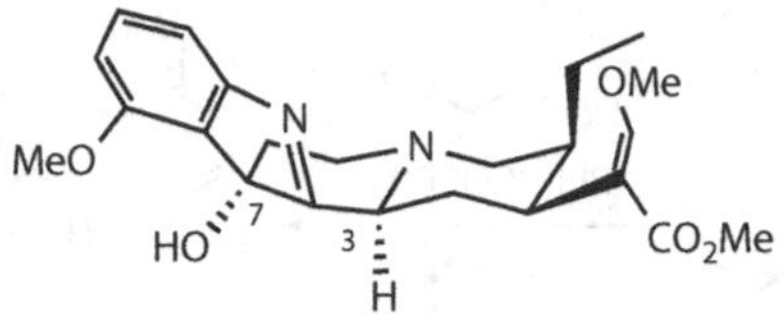

7-Hydroxymitragynine (17)

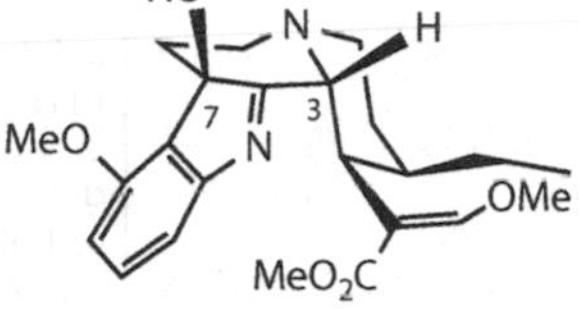

7-Hydroxyspeciociliatine (23)

FIGURE 9.5   Stereostructures of compounds **17** and **23**.

the injection. Mitragynine pseudoindoxyl (**15**), however, has lower activity than morphine (**4**). The EC_{50} value estimated for mitragynine pseudoindoxyl (**15**) was 6.51 nmol/mouse (95% confidence limit, 3.78–11.20 nmol/mouse), whereas that for morphine (**4**) was 3.20 nmol/mouse (95% confidence limit, 2.11–4.88 nmol/mouse). The antinociceptive activities of mitragynine (**1**), mitragynine pseudoindoxyl (**15**), and morphine (**4**) were completely inhibited by naloxone at 2 mg/kg, s.c. Therefore, these compounds induce analgesic activity via opioid receptors. Mitragynine pseudoindoxyl (**15**) exhibited lower analgesic activity than morphine (**4**) in spite of the very high opioid agonistic activity in the isolated guinea pig ileum test. It is speculated that the low analgesic activity of mitragynine pseudoindoxyl (**15**) is due to the instability of the compound in the brain and/or the low permeability of the blood–brain barrier.

The administration of 7-hydroxymitragynine (**17**) (2.5–10 mg/kg, s.c.) induced dose-dependent antinociceptive activities in both tail-flick and hot-plate tests in mice. Its activity was higher than that of morphine (**4**) in both tests. It is noteworthy that 7-hydroxymitragynine (**17**) exhibited potent antinociceptive activity after oral administration (5–10 mg/kg) in the tail-flick and hot-plate tests (Matsumoto et al. 2004). By contrast, orally administered morphine (**4**) (20 mg/kg) did not show antinociceptive activity in either test.

In order to develop more potent opioid receptor agonists based on 7-hydroxymitragynine (**17**), several compounds having various substituents on the benzene ring in **17** were synthesized, as follows (Takayama et al. 2006). Initially, mitragynine (**1**) (Takayama et al. 1995) was treated with one equivalent of phenyliodine bis(trifluoroacetate) (PIFA) in the presence of ethylene glycol (EG) in MeCN at 0°C to give 2,3-ethylene glycol-bridged indoline derivative (**24**) in a quantitative yield. By this transformation, the

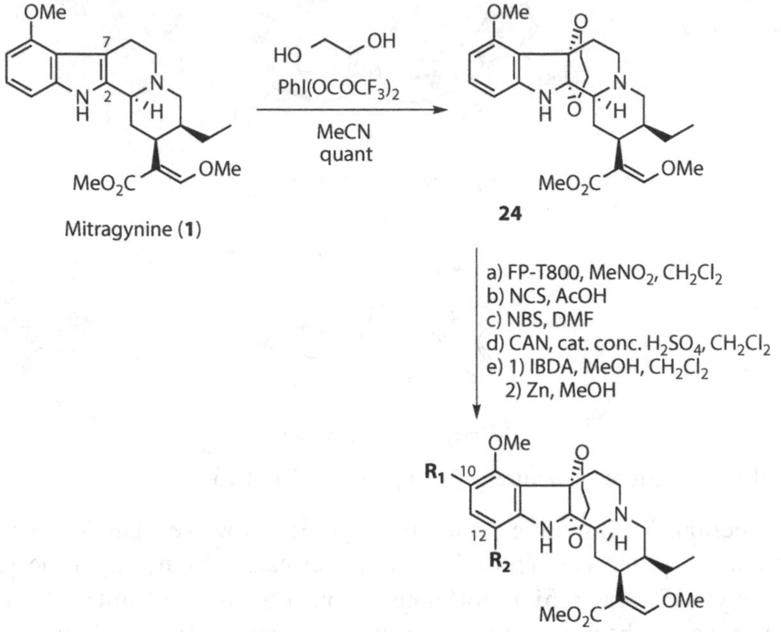

OMe
MeO2C
OMe
Mitragynine (1)
HO OH
PhI(OCOCF3)2
MeCN
quant
24
a) FP-T800, MeNO2, CH2Cl2
b) NCS, AcOH
c) NBS, DMF
d) CAN, cat. conc. H2SO4, CH2Cl2
e) 1) IBDA, MeOH, CH2Cl2
2) Zn, MeOH
R1
R2

a) **25**: R_1=F, R_2=H (53%)
b) **26a**: R_1=Cl, R_2=H (88%), **26b**: R_1=H, R_2=Cl (11%)
c) **27a**: R_1=Br, R_2=H (75%), **27b**: R_1=H, R_2=Br (24%)
d) **28a**: R_1=NO_2, R_2=H (52%), **28b**: R_1=H, R_2=NO_2 (21%)
e) **29**: R_1=OMe, R_2=H (64%, 2 steps)

SCHEME 9.4

reaction with electrophiles at the C3 position of the indole ring in **1** could be prevented. Using EG adduct **24**, various substituents were introduced onto the benzene ring, as shown in Scheme 9.4. Treatment of **24** with *N*-fluoro-2,6-dichloropyridinium triflate (FP-T800) gave compound **25** fluorinated at the C10 position in 53% yield. Exposure of **24** to NCS in AcOH afforded two chlorinated derivatives **26a** (10-chloro) and **26b** (12-chloro) in 88% and 11% yields, respectively. Using NBS in DMF, 10-bromo and 12-bromo derivatives (**27a** and **27b**) were obtained in 75% and 24% yields, respectively. To introduce a nitro group, a combination of CAN and concentrated H_2SO_4 in CH_2Cl_2 was used to generate **28a** in 52% yield together with its 12-isomer (**28b**) in 21% yield. 10-Methoxy derivative (**29**) was prepared in 64% yield by treatment of **24** with IBDA in MeOH/CH_2Cl_2, followed by reduction of the resulting iminoquinone intermediate with Zn in MeOH.

The C10-substituted derivatives thus obtained as the major product of each electrophilic aromatic substitution reaction were converted into their indole derivatives in good yields by reduction with $NaCNBH_3$ in AcOH. However, in the case of nitro derivative **28a**, a two-step procedure was required: **28a** was treated with TBSOTf in the presence of 2,6-lutidine,

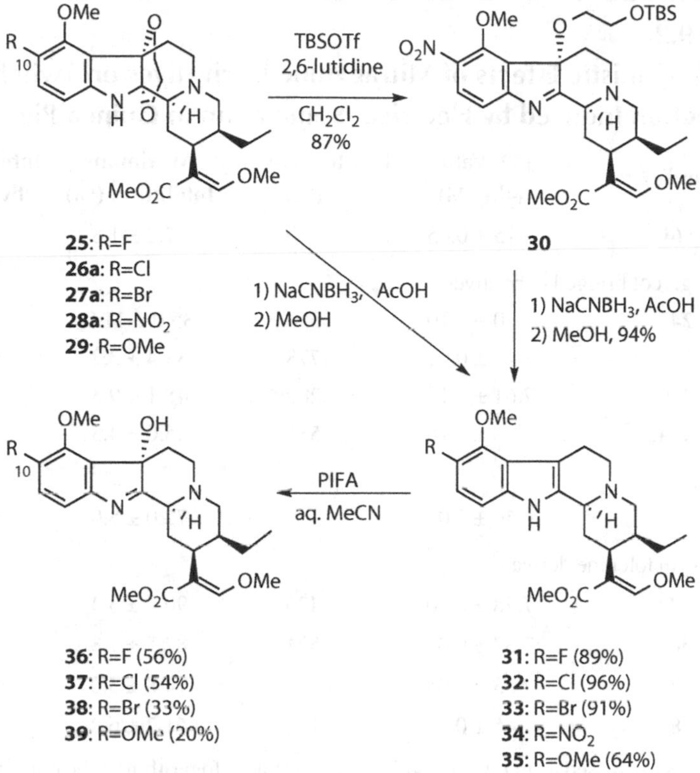

SCHEME 9.5

and resultant indolenine derivative **30** obtained in 87% yield was reduced with NaCNBH$_3$ to give indole derivative **34** in 94% yield (Scheme 9.5). The thus obtained indole derivatives (**31–35**) were respectively converted into 7-hydroxyindolenine derivatives (**36–39**) by oxidation with PIFA in aqueous MeCN.

The series of C10-substituted mitragynine derivatives obtained by the above reactions was subjected to pharmacological evaluation. Opioid agonistic effect was evaluated in an experiment involving twitch contraction induced by electrical stimulation of guinea pig ileum. The results are shown in Table 9.2. Among the EG bridged derivatives (**24**, **25**, **26a**, **27a**, **28a**, **29**) and the 7-hydroxyindolenine derivatives (**36–39**), the C10-fluorinated derivatives (**25**, **36**) showed the highest potency. Derivatives having a chloro or bromo group at C10 showed lower potency than the corresponding fluorinated derivatives. None of the indole derivatives (**31–35**) showed any opioid agonistic effect. These results suggest that the dimension or electronegativity of the functional group at the C10 position is important in eliciting the opioid agonistic effect. Compound **36** showed potent opioid agonistic

TABLE 9.2

Opioid Agonistic Effects of Mitragynine Derivatives on Twitch Contraction Induced by Electrical Stimulation in Guinea Pig Ileum[a]

Compound	pD$_2$ Value (–log M)	Relative Potency (%)	Maximum Inhibition (%)	Inhibitory Activity (%)
Morphine (**4**)	7.15 ± 0.05	100	87.2 ± 1.8	100
Ethylene glycol bridged derivatives				
24	7.70 ± 0.10	354	35.0 ± 11.0	40
25	8.40 ± 0.02	1778	83.4 ± 3.2	96
26a	7.61 ± 0.17	288	48.1 ± 9.3	55
28a	7.88 ± 0.18	537	65.0 ± 4.3	75
Mitragynine				
1	6.50 ± 0.06	22	72.0 ± 5.0	83
7-Hydroxyindolenine derivatives				
17	7.78 ± 0.10	426	90.8 ± 3.4	104
36	7.87 ± 0.04	524	82.5 ± 1.8	95
37	7.53 ± 0.08	239	74.8 ± 3.0	86
38	7.45 ± 0.04	199	61.7 ± 6.2	71

[a] Potency is expressed as pD$_2$ value, which is the negative logarithm of the concentration required to produce 50% of the maximum response to each compound (EC$_5$0). Relative potency is expressed as percentage of the pD$_2$ value of each compound relative to that of morphine. Maximum inhibition (%), which is elicited by a compound when the response reaches a plateau, is calculated by regarding the electrically stimulated contraction as 100%. Relative inhibitory activity, which means intrinsic activity toward opioid receptors, is expressed as percentage of the maximum inhibition by each compound relative to that by morphine. Compounds **27a–29**, **31–35**, and **39** did not show significant inhibition at 1 μM.

effect, but its potency was nearly equal to that of 7-hydroxymitragynine (**17**). On the other hand, compound **25** showed the most potent opioid agonistic effect among the derivatives tested in the present study. Its potency was 18- and 4-fold higher than that of morphine (**4**) and 7-hydroxymitragynine (**17**), respectively. Next, the contribution of opioid receptor subtypes to the pharmacological effects of **25** was examined. The μ-opioid receptor antagonist cyprodime (1 μM) and the κ-opioid receptor antagonist nor-binaltorphimine (30 nM) significantly reversed the inhibitory effect of **25** at 30 nM, suggesting that **25** activates not only μ-opioid receptors but also κ-opioid receptors. The

TABLE 9.3

Antinociceptive Effects Produced by Subcutaneous Administration of Morphine (4), 7-Hydroxymitragynine (17), and 25 in Mice[a]

Compound	Morphine (4)	7-Hydroxymitragynine (17)	25
Tail-flick	4.57 (3.12–6.69)	0.80 (0.43–1.33)	0.57 (0.36–0.90)
Hot-plate	4.08 (2.75–6.06)	0.93 (0.59–1.45)	0.70 (0.42–1.17)
Writhing	0.50 (0.31–0.80)	0.15 (0.09–0.24)	0.06 (0.03–0.09)

[a] ED_{50} represents the median effective dose (mg/kg) (95% confidence limit).

TABLE 9.4

Antinociceptive Effects Due to Oral Administration of Morphine (4), 7-Hydroxymitragynine (17), and 25 in Mice[a]

Compound	Morphine (4)	7-Hydroxymitragynine (17)	25
Tail-flick	63.0 (37.2–106.8)	4.43 (1.57–6.93)	2.84 (1.60–5.05)
Hot-plate	48.2 (27.5–84.5)	2.23 (1.38–3.60)	2.98 (1.79–4.92)
Writhing	4.60 (2.87–7.38)	1.05 (0.62–1.78)	0.63 (0.40–0.99)

[a] ED_{50} represents the median effective dose (mg/kg) (95% confidence limit).

antinociceptive activities of compound **25** in an in vivo experiment were more potent than those of 7-hydroxymitragynine (**17**) (Tables 9.3 and 9.4) (Matsumoto et al. 2008). The pharmacological profiles of compound **25** are described in Chapter 13 of this book in detail.

9.4 MGM-15 AND MGM-16

The above findings—that the opioid agonistic activity of the mitragynine derivatives increased in the order of mitragynine (**1**), 7-hydroxymitragynine (**17**), and compound **25**—point out that the presence of both the oxygen function at the C7 position and the tetrahedral sp^3 carbon at the C2 position would affect the affinity for opioid receptors. In particular, changing the spatial position of the benzene ring by the transformation of the C2 atom from sp^2 to sp^3 configuration would increase the activity (Figure 9.6). Based on this idea, compounds **40** and **41** having an sp^3 carbon at the C2 position were prepared by reduction of **17** and **36**, respectively, with sodium borohydride (Scheme 9.6).

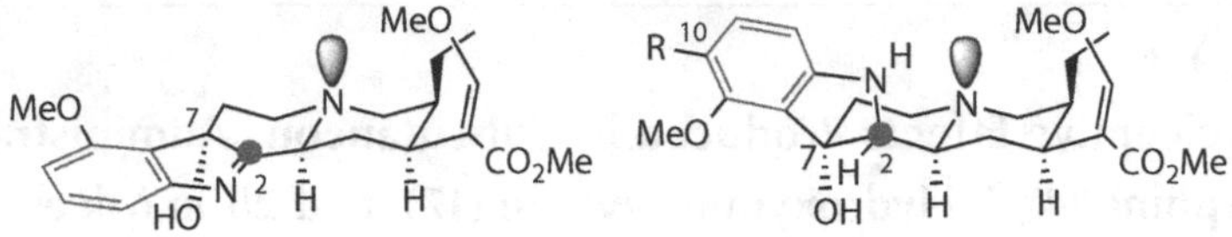

7-Hydroxymitragynine (17)

R=H: MGM-15 (40)
R=F: MGM-16 (41)

FIGURE 9.6 Stereostructures of compounds **17**, **40**, and **41**.

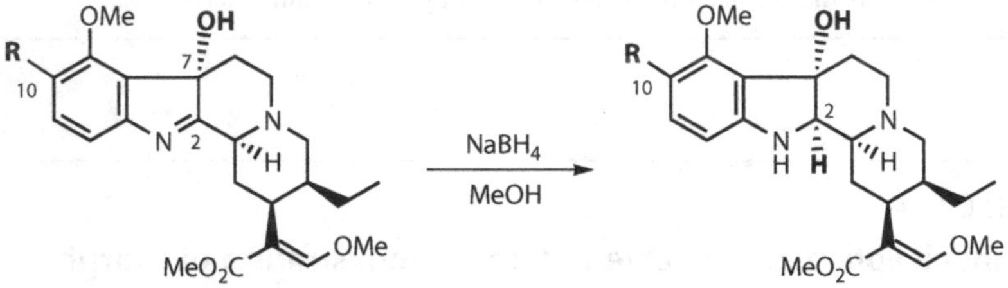

R=H: 7-Hydroxymitragynine (**17**)
R=F: **36**

R=H: MGM-15 (**40**)
R=F: MGM-16 (**41**)

SCHEME 9.6

TABLE 9.5

pD$_2$ Values and Maximum Inhibitory Effects of Mitragynine Derivatives in Guinea Pig Ileum

Compound	pD$_2$ Value (−log M)	Maximum Inhibition (%)
Morphine (**4**)	7.16 ± 0.05	79.4 ± 5.7
7-Hydroxymitragynine (**17**)	7.78 ± 0.10	90.8 ± 3.4
MGM-15 (**40**)	8.26 ± 0.05	78.2 ± 3.4
MGM-16 (**41**)	8.81 ± 0.09	89.0 ± 2.3

These compounds, tentatively called MGM-15 (**40**) and MGM-16 (**41**), showed, as expected, higher potency than 7-hydroxymitragynine (**17**), based on their pD$_2$ values (Table 9.5). The introduction of a fluoro group at the C10 position (MGM-16, **41**) led to a higher potency than MGM-15 (**40**). Further, MGM-15 (**40**) and MGM-16 (**41**) induced dose-related antinociceptive responses in the tail-flick tests after subcutaneous and oral administration (Table 9.6). The antinociceptive effect of MGM-15 (**40**) was approximately 15 and 50 times more potent than that of morphine (**4**) after subcutaneous and oral administration, respectively. The antinociceptive

TABLE 9.6
ED_{50} Values Indicating Antinociceptive Effects Produced by Subcutaneous and Oral Administration of Morphine (4), 7-Hydroxymitragynine (17), MGM-15 (40), and MGM-16 (41) in Mice (Tail-Flick)[a]

	ED_{50} (mg/kg)	
Compound	S.C.	P.O.
Morphine (4)	4.57 (3.12–6.69)	63.0 (37.2–106.8)
7-Hydroxymitragynine (17)	0.80 (0.48–1.33)	4.43 (1.57–6.93)
MGM-15 (40)	0.30 (0.18–0.50)	1.26 (0.84–1.88)
MGM-16 (41)	0.064 (0.040–0.10)	0.263 (0.165–0.420)

[a] ED_{50} represents the median effective dose (mg/kg) (95% confidence limit).

effect of MGM-16 (41) was approximately 71 and 240 times more potent than that of morphine (4) after subcutaneous and oral administration, respectively (Matsumoto et al. 2014). The pharmacological profiles of these compounds are described in Chapter 13 of this book in detail. These findings indicate that MGM-16 (41) could become a class of compounds with potential therapeutic utility for treating pain.

REFERENCES

Adkins, J.E., Boyer, E.W., and McCurdy, C.R. 2011. *Mitragyna speciosa*, a psychoactive tree from Southeast Asia with opioid activity. *Curr Top Med Chem* 11: 1165–75.

Beckett, A.H., and Casy, A.F. 1954. Synthetic analgesics: Stereochemical considerations. *J Pharm Pharmacol* 6: 986–1001.

Dhawan, B.N., Cesselin, F., Raghubir, R., Reisine, T., Bradley, P.B., Portoghese, P.S., and Hamon, M. 1996. International Union of Pharmacology. XII. Classification of opioid receptors. *Pharmacol Rev* 48: 567–92.

Horie, S., Yamamoto, L.T., Futagami, Y., Yano, S., Takayama, H., Sakai, S., Aimi, N., Ponglux, D., Shan, J., Pang, P.K.T., and Watanabe, K. 1995. Analgesic, neuronal Ca^{2+} channel-blocking and smooth muscle relaxant activities of mitragynine, an indole alkaloid, from the Thai folk medicine 'Kratom.' *J Traditional Med* 12: 366–67.

Horie, S., Koyama, F., Takayama, H., Ishikawa, H., Aimi, N., Ponglux, D., Matsumoto, K., and Murayama, T. 2005. Indole alkaloids of a Thai medicinal herb, *Mitragyna speciosa*, that has opioid agonistic effect in guinea-pig ileum. *Planta Med* 71: 231–36.

Kitajima, M., Misawa, K., Kogure, N., Said, I.M., Horie, S., Hatori, Y., Murayama, T., and Takayama, H. 2006. A new indole alkaloid, 7-hydroxyspeciociliatine, from the fruits of Malaysian *Mitragyna speciosa* and its opioid agonistic activity. *J Nat Med* 60: 28–35.

Matsumoto, K., Mizowaki, M., Thongpradichote, S., Murakami, Y., Takayama, H., Sakai, S., Aimi, N., and Watanabe, H. 1996a. Central antinociceptive effects of mitragynine in mice: Contribution of descending noradrenergic and serotonergic systems. *Eur J Pharmacol* 317: 75–81.

Matsumoto, K., Mizowaki, M., Thongpradichote, S., Takayama, H., Sakai, S., Aimi, N., and Watanabe, H. 1996b. Antinociceptive action of mitragynine in mice: Evidence for the involvement of supraspinal opioid receptors. *Life Sci* 59: 1149–55.

Matsumoto, K., Mizowaki, M., Takayama, H., Sakai, S., Aimi, N., and Watanabe, H. 1997. Suppressive effect of mitragynine on the 5-methoxy-*N,N*-dimethyltryptamine-induced head-twitch response in mice. *Pharmacol Biochem Behav* 57: 319–23.

Matsumoto, K., Horie, S., Ishikawa, H., Takayama, H., Aimi, N., Ponglux, D., and Watanabe, K. 2004. Antinociceptive effect of 7-hydroxymitragynine in mice: Discovery of an orally active opioid analgesic from the Thai medicinal herb *Mitragyna speciosa*. *Life Sci* 74: 2143–55.

Matsumoto, K., Horie, S., Takayama, H., Ishikawa, H., Aimi, N., Ponglux, D., Murayama, T., and Watanabe, K. 2005a. Antinociception, tolerance and withdrawal symptoms induced by 7-hydroxymitragynine, an alkaloid from the Thai medicinal herb *Mitragyna speciosa*. *Life Sci* 78: 2–7.

Matsumoto, K., Yamamoto, L.T., Watanabe, K., Yano, S., Shan, J., Pang, P.K.T., Ponglux, D., Takayama, H., and Horie, S. 2005b. Inhibitory effect of mitragynine, an analgesic alkaloid from Thai herbal medicine, on neurogenic contraction of the vas deferens. *Life Sci* 78: 187–94.

Matsumoto, K., Hatori, Y., Murayama, T., Tashima, K., Wongseripipatana, S., Misawa, K., Kitajima, M., Takayama, H., and Horie, S. 2006a. Involvement of μ-opioid receptors in antinociception and inhibition of gastrointestinal transit induced by 7-hydroxymitragynine, isolated from Thai herbal medicine *Mitragyna speciosa*. *Eur J Pharmacol* 549: 63–70.

Matsumoto, K., Takayama, H., Ishikawa, H., Aimi, N., Ponglux, D., Watanabe, K., and Horie S. 2006b. Partial agonistic effect of 9-hydroxycorynantheidine on μ-opioid receptor in the guinea-pig ileum. *Life Sci* 78: 2265–71.

Matsumoto, K., Takayama, H., Narita, M. et al. 2008. MGM-9 [(*E*)-methyl 2-(3-ethyl-7a,12a-(epoxyethanoxy)-9-fluoro-1,2,3,4,6,7,12,12b-octahydro-8-methoxyindolo[2,3-a]quinolizin-2-yl)-3-methoxyacrylate], a derivative of the indole alkaloid mitragynine: A novel dual-acting μ- and κ-opioid agonist with potent antinociceptive and weak rewarding effects in mice. *Neuropharmacology* 55: 154–65.

Matsumoto, K., Narita, M., Muramatsu, N. et al. 2014. Orally active opioid μ/δ dual agonist MGM-16, a derivative of the indole alkaloid mitragynine, exhibits potent analgesic effect on neuropathic pain in mice. *J Pharmacol Exp Ther*, 348:383–92.

Ponglux, D., Wongseripipatana, S., Takayama, H., Kikuchi, M., Kurihara, M., Kitajima, M., Aimi, N., and Sakai, S. 1994. A new indole alkaloid, 7α-hydroxy-7H-mitragynine, from *Mitragyna speciosa* in Thailand. *Planta Medica* 60: 580–81.

Portoghese, P.S. 1965. A new concept on the mode of interaction of narcotic analgesics with receptors. *J Med Chem* 8: 609–16.

Raffa, R.B., Beckett, J.R., Brahmbhatt, V.N. et al. 2013. Orally active opioid compounds from a non-poppy source. *J Med Chem* 56: 4840–48.

Saxton, J.E. (ed.) 1994. The Chemistry of Heterocyclic Compounds, Part 4, Suppl. Vol. 25. *Indoles: The Monoterpenoid Indole Alkaloids*. Chichester: John Wiley & Sons.

Takayama, H. 2004. Chemistry and pharmacology of analgesic indole alkaloids from the rubiaceous plant, *Mitragyna speciosa*. *Chem Pharm Bull* 52: 916–28.

Takayama, H., Maeda, M., Ohbayashi, S., Kitajima, M., Sakai, S., and Aimi, N. 1995. The first total synthesis of (–)-mitragynine, an analgesic indole alkaloid in *Mitragyna speciosa*. *Tetrahedron Lett* 36: 9337–40.

Takayama, H., Kurihara, M., Subhadhirasakul, S., Kitajima, M., Aimi, N., and Sakai, S. 1996. Stereochemical assignment of pseudoindoxyl alkaloids. *Heterocycles* 42: 87–92.

Takayama, H., Ishikawa, H., Kurihara, M. et al. 2002. Studies on the synthesis and opioid agonistic activities of mitragynine-related indole alkaloids: Discovery of opioid agonists structurally different from other opioid ligands. *J Med Chem* 45: 1949–56.

Takayama, H., Kitajima, M., and Kogure, N. 2005. Chemistry of Corynanthe-type related indole alkaloids from the *Uncaria, Nauclea,* and *Mitragyna* plants. *Curr Org Chem* 9: 1445–64.

Takayama, H., Misawa, K., Okada, N. et al. 2006. New procedure to mask the 2,3-π bond of the indole nucleus and its application to the preparation of potent opioid receptor agonists with a Corynanthe skeleton. *Org Lett* 8: 5705–708.

Thongpraditchote, S., Matsumoto, K., Tohda, M., Takayama, H., Aimi, N., Sakai, S., and Watanabe, H. 1998. Identification of opioid receptor subtypes in antinociceptive actions of supraspinally-administered mitragynine in mice. *Life Sci* 62: 1371–78.

Tohda, M., Thongpraditchote, S., Matsumoto, K. et al. 1997. Effects of mitragynine on cAMP formation mediated by δ-opiate receptors in NG108-15 cells. *Biol Pharm Bull* 20: 338–40.

Watanabe K., Yano S., Horie S., and Yamamoto L.T. 1997. Inhibitory effect of mitragynine, an alkaloid with analgesic effect from Thai medicinal plant, *Mitragyna speciosa*, on electrically stimulated contraction of isolated guinea-pig ileum through the opioid receptor. *Life Sci* 60: 933–42.

Watanabe, K., Yano, S., Horie, S. et al. 1999. Pharmacological properties of some structurally related indole alkaloids contained in the Asian herbal medicines, hirsutine and mitragynine, with special reference to their Ca^{2+} antagonistic and opioid-like effects. In *Pharmacological Research on Traditional Herbal Medicines*, Watanabe, H., and Shibuya, T. (eds.), 163–77. Amsterdam: Harwood Academic Publishers.

Winterfeldt, E. 1971. Reaktionen an Indolderivaten, XIII. Chinolon-Derivate durch Autoxydation. *Liebigs Annalen der Chemie* 745: 23–30.

Yamamoto, L.T., Horie, S., Takayama, H. et al. 1999. Opioid receptor agonistic characteristics of mitragynine pseudoindoxyl in comparison with mitragynine derived from Thai medicinal plant *Mitragyna speciosa*. *Gen Pharmacol* 33: 73–81.

Zarembo, J.E., Douglas, B., Valenta, J., and Weisbach, J.A. 1974. Metabolites of mitragynine. *J Pharm Sci* 63: 1407–15.

10 Chemical Structures of Opioids

Robert B. Raffa

CONTENTS

10.1 INTRODUCTION

A substance is defined as an opioid at the clinical level if it has a capability to produce effects that are produced by one or more of the poppy-derived opiates, and in particular morphine-like effects (or to block them if the substance is an antagonist). Since morphine mimics the pharmacology of endogenous compounds (such as endorphins and enkephalins), morphine-like effects can encompass a large variety of physiological processes. A more modern definition of opioid involves the molecular mechanism through which morphine produces its effect—activation of seven-transmembrane G-protein–coupled receptors (7-TM-GPCRs). Substances are said to be opioids if they possess an affinity for opioid receptors (this has an advantage in that it automatically includes antagonists).

There are three types of opioid receptors, designated mu (MOR), delta (DOR), and kappa (KOR), and subtypes of each of these. Compounds can display selective affinity for one receptor (sub)types or mixed affinity for multiple (sub)types, and the relative distribution of affinity across (sub) types gives rise to individualized pharmacologic characteristics. This chapter presents the varied chemical structures of substances that have affinity for opioid receptors and display opioid effects.

10.2   STRUCTURE TEMPLATES

I

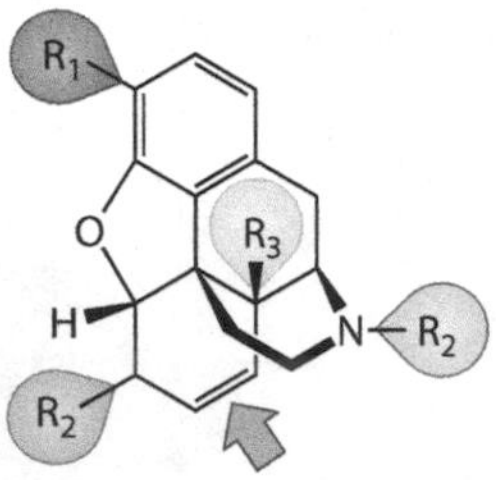

R$_1$	R$_2$	Bond	R$_3$	R$_4$	Name
OH	OH	2	H	CH$_3$	Morphine
OH	H	1	H	CH$_3$	Desmorphine
OH	OH	1	H	H	Dihydromorphine
OH	OH	1	OH	CH$_3$	Hydromorphinol
OH	CH$_3$	*	H	CH$_3$	Methyldesmorphine
HO	=CH$_2$	1	H	CH$_3$	6-MDDM
HO	N(CH$_2$CH$_2$Cl–)$_2$	1	OH		Chlornaltrexamine
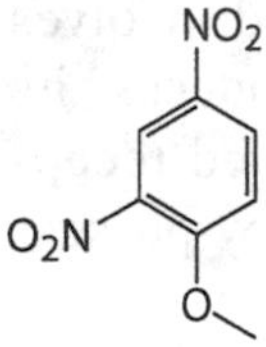	OH	2	H	CH$_3$	Dinitrophenylmorphine
OH	OH	2	H	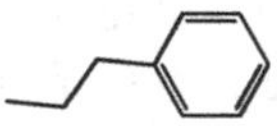	N-Phenethylnormorphine
OH	OH	1	OH	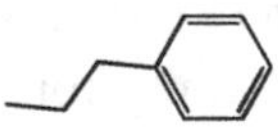	RAM-378
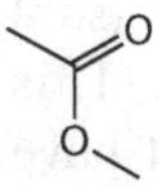	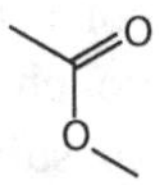	2	H	CH$_3$	Acetylpropionylmorphine
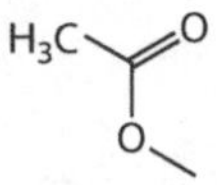	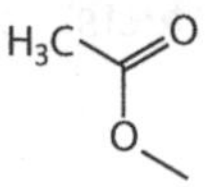	1	H	CH$_3$	Dihydroheroin
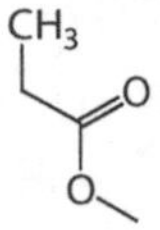	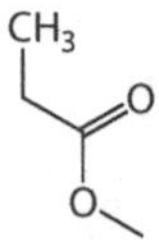	2	H	CH$_3$	Dipropanoylmorphine

(Continued)

R_1	R_2	Bond	R_3	R_4	Name
benzoyloxy (C$_6$H$_5$–CO–O–)	benzoyloxy (C$_6$H$_5$–CO–O–)	2	H	CH_3	Dibenzoyl-morphine
nicotinoyloxy (pyridine-3-CO–O–)	nicotinoyloxy (pyridine-3-CO–O–)	2	H	CH_3	Nicomorphine
OCH_3	OH	2	H	CH_3	Codeine
OCH_3	H_3C–CO–O–	2	H	CH_3	6-MAC
benzyloxy (C$_6$H$_5$CH$_2$–O–)	OH	2	H	CH_3	Benzylmorphine
OH	OCH_3	1	H	CH_3	Dihydrohet-erocodeine
ethoxy (CH$_3$CH$_2$–O–)	OH	2	H	CH_3	Ethylmorphine
OH	OCH_3	2	H	CH_3	Hetcrocodeine
morpholinoethoxy	OH	2	H	CH_3	Pholcodine
benzyloxy (C$_6$H$_5$CH$_2$–O–)	–O–CO–(CH$_2$)$_{13}$CH$_3$	2	H	CH_3	Myrophine
H_3C–CO–O–	=O	1	H	CH_3	Acetylmorphine
OCH_3	=O	2	H	CH_3	Codeinone
OCH_3	H_3C–CO–O–	2	H	CH_3	Thebacon
OCH_3	=O	1	H	CH_3	Hydrocodone
OH	=O	1	H	CH_3	Hydromorphone
OH	=O	2	H	CH_3	Morphinone
OCH_3	=O	1	OH	CH_3	Oxycodone
OH	=O	1	OH	CH_3	Oxymorphone
OH	=O	1	OH	–CH$_2$CH$_2$OCH$_3$	Semorphone
OH	Cl	2	H	CH_3	Chlormorphide
OCH_3	OH	1	OH	CH_3	14-Hydroxydi-hydrocodeine

(Continued)

R_1	R_2	Bond	R_3	R_4	Name
OCH_3	H_3C–C(=O)–O– (acetyl ester)	1	H	CH_3	Acetyldihydrocodeine
OCH_3	OH	1	H	CH_3	Dihydrocodeine
OH	OH	1	OH	(cyclobutylmethyl)	Nalbuphine
OCH_3	nicotinate (pyridine-3-carboxylate) ester	2	H	CH_3	Nicocodeine
OCH_3	nicotinate (pyridine-3-carboxylate) ester	1	H	CH_3	Nicodicodeine
OH	H_3C–C(=O)–O– (acetyl ester)	2	H	CH_3	6-MAM
OCH_3	OH	2	H	H	Norcodeine
OH	OH	2	H	H	Normorphine
OH	furan-acryloyl-N,N-dimethylamide	1	OH	cyclopropyl(methyl)	Nalfurafine
OH	=O	1	OH	cyclopropyl(methyl)	Nalmefene
OH	=O	1	OH	CH_2 (allyl)	Naloxone
OH	OH	2	H	CH_2 (allyl)	Nalorphine
OH	=O	1	OH	cyclopropyl(methyl)	Naltrexone
OH	glucuronide (carboxylic acid, trihydroxy pyranosyl)	2	H	CH_3	M6G
OCH_3	=N–O–CH_2–C(=O)OH (oxime acetic acid)	1	H	CH_3	Codoxime

* double-bond adjacent

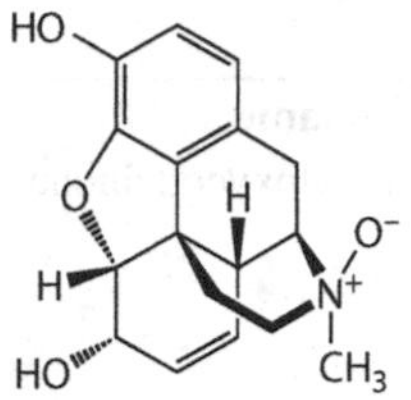

Morphine-*N*-oxide

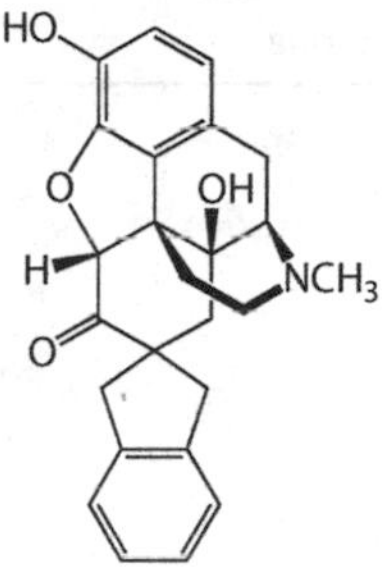

Spiroindanyloxymorphone

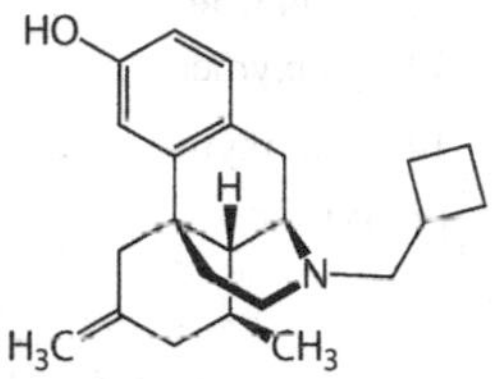

Xorphanol

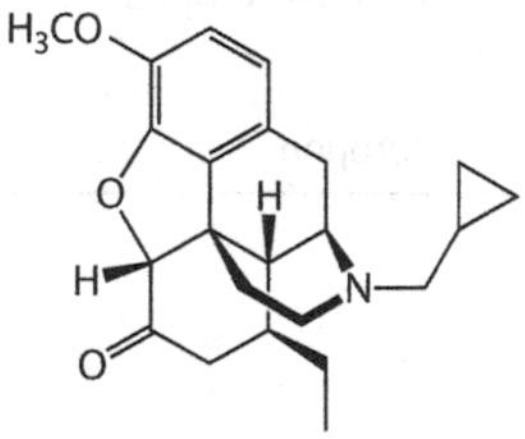

Conorphone

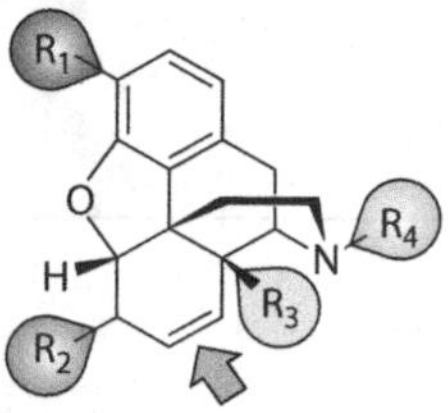

II

R_1	R_2	Bond	R_3	R_4	Name
OCH_3	$=O$	2	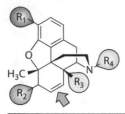	CH_3	14-Cinnamoyloxycodeinone
OH	$=N-NH_2$	1	OH	CH_3	Oxymorphazone
OH	$=O$	2	$NH(CH_2)_4CH_3$	CH_3	Pentamorphone

III

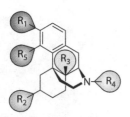

R_1	R_2	Bond	R_3	R_4	Name
OH	$=O$	1	$O–CH_2CH_3$	CH_3	14-Ethoxymetopon
OH	$=O$	1	$O–CH_3$	CH_3	14-Methoxymetopon
OH	$=O$	1		CH_3	PPOM
OH	$=O$	1	$O–CH_2CH_3$		N-Phenetyl-14-Ethoxymetopon
OH	$=O$	1	H	CH_3	Metopon

IV

R_1	R_2	R_3	R_4	R_5	Name
OH	H	H		H	Cyclorphan
OH	H	H		H	DXA

(Continued)

R₁	R₂	R₃	R₄	R₅	Name
OH	H	H	CH₃	H	Levorphanol
OCH₃	H	H	CH₃	H	Levomethorphan
OH	H	H		H	Levophenacylmorphan
OH	H	H	H	H	Norlevophanol
OH	H	OH		H	Oxilorphan
OH	H	H		H	Phenomorphan
OH	H	H		H	Furethylnorlevorphanol
OH	H	OH		H	Butorphanol
H	=O	OCH₃		OCH₃	Cyprodime
OCH₃	OH	OH	CH₃	OCH₃	Drotebanol

V

R₁	R₂	R₃	Bond	R₄	Name
OCH₃	OCH₃		2	CH₃	7-PET
	OCH₃		2	CH₃	Acetorphine
OH	OCH₃		1		Buprenorphine
OH	OCH₃		1		Cyprenorphine

(Continued)

R$_1$	R$_2$	Bond	R$_3$	R$_4$	Name
OH	OCH$_3$	CH$_3$ ⋯⟨C⟩—CH$_3$ OH	1	CH$_3$	Dihydroetorphine
OH	OCH$_3$	—CH$_3$ H$_3$C OH	2	CH$_3$	Etorphine
OH	OCH$_3$	HO CH$_3$ CH$_3$ CH$_3$ H$_3$C	1	H	Norbuprenorphine

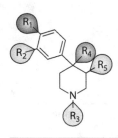

BU-48

VI

R$_1$	R$_2$	R$_3$	R$_4$	R$_5$	Name
F	H	CH$_3$	O= H$_3$C	H	4-Fluoromeperidine
H	H	CH$_2$	O= H$_3$C	H	Allylnorpethidine
H	H	NH$_2$	O= H$_3$C	H	Anileridine

(Continued)

R_1	R_2	R_3	R_4	R_5	Name
H	H			H	Benzethidine
H	H			H	Carperidine
H	H			H	Difenoxin
H	H			H	Diphenoxylate
H	H			H	Etoxcridine
H	H			H	Furethidine
H	OH	CH_3		H	Hydroxypethidine
H	H			H	Morpheridine
H	H			H	Oxpheneridine
H	H	CH_3		H	Pethidine

(*Continued*)

R$_1$	R$_2$	R$_3$	R$_4$	R$_5$	Name
H	H			H	Pheneridine
H	H			H	Phenoperidine
H	H			H	Piminodine
H	H	CH$_3$		H	Properidine
H	H			H	Prosidol
H	H			H	Sameridine
H	H	CH$_3$		CH$_2$	Allylprodine
H	H	CH$_3$		CH$_3$	α-Meprodine
H	H	CH$_3$		CH$_3$	α-Prodine
H	OH		CH$_3$	CH$_3$	Alvimopan

(*Continued*)

R$_1$	R$_2$	R$_3$	R$_4$	R$_5$	Name
H	OH	CH$_3$	–CH$_3$	CH$_3$	Picenadol
H	H	CH$_3$		H	MPPP
H	H			H	PEPAP
H		CH$_3$		H	Acetoxyketobemidone
H	H			H	Droxypropine
H	OH	CH$_3$		H	Ketobemidone
H	OH	CII$_3$		H	Methylketobemidone
H	OH	CH$_3$		H	Propylketobemidone
Cl	H		OH	H	Loperamide

Trimeperidine

VII

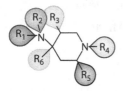

R₁	R₂	R₃	R₄	R₅	R₆	Name
phenyl	O=C–CH₂–CH₃ (H₃C)	CH₂ (allyl)	phenethyl	H	H	3-Allylfentanyl
phenyl	O=C–CH₂–CH₃ (H₃C)	CH₃	phenethyl	H	H	3-Methylfentanyl
phenyl	O=C–CH₂–CH₃ (H₃C)	CH₃	thienyl-ethyl	H	H	3-Methylthiofe-ntanyl
phenyl	O=C–CH₂–CH₃ (H₃C)	H	phenethyl	H	phenyl	4-Phenylfentanyl
phenyl	O=C–CH₂–CH₃ (H₃C)	H	tetrazolone–CH₃	H	CH₂OCH₃	Alfentanil
phenyl	O=C–CH₃ (CH₃)	H	phenyl–CH(CH₃)	H	H	α-Methylacetylfe-ntanyl
phenyl	O=C–CH₂–CH₃ (H₃C)	H	phenyl–CH(CH₃)	H	H	α-Methylfentanyl
phenyl	O=C–CH₂–CH₃ (H₃C)	H	thienyl–CH(CH₃)	H	H	α-Methylthiofe-ntanyl
phenyl	O=C–CH₂–CH₃ (H₃C)	H	HO–CH(phenyl)–CH₂	H	H	β-Hydroxyfe-ntanyl

(*Continued*)

R₁	R₂	R₃	R₄	R₅	R₆	Name
		H	HO	H	H	β-Hydroxythiofentanyl
		H		H	H	β-Methylfentanyl
		CH₃		H	H	Brifentanil
		H		H		Carfentanil
		H		H	H	Fentanyl
		CH₃		H		Lofentanil
		H		H	H	Mirfentanil
		H		H	H	Ocfentanil
		CH₃	HO	H	H	Ohmfentanil
		H		H	H	Parafluorofentanyl
		CH₃		CH₃	H	Phenaridine

(Continued)

R₁	R₂	R₃	R₄	R₅	R₆	Name
(phenyl)	(ethyl ketone)	H	(methyl butanoate)	H	(methyl carbonate)	Remifentanil
(phenyl)	(ethyl ketone)	H	(propyl thiophene)	H	H	Sufentanil
(phenyl)	(ethyl ketone)	H	(propyl thiophene)	H	H	Thiofentanyl
(2-fluorophenyl)	(ethyl ketone)	H	(tetrazolone with propyl and ethyl)	H	(phenyl)	Trefentanil

VIII

R₁	R₂	R₃	R₄	R₅	Name
(3-hydroxyphenyl)	(N,N-diethylbenzamide)	CH₂ (butenyl)	CH₃	CH₃	BW373U86
(phenyl)	(N,N-diethylbenzamide)	(3-fluorophenyl)	CH₃	CH₃	DPI-221
(3-hydroxyphenyl)	(N,N-diethylbenzamide)	(benzyl)	CH₃	CH₃	DPI-287
(3-hydroxyphenyl)	(N-(3-fluorophenyl)-N-methylbenzamide)	CH₂ (butenyl)	CH₃	CH₃	DPI-3290

(Continued)

R_1	R_2	R_3	R_4	R_5	Name
(4-methoxyphenyl)	(N,N-diethyl 4-benzamide)	CH_2 (allyl)	CH_3	CH_3	SNC-80
(phenyl)	(3-methyl-ethylbenzene)	CH_2 (allyl)	H	H	AD-1211

IX

R_1	R_2	R_3	R_4	Name
(acetyl-ethyl ester)	H	CH_3	H	Ethoheptazine
(acetyl-methyl ester)	CH_3	CH_3	H	Metheptazine
(acetyl-ethyl ester)	H	CH_3	CH_3	Metethoheptazine
(propyl ester)	H	CH_3	CH_3	Proheptazine

X

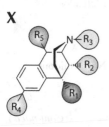

R₁	R₂	R₃	R₄	R₅	Name
CH_3	CH_3	cyclopropylmethyl	NH_2 (C=O)	H	8-CAC
CH_3	CH_3	CH_2 allyl	OH	H	Alazocine
CH_2CH_3	isopropyl (CH_3, CH_3)	ethyl hydroxycyclopropyl (OH)	OH	H	Bremazocine
CH_3	CH_3	cyclopropylmethyl	OH	=O	Ketazocine / Ketocyclazocine
CH_3	CH_3	CH_3	OH	H	Metazocine
CH_3	CH_3	CH_3, CH_3 (prenyl)	OH	H	Pentazocine
CH_3	CH_3	phenethyl	OH	H	Phenazocine
CH_3	CH_3	cyclopropylmethyl	OH	H	Cyclazocine

Dezocine

XI

R₁	R₂	R₃	R₄	Name
phenyl	$O=$ (H_3C) ethyl ketone	phenyl	piperidinyl (N)	Dipipanone
phenyl	$O=$ (H_3C) ethyl ketone	phenyl	H_3C-, $N-CH_3$, CH_3	Methadone
phenyl	$O=$ (H_3C) ethyl ketone	phenyl	$N-CH_3$, CH_3	Normethidone

(Continued)

R1	R2	R3	R4	Name
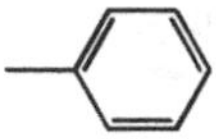	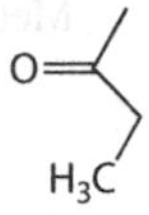	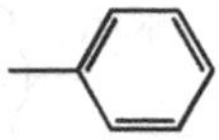	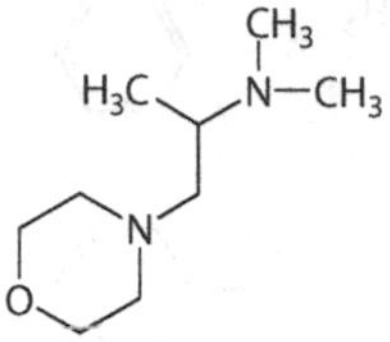	Phenadoxone
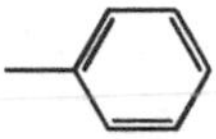	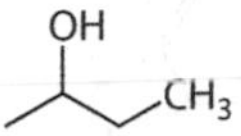	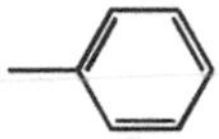	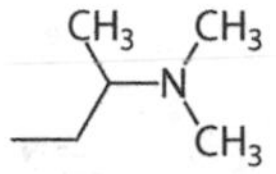	Dimepheptanol
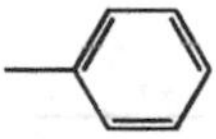	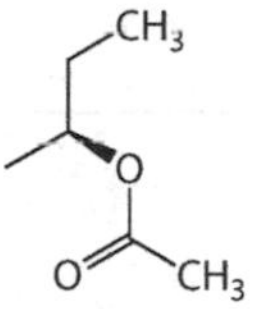	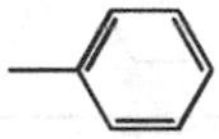	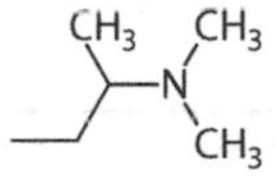	Levacetylmethadol
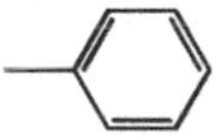	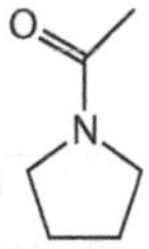	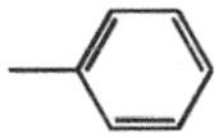	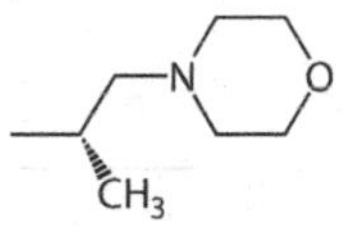	Dextromoramide
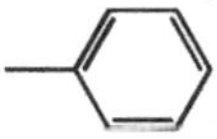	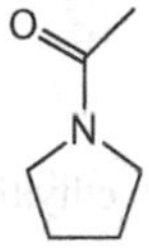	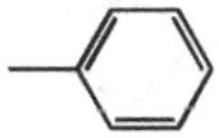	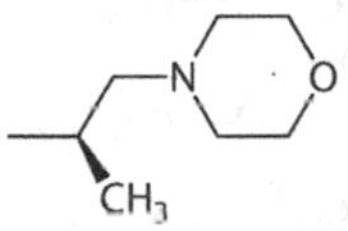	Levomoramide
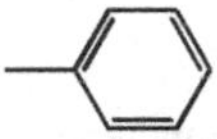	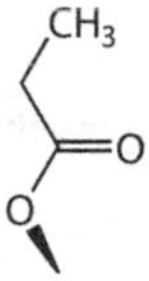	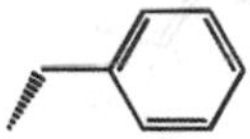	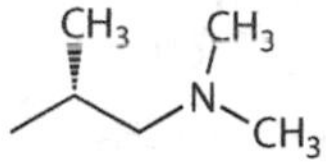	*d*-Propoxyphene
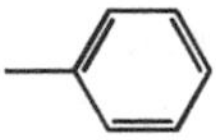	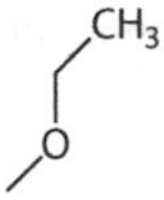	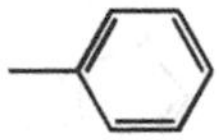	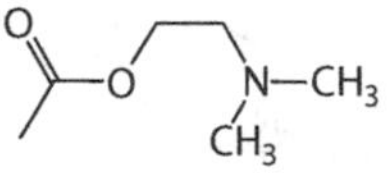	Dimenoxadol
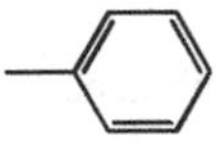	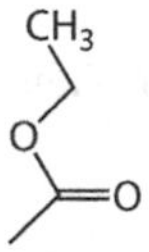	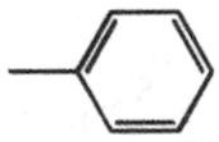	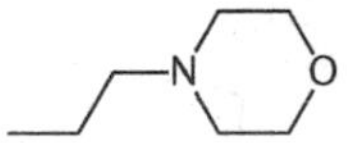	Dioxaphetyl butyrate
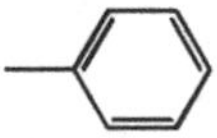	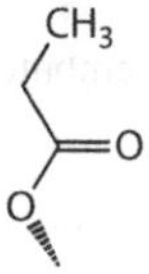	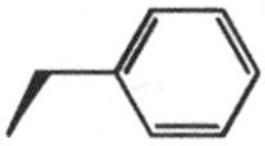	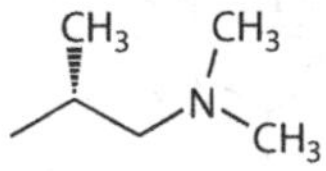	*l*-Propoxyphene
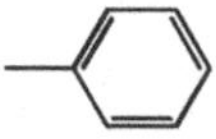	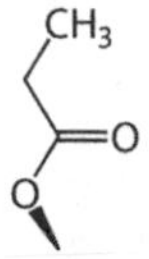	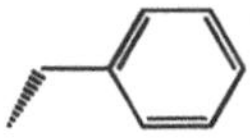	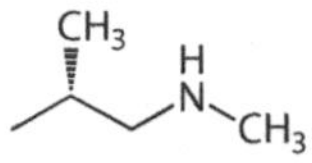	Norpropoxyphene

(*Continued*)

R₁	R₂	R₃	R₄	Name
phenyl	—S(=O)(=O)—CH₂CH₃	phenyl	CH₃ CH(CH₂CH₃)N(CH₃)CH₃	Methiodone
phenyl	O=C(CH₂CH₃) H₃C	phenyl	propyl-spiropiperidine tetrahydronaphthalene	R-4066
phenyl	CH(CH₃)CH₃ (isopropyl)	—≡N	CH₃ CH(CH₂CH₃)N(CH₃)CH₃	Isoaminile

XII

R₁	R₂	R₃	Name
thiophene	CH₃ CH-N(CH₂CH₃)(CH₂CH₃)	thiophene	Diethylthiambutene
thiophene	CH₃ CH₃ CH-N-CH₃	thiophene	Dimethylthiambutene
thiophene	CH₃ CH-N(CH₂CH₃)(CH₃)	thiophene	Ethylmethylthiambutene
thiophene	CH(CH₃)-N(piperidine)	thiophene	Piperidylthiambutene
thiophene	CH(CH₃)-N(pyrrolidine)	thiophene	Pyrrolidinylthiambutene
thiophene	CH(CH₃)-NH₂	thiophene	Thiambutene

Tipepidine

11 The Detection of Mitragynine and Its Analogs

Ruri Kikura-Hanajiri

CONTENTS

11.1 INTRODUCTION

The leaves of a plant endemic to tropical Southeast Asia, *Mitragyna speciosa* Korth. (Rubiaceae), are traditionally used as a substitute for opium in Thailand and Malaysia. The leaves are known as "kratom," "biak-biak," or "ketum" in these countries, and they contain several alkaloids that vary in composition (e.g., between geographic variants) (Shellard et al. 1978; Takayama 2004; Kitajima et al. 2006). *M. speciosa* leaves from adult plants in Thailand have been reported to contain as the main alkaloid over 60% mitragynine (MG), whereas the corresponding content from the leaves in Malaysia is only approximately 10%. Paynantheine (PAY) and the MG diastereomer speciogynine (SG) were reported as the second most abundant alkaloids, and the MG diastereomer speciociliatine (SC) was the third abundant alkaloid in both plants (Takayama 2004). Shellard et al. (1978) described the distribution and probable biogenetic route of *M. speciosa* in young and mature leaves from Thailand. The alkaloidal pattern is different in the young plants from that in the older trees. The leaves of young Thai plants contain, in addition to MG, PAY, SG, and SC, the MG diastereomer mitraciliatine (MC) and the PAY diastereomer isopaynantheine (ISO-PAY)

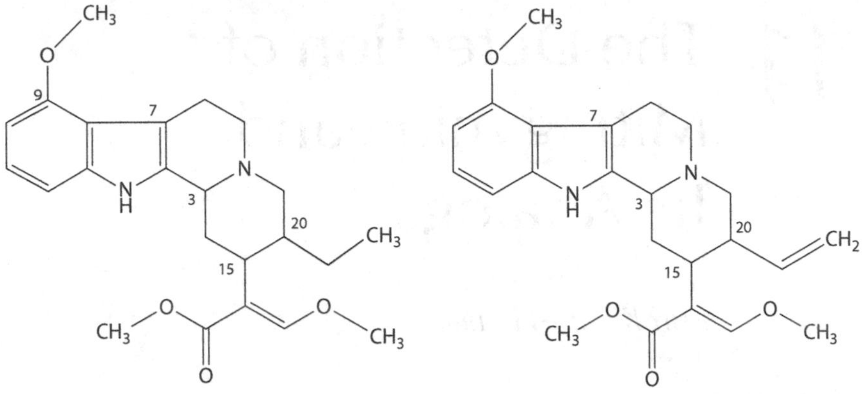

Mitragynine (MG): 3S, 15S, 20S
Speciogynine (SG): 3S, 15S, 20R
Speciociliatine (SC): 3R, 15S, 20S
Mitraciliatine (MC): 3R, 15S, 20R

Paynantheine (PAY): 3S, 15S, 20R
Isopaynantheine (ISO-PAY): 3R, 15S, 20R

FIGURE 11.1 Chemical structures of mitragynine (MG) and its related alkaloids in kratom.

(Figure 11.1). SG was found to be the main alkaloid in young plants. In the fruits of Malaysian plants, MG, PAY, and all of their diastereomers could be detected, with SC being the main alkaloid (Kitajima et al. 2006).

MG, a major constituent of *M. speciosa*, has an opioid agonistic activity (Matsumoto et al. 1996a; Thongpradichote et al. 1998; Takayama et al. 2000; Kitajima et al. 2006). Its derivative 7α-hydroxy-7*H*-mitragynine (7-OH MG) shows a much more potent antinociceptive effect in mice than does MG or morphine (Takayama et al. 2000; Takayama 2004; Matsumoto et al. 2004, 2005, 2006; Horie et al. 2005), even though it is a minor constituent (2.0% based on the crude base (Ponglux et al. 1994)) of the leaves. Moreover, 7-OH MG exhibits morphine-like tolerance and withdrawal effects in mice (Matsumoto et al. 2005). Kratom has been described as having dose-dependent opium-like sedative and coca-like stimulant effects due to its dual binding to opioid and alpha-adrenergic receptors (Matsumoto et al. 1996b).

Traditionally, kratom has been used in Malaysia and Thailand by laborers and farmers to enhance productivity, but also as a substitute for opium and in traditional medicine (e.g., as an antidiarrheal agent), owing to its morphine-like pharmacological effects. The possession of kratom has been illegal since 1943 in Thailand, and it is also illegal in other countries in the region (Malaysia and Myanmar) and outside the region (Australia, Bhutan, Finland, and Lithuania) (UNODC 2013). In Thailand, the growing and harvesting of the tree of kratom is prohibited, although this tree is found throughout the country, mostly in the southern region.

11.2 ABUSED SUBSTANCES OF KRATOM

In recent years, various products made from psychotropic plants—which are mostly not under regulation—have increased around the world. For a survey of current trends in the abuse of psychotropic plants in Japan, 127 types of plant products (purchased via Japanese-based websites from April 2004 to May 2007) were analyzed using liquid chromatography–mass spectrometry (LC-MS) (Kawamura et al. 2008). The subsequent analyses revealed typical hallucinogenic plant components in 51 products, as follows: *N,N*-dimethyltryptamine (the genus *Phalaris* and *Psychotria viridis* et al.); mescaline (San Pedro, *Trichocereus pachanoi*; and peyote, *Lophophoria williamsii* et al.); salvinorin A (*Salvia divinorum*); lysergamide (Hawaiian baby woodrose, *Argyreia nervosa*; and morning glory, *Ipomea violacea* et al.); and harmine and harmaline (*Banisteriopsis caapi* and the seeds of *Peganum harmala* et al.). Moreover, MG (kratom, *M. speciosa*), atropine and scoporamine (the genus *Datura* et al.) were detected in the products. One-third of the products were mixtures of several plant materials, and the compounds detected in some products were not consistent with the ingredients listed on the product labels. Moreover, there were products in which the contents of the active compounds were sufficient to produce hallucinogenic effects in humans.

According to a survey in the United Kingdom (Schmidt et al. 2010), the top five products marketed as "legal highs" on UK-based websites from April 2009 to June 2009 by frequency were "*Salvia divinorum*," "Kratom," "Hawaiian Baby Woodrose Seeds," "Fly Agaric" (*Amanita muscaria*, a mushroom containing ibotenic acid and muscimol), and an herbal mixture called "Genie" (containing CB_1/CB_2 cannabinoid receptors agonists, called "synthetic cannabinoids").

The use of kratom as a psychoactive substance has recently increased worldwide. According to the report of the United Nations Office on Drugs and Crime (UNODC 2013), a national household survey conducted in Thailand in 2007 suggested that kratom was the second most widely used drug after cannabis in terms of lifetime prevalence and the most widely used drug in terms of annual prevalence. The number of seizures of kratom quintupled in Thailand, from 1,100 in 2005 to 5,897 in 2011, and was far higher than those reported for heroin, opium, ketamine, "ecstasy," and cocaine. Kratom-related arrests more than doubled between 2007 and 2011 in both Myanmar and Thailand. In addition to the widespread consumption of kratom in Southeast Asia where the *Mitragyna speciosa* plant is native, in recent years kratom has become easy to obtain anywhere via the Internet and has begun to be used worldwide.

Internet surveys conducted by the European Monitoring Centre for Drugs and Drug Addiction (EMCDDA) in 2008 and 2011 revealed that kratom is one of the most widely offered new psychoactive substances (NPS) (EMCDDA 2013); the top three plant-based substances were khat (*Catha edulis*), *Salvia divinorum*, and kratom. Kratom leaves are usually consumed fresh, although dried leaves in powder form are also available. The fresh leaves are chewed whereas the powder form is often either swallowed or brewed into tea. More recently, plant products containing kratom have been sold as "incense" for their psychoactive effects; concentrations of the active components in these products differ depending on the variety of the plant used, the environment, and the time of harvesting.

Several cases of intoxication and fatalities related to products containing kratom have been reported around the world. Kapp et al. (2011) reported a case of intrahepatic cholestasis as an adverse effect after the intake of kratom for two weeks in the absence of any other causative agent. A fatal case possibly related to kratom toxicity was reported in the United States (Neerman et al. 2013). From 2009 to 2010, nine fatal cases of intoxication associated with the use of "Krypton," an herbal blend product containing MG and *O*-desmethyltramadol (an active metabolite of tramadol, a commonly prescribed analgesic), were reported in Sweden (Kronstrand et al. 2011). These fatalities might be attributed to the addition of the potent mu-receptor agonist *O*-desmethyltramadol to dried kratom leaves.

11.3   DETECTION OF MG AND ITS ANALOGS

11.3.1   MG AND ITS ANALOGS IN KRATOM

The alkaloid composition of botanical and forensic samples can be analyzed by regular chromatographic and spectroscopic methods (EMCDDA 2013). Kratom alkaloids can be separated by thin-layer chromatography (TLC) on silica gel plates with detection by UV (254 nm). Upon being sprayed with either modified Ehrlich's reagent or ferric chloride–perchloric acid reagent, MG gives purple or gray-to-brown spots, respectively. With the use of hexane/EA/25% ammonia solution (30:15:1, v/v/v) as the mobile phase, the Rf value of MG is 0.49. The UV spectrum of the methanol solution of MG shows a maximum at 225 nm with shoulders at 247, 285, and 293 nm. The characteristic absorption bands in the IR spectrum of MG are at 3365, 1690, and 1640 cm^{-1}.

Significant fragments in the electron impact ionization mass spectrum (*m/z*) have been observed: 398(M$^+$), 383, 366, 269, 214, 200, and 186. The UV spectrum of the ethanol solution of 7-hydroxy MG shows a maximum at 220 nm with shoulders at 245 and 305 nm. The characteristic absorption bands in the IR spectrum of 7-hydroxy MG in $CHCl_3$ are at 3590, 2850, 2820, 2750, 1700, 1645, 1630, 1600, 1490, 1465, and 1440 cm^{-1}. Significant fragments in the electron impact ionization mass spectrum (m/z) are 414(M^+), 397, 383, and 367.

To date, the identification of MG in plant materials has been achieved with the following methods: TLC and high-performance liquid chromatography (HPLC) (Vardakou et al. 2010), gas chromatography (GC) (Chan et al. 2005), liquid chromatography–electrospray ionization–mass spectrometry (LC-ESI-MS) (Kikura-Hanajiri et al. 2009), the identification of MG and other ingredients in a kratom cocktail by HPLC (Chittrakarn et al. 2012), and the qualitative and quantitative determination of MG by HPLC with diode array detection (HPLC-DAD) in plant material and a cocktail (Parthasarathy et al. 2013). Moreover, Kowalczuk et al. (2013) proposed a comprehensive procedure that involves the botanical analysis of leaf material on the basis of anatomical features and an MG identification using TLC and HPLC.

According to a survey of psychoactive plant products obtained via the Internet in 2006 and 2007 (Kikura-Hanajiri et al. 2009), many products of kratom distributed in Japan as "incense" claimed to have opioid-like effects. In that survey, the simultaneous analysis of MG, 7-OH MG, and other indole alkaloids (SG, SC, and PAY) in raw materials and 13 commercial products (six in the form of dried leaves, four in the form of powders, and three in the form of resins) of kratom was conducted using LC-MS. The reference raw materials of *M. speciosa* (dried leaves) were distinguishable as those with "big leaves" and those with "small leaves." There were obvious differences in the leaf sizes, although they were both recognized as *M. speciosa* in the area of Bangkok, Thailand, where the samples were collected.

For the LC-MS analysis of the target compounds, the protonated molecular ions ([M+H]$^+$) were observed as base peaks for all compounds except 7-OH MG. For 7-OH MG ($C_{23}H_{30}N_2O_4$), the ion corresponding to the [M+H+18]$^+$ (m/z 433) was detected mainly in the spectrum together with the protonated molecular ion (m/z 415). The elemental composition of the former ion (m/z 433) was estimated as $C_{23}H_{33}N_2O_6$ through exact mass measurements using a time-of-flight mass spectrometer (TOFMS), and it was found to be the protonated molecular ion adducted with a water molecule. Extraction was most effective when the sample was extracted with an 80% methanol aqueous solution by ultrasonication for 1 hour and kept at room temperature overnight.

MG and 7-OH MG in the solution were stable at least during the extraction procedure, and more than 95% of either compound remained in the solution throughout the procedure. As the results of the LC-MS analyses of the raw materials, all five targeted compounds were found in both the big and small leaves of *M. speciosa*. The concentrations of MG and 7-OH MG in the small leaves were significantly lower than those in the big leaves. The results of the DNA analyses (Maruyama et al. 2009) suggested that the plants with small leaves were a hybrid between *M. speciosa* and *M. hirsuta/diversifolia*. This could be the reason the small leaves had lower concentrations of the two compounds compared to the big leaves. Although *M. hirsuta* is a related species of *M. speciosa*, it was reported not to contain the major narcotic compounds found in *M. speciosa*, such as MG and 7-OH MG (Kitajima et al. 2007). Of the 13 commercial products, MG, 7-OH MG, and three other indole alkaloids could be detected in 11 products. The contents of MG in these products ranged from 1.2% to 6.3%, and those of 7-OH MG ranged from 0.01% to 0.04%. The concentrations of the targeted compounds in the resin samples were much higher than those in the dried leaves. Most of the samples of dried leaves contained the related alkaloids at the same levels as those in the big leaves of the reference raw material. The highest concentration of MG detected in the products was 62.6 mg/g (a resin form), followed by 47.4 mg/g (a powder form). The concentrations of 7-OH MG were less than 0.4 mg/g, and the sample forms might not be related to these concentrations.

MG, the major alkaloid of kratom, has been characterized as an opioid agonist, producing effects similar to those of morphine, but its weak potency cannot explain kratom's potent opium-like effect. Horie et al. (2005) reported that the opioid effect of *M. speciosa* was based mostly on the activity of the minor alkaloid 7-OH MG. This compound produced antinociceptive effects about 5.7 and 4.4 times more potent than those of morphine in tail-flick (ED_{50} = 0.80 mg/kg) and hot-plate (ED_{50} = 0.93 mg/kg) tests of mice, respectively (Matsumoto et al. 2006), when administered subcutaneously (s.c.). Moreover, oral doses of 7-OH MG to mice at 5–10 mg/kg showed antinociceptive activities in these tests, whereas the oral administration of morphine required a dose of 20 mg/kg to achieve similar activities (Matsumoto et al. 2004).

The potent antinociceptive effect of 7-OH MG was reported to be based on the activation of mu-opioid receptors, and 7-OH MG also exhibited morphine-like pharmacological characters, including tolerance and withdrawal (Matsumoto et al. 1996b). In the survey mentioned above (Kikura-Hanajiri et al. 2009), 0.1–0.4 mg/g of 7-OH MG was found in the products. As for mice (30 g weight), approximately 0.1 mg (s.c. administration) or 1 g (oral administration) of the products might be enough to show potent antinociceptive activities, possibly including other opium-like effects.

11.3.2 MG AND ITS ANALOGS IN BIOLOGICAL SPECIMENS

Kratom has been used for centuries for its medicinal and psychoactive qualities, which are comparable to that of opiate-based drugs. The current widespread availability of kratom on the Internet indicates an emerging trend of using kratom as a non-controlled drug substance to replace morphine-like drugs. Despite the interesting pharmacological properties of kratom, few researchers have analyzed the effects of MG and its analogues on human samples. The development of analytical methods that can be used for these compounds in biological samples is required for emergency and forensic cases.

De Moraes et al. (2009) reported a method for analyzing MG in rat plasma using HPLC coupled with electrospray tandem MS (LC-MS/MS). MG was extracted from rat plasma with hexane-isoamyl alcohol and separated on a LiChrospher RP-Select B column. The analysis was carried out in the multiple reaction monitoring (MRM) mode. Protonated ions [M+H]+ and their respective ion products were monitored at the transition 398 > 174 for MG. The quantification limit was 0.2 ng/mL, within a linear range of 0.2–1000 ng/mL. The method was applied to quantify MG in plasma samples of rats treated after an oral administration of MG.

Lu et al. reported a method for analyzing MG in human urine using LC-MS/MS (Lu et al. 2009). MG was extracted by methyl t-butyl ether and separated on an HILIC column. The regression linearity of MG calibration ranged from 0.01 to 5.0 ng/mL, achieved with a correlation coefficient greater than 0.995; the detection limit was 0.02 ng/mL. By this LC-MS/MS analysis, an amount of 167 ± 15 ng/mL MG was found in human urine samples in the case of serious toxicity following kratom use.

Le et al. (2012) conducted a quantitative analysis of MG and qualitative analyses of two metabolites of MG (5-desmethyl MG and 17-desmethyldihydro MG) and the related active alkaloid 7-OH MG in human urine following kratom use by LC-MS/MS. In a fatal case related to the use of propylhexedrine (a potent α-adrenergic sympathomimetic amine found in nasal decongestant inhalers) and MG, 1.7 mg/L of propylhexedrine and 0.39 mg/L of MG were detected in the user's blood, and both drugs as well as acetaminophen, morphine, and promethazine were detected in the urine. GC-MS and LC-MS/MS were used to obtain the quantitative results for the propylhexedrine heptafluorobutyryl derivative and MG, respectively (Holler et al. 2011).

As mentioned earlier, from 2009 to 2010, nine fatal cases of intoxication associated with the use of "Krypton," an herbal blend product containing kratom and O-desmethyltramadol (an active metabolite of tramadol, a commonly prescribed analgesic), were reported in Sweden (Kronstrand et al. 2011). The blood concentrations of MG, determined by ultra-high performance LC (UHPLC)-MS/MS, ranged from 0.02 to 0.18 μg/g, and the O-desmethyltramadol concentrations, determined by GC with nitrogen-specific detection, ranged from 0.4 to 4.3 μg/g. Neither tramadol nor N-desmethyltramadol was detected in these samples, which implies that the ingested drug was O-desmethyltramadol. The addition of the potent mu-receptor agonist O-desmethyltramadol to powdered leaves from kratom might have contributed to the nine fatalities. In Germany, health damage caused by Krypton was also reported (Arndt et al. 2011). Arndt et al. (2011) reported that kratom alkaloids (MG, speciociliatine, speciogynine, mitraciliatine, and paynantheine) and approximately 9 ng/mL O-desmethyltramadol, but no tramadol and N-desmethyltramadol, were detected in the urine of a Krypton consumer.

H. H. Mauler's group investigated systematic analytical methods for MG, its related compounds, and their metabolites in human biological specimens (Philipp et al. 2009, 2011a,b,c). Philipp et al. (2009) investigated the phase I and II metabolites of MG in rat and human urine after solid-phase extraction using LC-linear ion trap MS, providing detailed structure information in the MS^n mode, particularly with high resolution. The seven identified phase I metabolites indicated that MG was metabolized by hydrolysis of the methylester in position 16, O-demethylation of the 9-methoxy group and the 17-methoxy group, followed via the intermediate aldehydes by oxidation to carboxylic acids or reduction to alcohols and the combinations of some steps. In rats, four metabolites were additionally conjugated to glucuronides and one to sulfate, but in humans, three metabolites were conjugated to glucuronides and three to sulfates.

The metabolites of other related alkaloids contained in kratom (PAY, the MG diastereomers SG, SC, and MC, and the PAY diastereomer ISO-PAY) were also investigated in rat urine after the administration of each pure alkaloid (Philipp et al. 2011a,c). PAY was metabolized via the same pathways as MG. Several metabolites were excreted as glucuronides or sulfates. After the administration of SC, its phase I and II metabolites were identified in rat urine. Considering the mass spectra and retention times, SC and its metabolites identified in the rat could also be confirmed in human urine. The metabolic pathways of MC and ISO-PAY in rats were comparable to those of their corresponding diastereomers. In the human urine samples tested, not all metabolites found in rats could be detected because of the much lower amounts of MC and ISO-PAY in kratom.

Philipp et al. (2011b) developed a full-scan GC-MS procedure for monitoring kratom or Krypton intake in urine after the enzymatic cleavage of conjugates, solid-phase extraction, and trimethylsilylation, focusing on the metabolites of the kratom alkaloids MG, PAY, SG, and SC. The limits of detection (signal-to-noise ratio of 3) were 100 ng/mL for the parent alkaloids and 50 ng/mL for O-demethyltramadol. In most of the human urine samples that tested positive, MG, 16-carboxy-MG, 9-O-demethyl-MG, and/or 9-O-demethyl-16-carboxy-MG could be detected.

11.4 IDENTIFICATION OF *M. SPECIOSA* USING DNA ANALYSES

The variety of combinations among *M. speciosa* products and the cases of plant or chemical adulteration give rise to a need to develop methods for the identification of both the plant materials and the main constituent, MG. The phylogenetic characterization of kratom samples by specific DNA nucleotide sequences can complement the phytochemical analyses.

The genus *Mitragyna* consists of ten species located around the world. A phylogenetic analysis of the *Mitragyna* plants was performed using nuclear rDNA, internal transcribed spacer (ITS) and chloroplast DNA, and rbcL and trnT-F sequences (Sukrong et al. 2007; Razafimandimbison and Bremer 2002), and it was shown that *M. speciosa* is well discriminated from other related plants based on the ITS sequence (Sukrong et al. 2007). Maruyama et al. (2009) investigated the botanical origin of 13 commercial kratom products obtained via the Internet in Japan using the ITS sequence analysis of rDNA. In addition, the polymerase chain reaction–restriction fragment length polymorphism (PCR-RFLP) method was applied to the same products in order to estimate the method's accuracy and utility (Maruyama et al. 2009; Sukrong et al. 2007). The ITS sequence analysis of the commercial products showed that most of them were derived from *M. speciosa* or closely related plants, while the others were made from a plant of the same tribe as *M. speciosa*. The reported PCR-RFLP method could clearly distinguish kratom from the other psychoactive plants available in Japanese markets as well as from related plants.

In addition to plant materials with psychoactive effects such as kratom, the distribution of herbal blend products that contain psychoactive synthetic compounds as adulterants has increased remarkably worldwide. In particular, ever since cannabinoid receptor agonists (synthetic cannabinoids) were first detected in herbal blend products in 2008, the situation in the global drug market has changed dramatically. Ogata et al. (2013) investigated the origins of botanical materials in 62 herbal blend products by DNA sequence analyses and BLAST searches. These products, in which various synthetic cannabinoids were detected, were obtained via the Internet in Japan from 2008 to 2011. The nucleotide sequences of four regions were analyzed to identify the origins of each plant species in the herbal mixtures. The sequences of damiana (*Turnera diffusa*) and Lamiaceae herbs (*Melissa*, *Mentha*, and *Thymus*) were frequently detected in a number of products. However, the sequences of other plant species indicated on the packaging labels were not detected.

For most of the plants identified, no reliable psychoactive effects have been reported. In only a few products were DNA fragments of potent psychotropic plants such as marijuana (*Cannabis sativa* L.), diviner's sage (*S. divinorum*), and kratom (*M. speciosa*) detected, and their active constituents were confirmed through GC-MS and LC-MS. The names of these plants were never indicated on the labels. With the exception of some psychotropic plants, the plant materials of recent herbal products, which were enhanced with psychoactive synthetic compounds, would be used mainly as diluents for the synthetic compounds.

11.5 GENERAL CONCLUSIONS

The use of kratom as a psychoactive substance, as well as the use of other psychoactive plant products such as khat and *Salvia divinorum*, has increased not only in Southeast Asia where the *M. speciosa* plant is native but also in European countries and the United States, because of their easy availability via the Internet. These naturally occurring alkaloids give a useful key to the development of new types of analgesic drugs with structures quite different from that of morphine, but the abuse of these plant products is of serious concern. In addition, the distribution of herbal blend products augmented with potent synthetic psychoactive compounds (e.g., synthetic cannabinoids) has increased dramatically in the past five years. To prevent the abuse of these psychoactive substances, continuous and dedicated monitoring of the distribution of these substances is essential.

REFERENCES

Arndt, T., Claussen, U., Güssregen, B., Schröfel, S., Stürzer, B., Werle, A. and Wolf, G. 2011. Kratom alkaloids and *O*-desmethyltramadol in urine of a 'Krypton' herbal mixture consumer. *Forensic Sci Int* 208: 47–52.

Chan, K.B., Pakiam, C. and Rahim, R.A. 2005. Psychoactive plant abuse: identification of mitragynine in ketum and in ketum preparations. *Bull Narc* 57: 249–56.

Chittrakarn, S., Penjamras, P. and Keawpradub, N. 2012. Quantitative analysis of mitragynine, codeine, caffeine, chlorpheniramine and phenylephrine in a kratom (*Mitragyna speciosa* Korth.) cocktail using high-performance liquid chromatography. *Forensic Sci Int* 217: 81–86.

de Moraes, N.V., Moretti, R.A. C., Furr E.B. III, McCurdy, C.R. and Lanchote, V.L. 2009. Determination of mitragynine in rat plasma by LC–MS/MS: Application to pharmacokinetics. *J Chromatogr B* 877: 2593–97.

EMCDDA European Monitoring Centre for Drugs and Drug Addiction. 2013. Drug profiles, Kratom (*Mitragyna speciosa*). http: //www.emcdda.europa .eu/publications/drug-profiles/kratom (accessed December 16, 2013).

Holler, J.M., Vorce, S.P., McDonough-Bender, P.C., Magluilo Jr., J., Solomon, C.J. and Levine, B. 2011. A drug toxicity death involving propylhexedrine and mitragynine. *J Anal Toxicol* 35: 54–59.

Horie, S., Koyama, F., Takayama, H., Ishikawa, H., Aimi, N., Ponglux, D., Matsumoto, K. and Murayama, T. 2005. Indole alkaloids of a Thai medicinal herb, *Mitragyna speciosa*, that has opioid agonistic effect in guinea-pig ileum. *Planta Med* 71: 231–36.

Kapp, F.G., Maurer, H.H., Auwärter, V., Winkelmann, M. and Hermanns-Clausen, M. 2011. Intrahepatic cholestasis following abuse of powdered kratom (*Mitragyna speciosa*). *J Med Toxicol.* 7: 227–31.

Kawamura, M., Kikura-Hanajiri, R. and Goda, Y. 2008. Survey of current trends in the abuse of psychotropic plants using LC-MS. *Jpn J Food Chemistry* 15: 73–78.

Kikura-Hanajiri, R., Kawamura, M., Maruyama, T., Kitajima, M., Takayama, H. and Goda, Y. 2009. Simultaneous analysis of mitragynine, 7-hydroxymitragynine and other alkaloids in the psychotropic plant kratom (*Mitragyna speciosa*) by LC–ESI-MS. *Forensic Toxicol* 27: 67–74.

Kitajima, M., Misawa, K., Kogure, N., Said I. M., Horie, S., Hatori, Y., Murayama, T. and Takayama, H. 2006. A new indole alkaloid, 7-hydroxyspeciociliatine, from the fruits of Malaysian *Mitragyna speciosa* and its opioid agonistic activity. *J Nat Med* 60: 28–35.

Kitajima, M., Nakayama, T., Kogure, N., Wongseripipatana, S. and Takayama, H. 2007. New heteroyohimbine-type oxindole alkaloid from the leaves of Thai *Mitragyna hirsute*. *J Nat Med* 61: 192–5.

Kowalczuk, A.P., Łozak, A. and Zjawiony, J.K. 2013. Comprehensive methodology for identification of Kratom in police laboratories. *Forensic Sci Int* 233: 238–43.

Kronstrand, R., Roman, M., Thelander, G. and Eriksson, A. 2011. Unintentional fatal intoxications with mitragynine and O-desmethyltramadol from the herbal blend krypton. *J Anal Toxicol* 35: 242–47.

Le, D., Goggin M.M. and Janis, G.C. 2012. Analysis of mitragynine and metabolites in human urine for detecting the use of the psychoactive plant kratom. *J Anal Toxicol* 36: 616–25.

Lu, S., Trana, B.N., Nelsen, J L. and Aldous, K. M. 2009. Quantitative analysis of mitragynine in human urine by high performance liquid chromatography-tandem mass spectrometry. *J Chromatogr B* 877: 2499–2505.

Maruyama, T., Kawamura, M., Kikura-Hanajiri, R., Takayama, H., Goda, Y. 2009. The botanical origin of kratom (*Mitragyna speciosa*; Rubiaceae) available as abused drugs in the Japanese markets. *J Nat Med* 63: 340–44.

Matsumoto, K., Mizowaki, M., Suchitra, T., Takayama, H., Sakai, S., Aimi, N. and Watanabe, H. 1996a. Antinociceptive action of mitragynine in mice: Evidence for the involvement of supraspinal opioid receptors. *Life Sci* 59: 1149–55.

Matsumoto, K., Mizowaki, M., Suchitra, T., Murakami, Y., Takayama, H., Sakai, S. Aimi, N. and Watanabe, H. 1996b. Central antinociceptive effects of mitragynine in mice: Contribution of descending noradrenergic and serotonergic systems. *Eur J Pharmacol* 317: 75–81.

Matsumoto, K., Horie, S., Ishikawa, H., Takayama, H., Aimi, N., Ponglux, D. and Watanabe, K. 2004. Antinociceptive effect of 7-hydroxymitragynine in mice: Discovery of an orally active opioid analgesic from the Thai medicinal herb *Mitragyna speciosa*. *Life Sci* 74: 2143–55.

Matsumoto, K., Horie, S., Takayama, H., Ishikawa, H., Aimi, N., Ponglux, D., Murayama, T. and Watanabe, K. 2005. Antinociception, tolerance and withdrawal symptoms induced by 7-hydroxymitragynine, an alkaloid from the Thai medicinal herb *Mitragyna speciosa*. *Life Sci* 78: 2–7.

Matsumoto, K., Hatori, Y., Murayama, T., Tashima, K., Wongseripipatana, S., Misawa, K., Kitajima, M., Takayama, H. and Horie, S. 2006. Involvement of mu-opioid receptors in antinociception and inhibition of gastrointestinal transit induced by 7-hydroxymitragynine, isolated from Thai herbal medicine *Mitragyna speciosa*. *Eur J Pharmacol* 549: 63–70.

Neerman, M.F., Frost, R.E. and Deking, J. 2013. A drug fatality involving kratom. *J Forensic Sci* 58: S280–84.

Ogata, J., Uchiyama, N., Kikura-Hanajiri, R. and Goda, Y. 2013. DNA sequence analyses of blended herbal products including synthetic cannabinoids as designer drugs. *Forensic Sci Int* 227: 33–41.

Parthasarathy, S., Ramanathan S., Murugaiyah, V., Hamdan, M.R., Said M.I.M., Lai C. S. and Mansor, S.M. 2013. A simple HPLC-DAD method for the detection of psychotropic mitrahynine in *Mitragyna speciosa* (ketum) and its products for the application in forensic investigation. *Forensic Sci Int* 226: 183–87.

Philipp, A.A., Wissenbach, D.K., Zoerntlein, S.W., Klein, O.N., Kanogsunthornrat, J. and Maurer, H.H. 2009. Studies on the metabolism of mitragynine, the main alkaloid of the herbal drug Kratom, in rat and human urine using liquid chromatography–linear ion trap mass spectrometry. *J Mass Spectrom* 44: 1249–61.

Philipp, A.A., Wissenbach, D.K., Weber, A.A., Zapp, J. and Maurer, H.H. 2011a. Metabolism studies of the Kratom alkaloid speciociliatine, a diastereomer of the main alkaloid mitragynine, in rat and human urine using liquid chromatography–linear ion trap mass spectrometry. *Anal Bioanal Chem* 399: 2747–53.

Philipp, A.A., Meyer, M.R., Wissenbach, D.K., Weber, A.A., Zoerntlein, S.W., Zweipfenning, P.G.M. and Maurer, H.H. 2011b. Monitoring of kratom or Krypton intake in urine using GC-MS in clinical and forensic toxicology. *Anal Bioanal Chem* 400: 127–35.

Philipp, A.A., Wissenbach, D.K., Weber, A.A., Zapp, J. and Maurer, H.H. 2011c. Metabolism studies of the Kratom alkaloids mitraciliatine and isopaynantheine, diastereomers of the main alkaloids mitragynine and paynantheine, in rat and human urine using liquid chromatography–linear ion trap-mass spectrometry, *J Chromatogra B* 879: 1049–55.

Ponglux, D., Wongseripipatana, S., Takayama, H., Kikuchi, M., Kurihara, M., Kitajima, M., Aimi, N. and Sakai, S. 1994. A new indole alkaloid, 7 alpha-hydroxy-7*H*-mitragynine, from *Mitragyna speciosa* in Thailand. *Planta Med* 60: 580–1.

Razafimandimbison, S.G. and Bremer, B. 2002. Phylogeny and classification of Naucleeae s.l. (Rubiaceae) inferred from molecular (ITS, rbcL, and trnT-F) and morphological data. *Am J Bot* 89: 1027–41.

Schmidt, M.M., Sharma, A., Schifano, F. and Feinmann, C. 2010. 'Legal highs' on the net-evaluation of UK-based websites, products and product information. *Forensic Sci Int* 206: 92–97.

Shellard, E.J., Houghton, P.J. and Resha, M. 1978. The *Mitragyna* species of Asia. *Planta Med* 34: 253–63.

Sukrong, S., Zhu, S., Ruangrungsi, N., Phadungcharoen, T., Palanuvej, C., Komatsu, K. 2007. Molecular analysis of the genus *Mitragyna* existing in Thailand based on rDNA ITS sequences and its application to identify a narcotic species: *Mitragyna speciosa*. *Biol Pharm Bull* 30: 1284–88.

Takayama, H. 2004. Chemistry and pharmacology of analgesic indole alkaloids from the rubiaceous plant, *Mitragyna speciosa*. *Chem Pharm Bull* 52: 916–28.

Takayama, H., Aimi, N. and Sakai, S. 2000. Chemical studies on the analgesic indole alkaloids from the traditional medicine (*Mitragyna speciosa*) used for opium substitute. *Yakugaku Zasshi* 120: 959–67.

Thongpradichote, S., Matsumoto, K., Tohda, M., Takayama, H., Aimi, N., Sakai, S. and Watanabe, H. 1998. Identification of opioid receptor subtypes in antinociceptive actions of supraspinally administered mitragynine in mice. *Life Sci* 62: 1371–78.

UNODC. 2013. United Nations Office on Drugs and Crime, World Drug Report 2013, United Nations publication. http: //www.unodc.org/unodc/secured /wdr/wdr2013/World_Drug_Report_2013.pdf.

Vardakou, I., Pistos, C. and Spiliopoulou, C. 2010. Spice drugs as a new trend: Mode of action, identification and legislation. *Toxicol Lett* 197: 157–62.

12 The ADME of Mitragynine and Analogs

Annette Cronin, Ines A. Ackerman, and Jochen Beyer

CONTENTS

12.1 INTRODUCTION

ADME is the abbreviation used in pharmacology to describe the processes of absorption, distribution, metabolism, and excretion of substances introduced into an organism. The four steps influence the movement and concentration of a xenobiotic (exogenous) substance in the organism, and therefore the pharmacological effects. ADME can also be used to describe and determine the fate of a substance from the moment of entering the body until its complete elimination.

12.2 ABSORPTION

Absorption is the process of the movement of a xenobiotic substance from the site of its administration into the bloodstream of a body. Substances can be administered into a body through the mucosa of the digestive system after oral, sub-lingual, or rectal ingestion. Further epithelial surfaces as site of administration include skin, cornea, vagina, and nasal mucosa. Additionally, substances can be inhaled or injected.

After oral ingestion of xenobiotics such as mitragynine contained in kratom leaves, the mechanism of the transfer across the intestinal epithelial barrier is a passive diffusion. The rate of transfer of xenobiotics is determined by substance specific properties as well as physiological factors. The main substance-specific properties are lipid solubility, ionization, and molecule size. High-lipid solubility and small molecule size result in a high absorption of the ingested substance.

Physiological factors that influence the rate of absorption include the splanchnic blood flow (rate of blood circulation around the internal organs) and the gastrointestinal motility. These physiological factors can be influenced by the drug itself when it pharmacologically reduces the splanchnic blood flow or the gastrointestinal motility.

Bioavailability is a measurement of the extent of absorption. By definition, bioavailability of a xenobiotic substance is the fraction of an administered substance that reaches the bloodstream unchanged. The bioavailability of a substance that is administered intravenously is therefore 100%.

12.3   DISTRIBUTION

The distribution of a xenobiotic substance is the reversible transfer of a substance from one body compartment to another. After absorption into the bloodstream, substances are carried throughout the body and distributed from the blood into muscle and organs. The different characteristics of the tissues thereby result in different concentrations of the substance in each compartment. These distribution processes usually lower the concentration of the ingested substance in the blood.

To describe these distribution processes, the pharmacokinetic parameter *volume of distribution* has been introduced. The volume of distribution is the apparent volume of liquid required to represent the measured plasma concentration of a drug after the application of a certain concentration. For drugs that distribute outside the plasma compartment, the volume of distribution can therefore be much larger than the total body volume.

12.4   XENOBIOTIC METABOLISM

Xenobiotic metabolism is the chemical modification of foreign compounds by the organism using specialized enzymatic systems. During this process, lipophilic chemical substances are often converted into more hydrophilic derivatives and therefore readily excretable metabolites. Xenobiotic metabolism is generally considered a detoxifying process. However, in some cases substances are bio-activated by the formation of reactive intermediates, which may lead to undesired toxicity. For this reason, the analysis of the metabolic pathway of a chemical compound/drug is vital for understanding the reaction mechanism and/or toxicity of the substance.

Xenobiotic metabolism is divided into distinct phases. In phase I, known as functionalization, several enzymatic systems introduce polar functional groups into lipophilic substrates. Next, in phase II (conjugation), functionalized metabolites are conjugated to polar endogenous compounds such as glutathione, glucuronic acid, or sulfate. Conjugated metabolites may then be transported out of the cell by various efflux transporters in phase III (excretion).

12.4.1 PHASE I

The most important enzymatic systems of phase I are the cytochrome P450-dependent monooxygenases (CYPs). CYPs are localized in the endoplasmic reticulum membrane of the liver and many other organs. CYPs form a complex with cytochrome P450 reductase that is required for activity. The enzymes contain the prosthetic group hemin, which is reduced to the oxygen-binding heme during the catalytic cycle. CYPs introduce oxygen into their substrates or—in few cases—abstract electrons. CYPs accept a broad spectrum of substrates and CYP-catalyzed reactions result in hydroxylation, epoxidation, dealkylation, desaturation, heteroatom oxidation, or heteroatom replacement. CYP enzymes are found in nearly every living organism; in humans more than 60 CYP isoforms with varied roles in xenobiotic as well as endogenous metabolism are identified (Guengerich 2010).

Other important phase I enzymes include oxidases (flavin-containing monooxygenases (FMO), monoamine oxidases (MAO)), dehydrogenases (alcohol and aldehyde dehydrogenases (ADH, ALDH)), and hydrolases (esterases, amidases and epoxide hydrolases (EH)). FMOs contain flavin as their prosthetic group and use NADPH as cofactor to oxygenate mainly nucleophilic nitrogen and sulfur in a wide range of substrates. MAOs are localized in the mitochondria membrane and catalyze the oxidation of endogenous (serotonin, dopamine) and exogenous monoamines (tyramine, phenylethylamines). ADH and ALDH metabolize ethanol and other alcohols to the corresponding aldehyde and carboxylic acid, respectively. Esterases catalyze the hydrolysis of esters (such as acetylcholine) to an alcohol and carboxylic acid. Finally, EHs convert compounds containing an epoxy group to the corresponding diols (Decker et al. 2009).

12.4.2 PHASE II

Several transferase systems are involved in phase II of xenobiotic metabolism. Electrophilic substrates (α,β-unsaturated carbonyl compounds or epoxides) are conjugated by glutathion transferases (GST) (or epoxide hydrolases), while nucleophilic substrates are conjugated mainly by UDP-glucuronosyl transferases (UGT), sulfotransferases (SULT), and N-acetyl transferases (NAT). Normally, a conjugation with hydrophilic endogenous intermediates results in a loss of chemical reactivity of electrophilic compounds as well as loss of activity of nucleophilic substrates. The water solubility of metabolite conjugates is increased, and therefore the excretion is facilitated. In an alternative mechanism relevant for the detoxification of plant alkaloids, compounds can be acetylated by acetyltransferases, leading to less hydrophilic derivatives.

12.5 PHARMACOKINETIC PROFILE OF MITRAGYNINE

Common routes of administration of kratom are oral administration of fresh or dried leaves (chewing the leaves, drinking the tea) and pulmonal inhalation of the smoke of burned dried leaves (smoking).

Parthasarathy et al. (2010) presented a detailed pharmacokinetic profile of mitragynine in rats after oral and intravenous administration. After intravenous administration of mitragynine (1.5 mg/kg) the concentration peaked at 1.2 hours with 2.3 µg/mL, followed by a biphasic elimination with a half-life of 2.9 hours and a clearance of 0.09 L/h per kg. The volume of distribution was relatively small with 0.79 L/kg, indicating that mitragynine is not widely distributed into tissue compartments (Parthasarathy et al. 2010). In contrast to the intravenous application, the oral absorption of mitragynine was shown to be prolonged and incomplete, with an absolute oral bioavailability of around 3%. Several reports show that after oral application (20–50 mg mitragynine), clearance (1.6–7 L/h per kg) and volume distribution (37–89 L/kg) show much higher values than in intravenous application (Janchawee et al. 2007; deMoraes et al. 2009; Parthasarathy et al. 2010), thus evidencing the poor absorption/low bioavailability of mitragynine. A comparison of the published pharmacokinetic parameters of mitragynine is summarized in Table 12.1.

The bioavailability of mitragynine following inhalation and the bioavailability of other alkaloids found in kratom have not been studied in detail so far.

12.6 METABOLISM OF MITRAGYNINE

Of the several indole alkaloids found in the rubiaceous plant *Mitragyna speciosa*, mitragynine is the most abundant. A number of mitragynine analogs are found in *M. speciosa* leaves to various extents. As illustrated

TABLE 12.1

Summary of the Published Pharmacokinetic Data on Mitragynine

Application	Parthasarathy et al. (2010) Intravenous (1.5 mg/kg)	Oral (50 mg/kg)	Janchawee et al. (2007) Oral (40 mg/kg)	de Moraes et al. (2009) Oral (20 mg/kg)
Bioavailability [%]	100	3.03 ± 1.47		
C_{max} [µg/mL]	2.3 ± 1.2	0.70 ± 0.21	0.63 ± 0.18	0.42 ± 0.06
T_{max} [h]	1.2 ± 1.1	4.5 ± 3.6	1.83 ± 1.25	1.26 ± 0.2
$t_{1/2\,a}$ [h]		1.72 ± 0.90	0.48 ± 0.36	0.28 ± 0.095
$t_{1/2}$ [h]	2.9 ± 2.1	6.6 ± 1.3	9.43 ±1.74	3.85 ± 0.51
AUC $^{0-\infty}$ [µg h/mL]	9.2 ± 6.5	8.2 ± 3.0	6.99 ± 2.93	3.15
V_d/F [L/kg]	0.79 ± 0.42	64 ± 23	89.50 ± 30.30	37.90 ± 5.41
Cl/F [L/h kg]	0.29 ± 0.27	7.0 ± 3.0	1.60 L/h ± 0.58 L/h	6.35 ± 0.43

C_{max}: peak plasma concentration; T_{max}: time to reach Cmax; $t_{1/2a}$: absorption half-life; $t_{1/2}$: elimination half-life; $AUC^{0-\infty}$: area under the plasma concentration-time curve; V_d/F: apparent volume of distribution; CL/F: apparent total clearance

in Figure 12.1, speciogynine, mitraciliantine, and speciociliantine are diastereomers of mitragynine. Paynantheine and isopaynantheine represent dehydrated analogs of speciogynine and mitraciliantine, respectively (Takayama 2004; Philipp et al. 2011b).

12.6.1 MITRAGYNINE

Only limited information is currently available on the metabolism of kratom alkaloids in rodents and humans. Phillip et al. provided data on the metabolism of mitragynine and its analogs in rat urine. The single compounds were isolated and purified from kratom leaves. Metabolism was analyzed using LC-MS/MS and high-resolution mass spectrometry after administration of a single oral dose of 40 mg/kg of each pure compound. These results were compared to human patient urine samples with suspected kratom consumption that were submitted to the authors' laboratory for toxicological testing (Philipp et al. 2009, 2010a,b, 2011a,b).

In their first study Phillip et al. identified seven phase I metabolites as well as the corresponding phase II metabolites of mitragynine in rat and human urine. The authors therefore postulated a metabolic pathway of mitragynine, which is shown in detail for humans in Figure 12.2. The identification of metabolites in human urine using LC/MS/MS and high-resolution mass spectrometry indicated an O-demethylation of the methoxy group in positions 9 and 17. The presence of metabolite 3 further indicated a hydrolysis of the

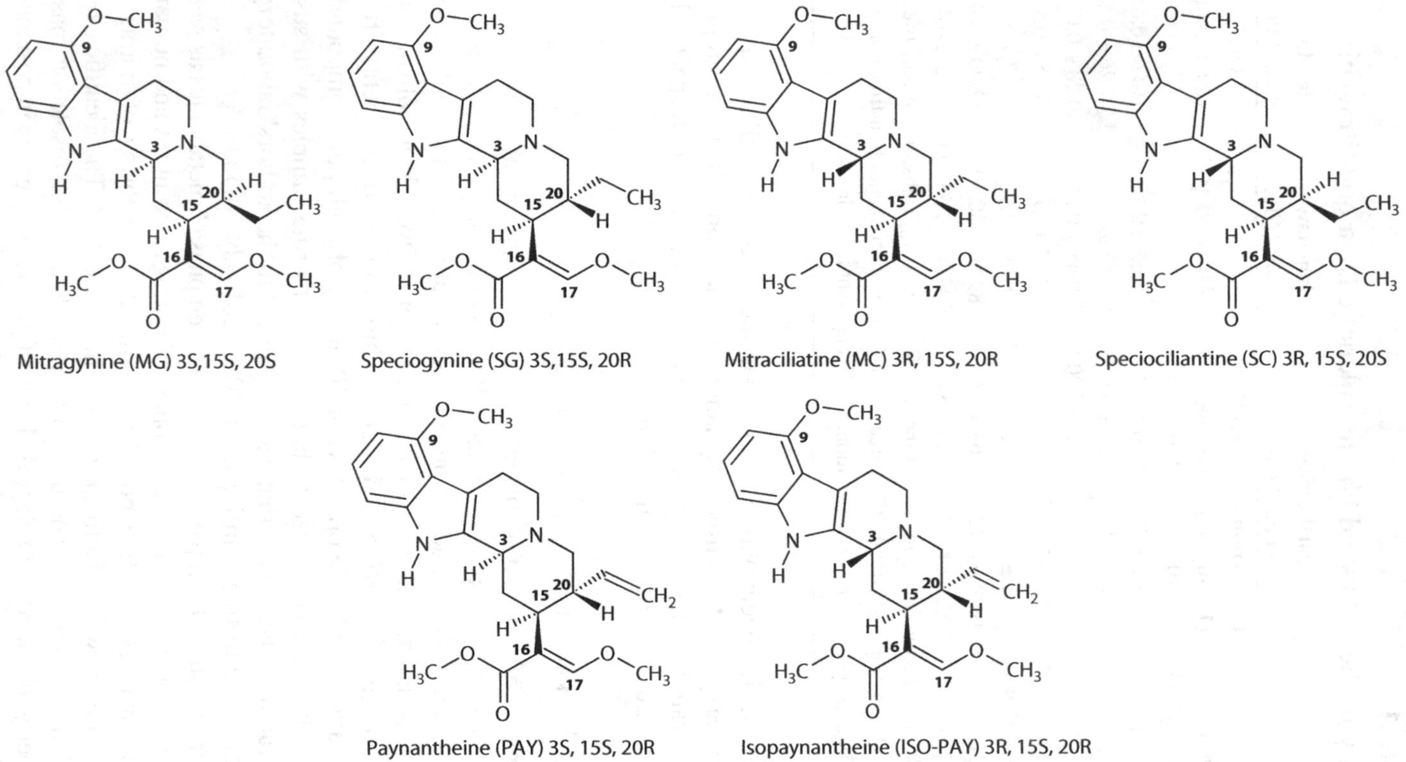

FIGURE 12.1 Structures and sterical configurations of the alkaloid mitragynine and its analogues from *Mitragyna speciosa*. (Adapted from Philipp et al. 2011b.)

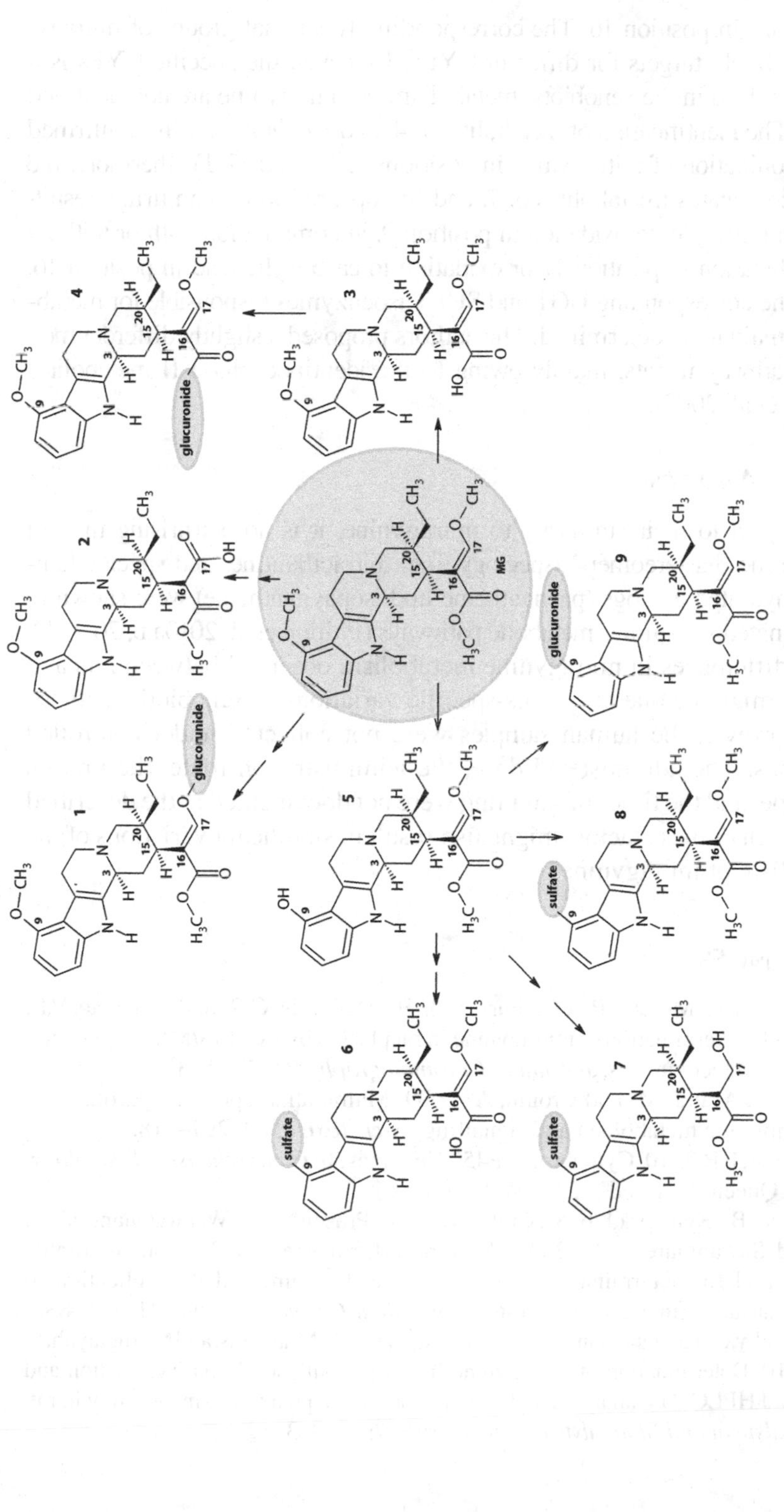

FIGURE 12.2 Postulated metabolic pathway of mitragynine (highlighted in a gray circle) in humans. The illustration shows several phase I and phase II metabolites (highlighted in gray) identified in human urine of assumed kratom users. (Adapted from Philipp et al. 2011b.)

methylester in position 16. The corresponding functional groups of mitragynine are likely targets for different CYPs. However, the specific CYPs isoforms involved in the xenobiotic metabolism of mitragynine are not identified to date. The identification of metabolites 1, 4, and 9 in human urine confirmed a glucuronidation of mitragynine in positions 9, 16, and 17. Furthermore, the sulfate conjugates (metabolites 6, 7, and 8) appeared in human urine, resulting from mitragynine oxidation in position 9, in combination with or without a demethylation in position 17 or oxidation to carboxylic acid in position 16. Again, the corresponding UGT and SULT isoenzymes responsible for metabolism remain to be determined. The authors proposed a slightly different metabolic pathway in rats, mainly owing to the identified phase II metabolites (Philipp et al. 2009).

12.6.2 ANALOGS

With respect to their similarity to mitragynine, it is not surprising that all mitragynine diastereomers' (speciogynine, mitraciliantine, and speciociliantine) dehydrated analogs (paynantheine and isopaynantheine) were shown to be eliminated via similar metabolic pathways (Philipp et al. 2010a,b, 2011a,b).

The differences in mitragynine metabolism observed between rats and humans might be due to species-specific variations in xenobiotic metabolism. However, the human samples were not collected under controlled conditions. The administered dose, the administration route, the kratom plant type, and the times of sampling were not documented in the described study. Each of these factors might also result in significant variations of the metabolism of mitragynine.

REFERENCES

de Moraes, N.V., Moretti, R.A.C., Furr III, E.B., McCurdy, C.R. and Lanchote, V.L. 2009. Determination of mitragynine in rat plasma by LC–MS/MS: Application to pharmacokinetics. *Journal of Chromatography B* 877: 2593–97.
Decker, M., Arand, M. and Cronin, A. 2009. Mammalian epoxide hydrolases in xenobiotic metabolism and signalling. *Arch Toxicol* 83: 297–318.
Guengerich, F.P. 2010. Cytochrome P450 Enzymes. In *Comprehensive Toxicology*, McQueen, C.A. (ed.), 275–94. Elsevier.
Janchawee, B., Keawpradub, N., Chittrakarn, S., Prasettho, S., Wararatananurak, P. and Sawangjareon, K. 2007. A high-performance liquid chromatographic method for determination of mitragynine in serum and its application to a pharmacokinetic study in rats. *Biomedical Chromatography* 21: 176–83.
Parthasarathy, S., Ramanathan, S., Ismail, S., Adenan, M., Mansor, S. and Murugaiyah, V. 2010. Determination of mitragynine in plasma with solid-phase extraction and rapid HPLC–UV analysis, and its application to a pharmacokinetic study in rat. *Analytical and Bioanalytical Chemistry* 397: 2023–30.

Philipp, A., Wissenbach, D., Weber, A., Zapp, J. and Maurer, H. 2011a. Metabolism studies of the Kratom alkaloid speciociliatine, a diastereomer of the main alkaloid mitragynine, in rat and human urine using liquid chromatography-linear ion trap mass spectrometry. *Analytical and Bioanalytical Chemistry* 399: 2747–53.

Philipp, A., Wissenbach, D., Weber, A.A., Zapp, J., Zoerntlein, S., Kanogsunthornrat, J. and Maurer, H. 2010a. Use of liquid chromatography coupled to low- and high-resolution linear ion trap mass spectrometry for studying the metabolism of paynantheine, an alkaloid of the herbal drug Kratom in rat and human urine. *Analytical and Bioanalytical Chemistry* 396: 2379–91.

Philipp, A.A., Wissenbach, D.K., Weber, A.A., Zapp, J. and Maurer, H.H. 2010b. Phase I and II metabolites of speciogynine, a diastereomer of the main Kratom alkaloid mitragynine, identified in rat and human urine by liquid chromatography coupled to low- and high-resolution linear ion trap mass spectrometry. *Journal of Mass Spectrometry* 45: 1344–57.

Philipp, A.A., Wissenbach, D.K., Weber, A.A., Zapp, J. and Maurer, H.H. 2011b. Metabolism studies of the Kratom alkaloids mitraciliatine and isopaynantheine, diastereomers of the main alkaloids mitragynine and paynantheine, in rat and human urine using liquid chromatography–linear ion trap-mass spectrometry. *Journal of Chromatography B* 879: 1049–55.

Philipp, A.A., Wissenbach, D.K., Zoerntlein, S.W., Klein, O.N., Kanogsunthornrat, J. and Maurer, H.H. 2009. Studies on the metabolism of mitragynine, the main alkaloid of the herbal drug Kratom, in rat and human urine using liquid chromatography–linear ion trap mass spectrometry. *Journal of Mass Spectrometry* 44: 1249–61.

Takayama, H. 2004. Chemistry and pharmacology of analgesic indole alkaloids from the rubiaceous plant, *Mitragyna speciosa*. *Chem Pharm Bull* 52: 916–28.

13 Analgesic Effects of Mitragynine and Analogs

Kenjiro Matsumoto and Syunji Horie

CONTENTS

13.1 INTRODUCTION

Substances derived from natural products have been utilized since the beginning of history for various medical purposes, including the treatment of pain. A prototypical example of such natural products is the opium poppy (*Papaver somniferum*). Morphine, an alkaloid component of the opium poppy, is the most widely used compound among narcotic analgesics and remains the gold standard. Recently, analogs have been produced from natural compounds, and completely synthetic compounds based on natural pharmacophores have been introduced to the market. However, the research and medical fields still struggle with the undesirable side effects of these analgesic substances (McCurdy and Scully 2005).

The traditional herbal medicine *Mitragyna speciosa* has long been used in Thailand for its opioid-like effects and as a replacement for opium (Burkill 1935; Suwanlert 1975; Adkins et al. 2011). This medicinal herb contains many indole alkaloids (Takayama 2004). Mitragynine, a main constituent of this plant, is an indole alkaloid that is structurally different from morphine. Pharmacological activities of mitragynine have been studied, and we have found that mitragynine has agonistic effects on opioid

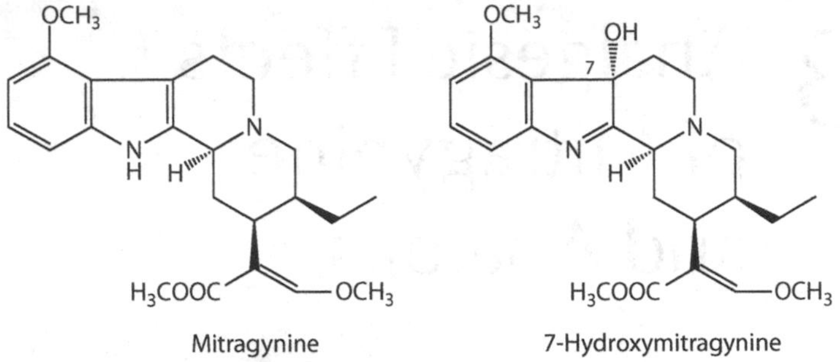

Mitragynine   7-Hydroxymitragynine

FIGURE 13.1 Chemical structure of mitragynine and 7-hydroxymitragynine.

receptors, but its antinociceptive effect was less potent than that of the crude extract of *M. speciosa* (Watanabe et al. 1997). That is, the opium-like effect of *M. speciosa* cannot be fully explained by that of mitragynine. Thus the opioid agonistic effects of other constituents of *M. speciosa* were investigated using in vitro assays. Among them, 7-hydroxymitragynine, which has a hydroxyl group at the C7 position of mitragynine (Figure 13.1), produced the most potent antinociceptive effect, suggesting that the opioid effect of *M. speciosa* is based mostly on the activity of 7-hydroxymitragy-nine (Horie et al. 2005). This chapter covers published work on the analgesic activities of mitragynine, 7-hydroxymitragynine, and synthesized mitragynine derivatives (Figure 13.2).

13.2   ANTINOCICEPTIVE EFFECT OF MITRAGYNINE AND ITS METABOLITE

The pharmacology of *M. speciosa* and mitragynine was first explored by K.S. Grewal at the University of Cambridge in 1932 (Grewal 1932). He performed a series of experiments on animal tissues and a group of five male volunteers. He described mitragynine as having a central nervous system stimulant effect resembling that of cocaine. Macko et al. (1972) reported that mitragynine exhibited antinociceptive and antitussive actions in mice comparable to those of codeine. Their findings were that, unlike opioid analgesics at equivalent doses, mitragynine did not possess the side effects common to opioids. In addition, the absence of an antagonistic effect of nalorphine on mitragynine-induced antinociception led them to postulate non-involvement of the opioid system in the action of mitragynine.

We have studied the pharmacological effects of mitragynine on guinea pig ileum, mouse vas deferens, radioligand binding, and the tail-flick test in mice, and found that mitragynine acts on opioid receptors and possesses antinociceptive effects (Watanabe et al. 1997; Yamamoto et al. 1999; Takayama et al. 2002). But the effect of mitragynine was less potent than

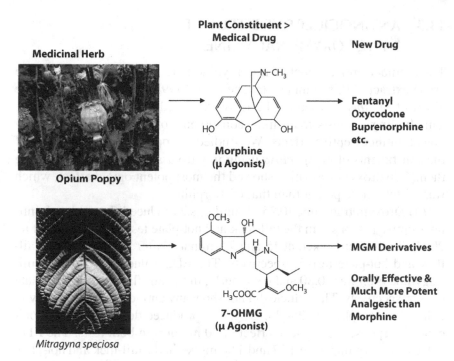

FIGURE 13.2 Development of an MGM derivative that exerts an analgesic effect from an alkaloid of *Mitragyna speciosa*. Our development research on *Mitragyna* alkaloids is similar to that of morphine.

that of morphine. Some pharmacological investigations of mitragynine have also revealed that it has an antinociceptive action through the supraspinal opioid receptors, and that its action is dominantly mediated by μ- and δ-opioid receptors in in vivo and in vitro studies (Matsumoto et al. 1996a,b; Tohda et al. 1997; Thongpraditchote et al. 1998).

Another alkaloid of interest is mitragynine pseudoindoxyl, which was at first isolated as a metabolite of mitragynine by microbial biotransformation. Macko et al. (1972) reported that oral administration of mitragynine was more effective than subcutaneous administration. This finding suggested that the antinociceptive effect of mitragynine exists predominantly in its derivatives. Our previous study demonstrated a potent opioid agonistic property of the compound mitragynine pseudoindoxyl in in vitro experiments (Matsumoto et al. 1996a). In guinea pig ileum, mitragynine and mitragynine pseudoindoxyl inhibit the twitch contraction through opioid receptors. The effect of mitragynine pseudoindoxyl was 20-fold more potent than that of morphine. In mouse vas deferens, the effect of mitragynine pseudoindoxyl was 35-fold more potent than that of morphine. In spite of its potent opioid effect, mitragynine and mitragynine pseudoindoxyl induced only a weak antinociceptive effect in the mouse tail-flick test in comparison with morphine (Takayama et al. 2002).

13.3   ANTINOCICEPTIVE EFFECT OF 7-HYDROXYMITRAGYNINE

The antinociceptive effect of mitragynine is less potent than that of the crude extract of this plant (Watanabe et al. 1999). That is, the opium-like effect of *M. speciosa* cannot be fully explained by that of mitragynine. This finding suggests that minor constituents of *M. speciosa* have very potent antinociceptive effects. We studied the opioid agonistic effects of the constituents of *M. speciosa* using in vitro assays (Chapter 9). Among them, 7-hydroxymitragynine showed the most potent opioid effect, which was 17-fold more potent than that of morphine.

7-Hydroxymitragynine (0.25–2 mg/kg, s.c.) induced dose-related antinociceptive responses in the tail-flick and hot-plate tests (Matsumoto et al. 2004). The effect peaked at 15 and 7.5 minutes after injection in the tail-flick and hot-plate tests, respectively. The ED_{50} values for 7-hydroxymitragynine (s.c.) were 0.80 and 0.93 mg/kg in the tail-flick and the hot-plate tests, respectively. The vehicle did not show any antinociceptive activity in either test. Morphine (1.25–8 mg/kg, s.c.) produced dose-related antinociceptive response with a peak effect at 30 minutes in both tests. The ED_{50} values for morphine were 4.57 and 4.08 mg/kg in the tail-flick and hot-plate tests, respectively. Compared to morphine on mg/kg basis, 7-hydroxymitragynine was 5.7 and 4.4 times more potent in the tail-flick and hot-plate tests, respectively. 7-Hydroxymitragynine affected behavioral responses: 2 mg/kg of 7-hydroxymitragynine elicited an increase spontaneous locomotor activity and Straub tail, as did 8 mg/kg of morphine.

In order to determine the opioid receptor-type selectivity of 7-hydroxymitragynine antinociception, mice were pretreated with selective opioid receptor antagonists (Matsumoto et al. 2005). In the tail-flick test, the antinociceptive effect of 7-hydroxymitragynine was significantly blocked by the non-selective opioid antagonist naloxone, the irreversible μ_1/μ_2-opioid receptor selective antagonist β-funaltrexamine (β-FNA), and the μ_1-opioid receptor selective antagonist naloxonazine. The selective δ-opioid antagonist naltrindole (NTI) and the selective κ-opioid antagonist nor-binaltorphimine (norBNI) were ineffective against 7-hydroxymitragynine-induced antinociception. In the hot-plate test, the effect of 7-hydroxymitragynine was completely blocked by naloxone and β-FNA, and partially (38%) blocked by NTI. The κ-opioid receptor antagonist norBNI was ineffective against 7-hydroxymitragynine–induced antinociception.

Selective antagonists were employed in order to clarify the involvement of the opioid receptor subtypes in the antinociceptive effect of 7-hydroxymitragynine. μ-Opioid receptors are divided into two distinct subtypes that mediate antinociception at the spinal and supraspinal levels: the μ_1-opioid receptor, important for supraspinal antinociception, and the μ_2-opioid receptor, which is involved in spinal antinociception (Ling and Pasternak 1983; Bodner et al. 1988; Paul et al. 1989). To investigate the relative involvement of μ_1- and μ_2-opioid receptors in spinal and supraspinal antinociception of 7-hydroxymitragynine, the μ_1/μ_2-opioid receptor antagonist β-FNA and the μ_1-opioid antagonist naloxonazine were used. It was found that the antinociceptive effects of 7-hydroxymitragynine are mediated primarily through the μ-opioid receptors because the μ_1/μ_2-opioid receptor antagonist β-FNA almost completely blocked the effect in the tail-flick and hot-plate tests. In addition, naloxonazine has been shown to preferentially block μ_1-opioid receptors rather than μ_2-opioid receptors (Sakurada et al. 1999). Naloxonazine significantly blocked the antinociceptive effect of 7-hydroxymitragynine in the tail-flick and hot-plate tests, suggesting that the antinociception induced by 7-hydroxymitragynine i s highly involved in the μ_1-receptors. However, it was also found that the effect of 7-hydroxymitragynine was partially blocked by the δ-selective antagonist naltrindole in the hot-plate test, suggesting partial involvement of the supraspinal δ-opioid receptors. In addition, Thongpradichote et al. (1988) revealed that mitragynine, which is a main constituent of *M. speciosa* and has structural similarities to 7-hydroxymitragynine, has an antinociceptive activity through the supraspinal μ- and δ-opioid receptors. These results suggest that the supraspinal δ-opioid receptors are also involved in the antinociceptive effect of 7-hydroxymitragynine.

When orally administered, 7-hydroxymitragynine (1–8 mg/kg, p.o.) induced dose-related antinociceptive response in the tail-flick and the hot-plate tests. The effect peaked at 15 and 7.5–15 minutes after injection in the tail-flick and the hot-plate tests, respectively. The ED_{50} values for 7-hydroxymitragynine were 4.43 and 2.23 mg/kg in the tail-flick and the hot-plate test, respectively.

Morphine (25–100 mg/kg, p.o.) produced dose-related antinociceptive response, with a peak effect at 60 and 30 minutes after injection in the tail-flick and the hot-plate tests, respectively. The ED_{50} values for morphine were 63.0 and 48.2 mg/kg in the tail-flick and the hot-plate tests, respectively. Compared to morphine on mg/kg, 7-hydroxymitragynine (p.o.) was 14.2 and 21.6 times more potent in the tail-flick and hot-plate tests, respectively (Table 13.1).

The higher potency and more rapid effect of 7-hydroxymitragynine than morphine may be a result of its high lipophilicity and its ease in penetrating the blood–brain barrier. Indeed, it has been shown that analgesics with high lipophilicity, such as fentanyl, rapidly penetrate the blood–brain barrier, and thus fentanyl produces more potent and rapid antinociception than morphine does (Narita et al. 2002).

TABLE 13.1

**Antinociceptive Effect (ED$_{50}$) of Morphine and
7-Hydroxymitragynine after s.c. or p.o. Administration in Mice
Tail-Flick and Hot-Plate Tests**

	Tail-Flick			Hot-Plate		
	ED$_{50}$ (s.c.)	ED$_{50}$ (p.o.)	p.o./s.c.	ED$_{50}$ (s.c.)	ED$_{50}$ (p.o.)	p.o./s.c.
Morphine	4.57	63.0	13.8	4.08	48.2	11.8
7-Hydroxymitragynine	0.80	4.43	5.54	0.93	2.23	2.40

ED$_{50}$ value represents effective dose (mg/kg) 50%.

We isolated a new analgesic compound 7-hydroxymitragynine, a minor constituent of the Thai medicinal herb *M. speciosa*. The compound induced a potent antinociceptive effect via the μ-opioid receptor mechanism (Matsumoto et al. 2004, 2006), and its effect was more potent than that of morphine. On the other hand, it was found to inhibit gastrointestinal transit less potently than morphine at each equi-antinociceptive dose (Matsumoto et al. 2006). We investigated the structural similarities between morphine and 7-hydroxymitragynine using molecular modeling techniques (Matsumoto et al. 2005), but we could not superimpose all three functional groups—a nitrogen atom, a benzene residue, and an oxygen atom—on the benzene ring in the structures of morphine and 7-hydroxymitragynine. These functional groups of the structure play an important role in producing analgesic activity (Dhawan et al. 1996). Therefore it is speculated that 7-hydroxymitragynine binds to opioid receptor sites other than those that morphine binds to. 7-Hydroxymitragynine may be a seed for novel analgesics because of its unique structure and strong potency.

13.4   MGM-9 DERIVATIVE OF MITRAGYNINE

13.4.1   NOVEL μ/κ-OPIOID AGONIST

For the clinical treatment of acute and chronic severe pain, morphine is the standard analgesic. Morphine-related derivatives have been synthesized by simplification and introduction of substituents into the morphine structure in order to develop powerful analgesics without side effects (Matsumoto et al. 2004). Analgesics such as fentanyl and buprenorphine have been consequently derived from morphine. Most of those used clinically have μ-receptor agonist profiles. Despite their profound utility in the management of pain, they have undesirable side effects such as constipation, development of dependence, and tolerance.

We have studied the opioid agonistic effects of the constituents of *M. speciosa* using in vitro assays. Among them, 7-hydroxymitragynine, which has a hydroxyl group at the C7 position of mitragynine, produced the most potent effect, suggesting that the opioid effect of *M. speciosa* is based mostly on the activity of 7-hydroxymitragynine. 7-Hydroxymitragynine induced a potent antinociceptive effect in mice, and its effect was more potent than that of morphine when subcutaneously or orally administered and mediated by the μ-opioid receptor mechanism (Matsumoto et al. 2004, 2006).

These pharmacologically and chemically interesting properties of 7-hydroxymitragynine encouraged us to pursue further investigation for the development of novel analgesics, and we have synthesized a large number of mitragynine derivatives. Among them, an ethylene- glycol-bridged and C10-fluorinated derivative of mitragynine, MGM-9 (Figure 13.3), induced more potent opioid agonistic effects than morphine and 7-hydroxymitragynine in the electrical stimulation assay using a guinea pig isolated ileum preparation (Takayama et al. 2006).

It is well known that μ-opioids induce potent antinociception, but they also induce psychological dependence during chronic administration. Activation of dopaminergic systems after administration of the μ-opioid agonist induces the development of rewarding effects. In contrast, κ-opioid receptors negatively modulate the activity of dopaminergic neurons and inhibit the rewarding effects mediated by μ-opioid receptors (Narita et al. 2001). Therefore, we hypothesized that a dual-acting μ- and κ-opioid agonist would induce potent antinociceptive effects and fewer rewarding effects than μ-agonists such as morphine.

Antinociceptive effects of MGM-9, 7-hydroxymitragynine, and morphine were investigated in acute thermal pain tests in mice. MGM-9 (0.25-2 mg/kg) induced dose-related antinociceptive responses in the tail-flick and hot-plate tests after subcutaneous administration. The effect peaked at 15 and 7.5 minutes after injection in the tail-flick and hot-plate tests, respectively. The ED_{50}

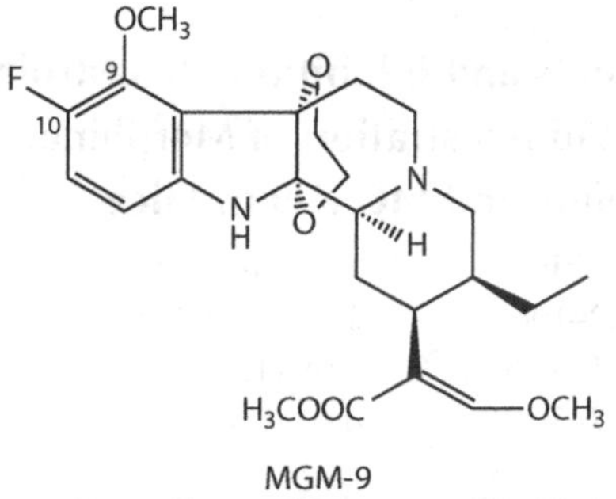

MGM-9

FIGURE 13.3 Chemical structure of MGM-9.

values for MGM-9 were 0.57 and 0.70 mg/kg in the tail-flick and the hot-plate tests, respectively. Compared to morphine, MGM-9 was 8 and 6 times more potent in the tail-flick and hot-plate tests, respectively (Table 13.2). Potent and dose-related antinociceptive responses were exhibited in both tests in response to oral administration of MGM-9 (1–8 mg/kg) (Matsumoto et al. 2008). The effect peaked at 15 and 7.5 min after injection in the tail-flick and hot-plate tests, respectively. The ED_{50} values for MGM-9 were 2.84 and 2.98 mg/kg in the tail-flick and hot-plate tests, respectively. Compared to morphine, MGM-9 was 22 and 16 times more potent in the tail-flick and hot-plate tests, respectively (Table 13.3).

In the acetic acid writhing test, MGM-9 induced potent and dose-related antinociceptive responses in mice after subcutaneous and oral administration. The ED_{50} values (95% confidence limits) for MGM-9 were 0.06 and 0.63 mg/kg after subcutaneous and oral administration, respectively. MGM-9 was about 2 times more potent than 7-hydroxymitragynine (Tables 13.2 and 13.3). Compared to morphine, MGM-9 was 8 and 7 times more potent after subcutaneous and oral administrations, respectively (Tables 13.2 and 13.3).

TABLE 13.2

Antinociceptive Effects Produced by Subcutaneous Administration of Morphine, 7-Hydroxymitragynine and MGM-9 in Mice

	Morphine	MGM-9	7-Hydroxymitragynine
Tail-flick	4.57 (3.12–6.69)	0.57 (0.36–0.90)	0.80 (0.48–1.33)
Hot-plate	4.08 (2.75–6.06)	0.70 (0.42–1.17)	0.93 (0.59–1.45)
Writhing	0.50 (0.31–0.80)	0.06 (0.03–0.09)	0.15 (0.09–0.24)

ED_{50} represents the median effective dose (mg/kg) (95% confidence limits).

TABLE 13.3

Antinociceptive Effects and Inhibition of Gastrointestinal Transit (IGIT) Due to Oral Administration of Morphine, 7-Hydroxymitragynine, and MGM-9 in Mice

	Morphine	MGM-9	7-Hydroxymitragynine
Tail-flick	63.0 (37.2–106.8)	2.84 (1.60–5.05)	4.43 (1.57–6.93)
Hot-plate	48.2 (27.5–84.5)	2.98 (1.79–4.92)	2.23 (1.38–3.60)
Writhing	4.60 (2.87–7.38)	0.63 (0.40–0.99)	1.05 (0.62–1.78)

ED_{50} represents the median effective dose (mg/kg) (95% confidence limits).

In order to determine the opioid receptor-type selectivity of MGM-9 antinociception, mice were pretreated with selective opioid receptor antagonists in the tail-flick, hot-plate, and writhing tests (Matsumoto et al. 2008). We chose a dose of MGM-9 that produces a response of 80–90% to detect the effects of the antagonists easily. In the tail-flick test, the antinociceptive effect of MGM-9 was completely blocked by the non-selective opioid antagonist naloxone, and it was significantly blocked by the irreversible μ-opioid receptor selective antagonist β-FNA and the κ-opioid receptor selective antagonist norBNI in the tail-flick, hot-plate, and writhing tests. The antinociceptive effect of MGM-9 was completely blocked by coadministration of β-FNA and norBNI in the three tests. To confirm the involvement of κ-opioid receptor in the antinociceptive effect of MGM-9, the effect of GNTI, a selective κ-opioid antagonist, was investigated in the tail-flick assay. GNTI (3 mg/kg, s.c.) significantly blocked the antinociceptive effects of MGM-9 ($27.4 \pm 4.9\%$ MPE, $P < 0.01$, $n = 7$). The antinociceptive effect of U50,488 (10 mg/kg, s.c., $86.6 \pm 6.2\%$ MPE, $n = 6$) was completely blocked by the same dose of GNTI ($6.1 \pm 3.4\%$ MPE, $P < 0.01$, $n = 6$). The selective δ-antagonist naltrindole was ineffective against MGM-9–mediated antinociception in the three tests.

The in vivo studies revealed that MGM-9 produced a dose-dependent and strong antinociceptive effect when subcutaneously and orally administered to mice in tail-flick, hot-plate, and writhing tests. The antinociceptive effect of MGM-9 was about 8 and 7–22 times more potent than morphine after subcutaneous and oral administration, respectively. In the receptor-binding assay, MGM-9 showed similar binding affinities for μ-opioid receptors, as compared with morphine, but the antinociceptive effect of MGM-9 was much stronger than that of morphine (Matsumoto et al. 2008). In order to elucidate the reason for this inconsistency, we performed a functional assay using the isolated guinea pig ileum and compared the potency of MGM-9 with that of morphine. MGM-9 induced opioid agonistic effects about 18 times more potent than morphine in isolated guinea pig ileum. Then we found that the strong antinociceptive effect of MGM-9 correlates not with the binding affinity in the receptor binding tests, but with its potency in the experiments with isolated guinea pig ileum. A possible explanation is the spare receptors in classical receptor theory. Full agonists do not need to bind all the specific receptors in order to induce full intrinsic activity. Therefore the affinity in the receptor binding assay is somewhat lower than that obtained from the potency in the functional assay. On the other hand, the antinociceptive effect of MGM-9 correlated with the affinity observed in the in vitro functional assay. This may be the reason MGM-9 induces a more potent antinociceptive effect than expected from the receptor-binding assay data.

Natives of Thailand and Malaysia use the leaves of the *M. speciosa* for pharmacologic effects. When taken orally, the leaves are very effective in increasing work endurance and as a substitute for opium in treating addicts. It is reported that naturally derived indole alkaloids such as 7-hydroxymitragynine from *M. speciosa* and pseudo-akuammigine from *Picralima nitida* exert antinociceptive activity when administered orally (Duwiejua et al. 2002; Matsumoto et al. 2004). Thus we investigated the antinociceptive effects of MGM-9 via the oral route, based on the traditional usage of *M. speciosa* and the clinical relevance of this route for administration to human patients. In the present study, MGM-9 induced potent antinociceptive effects especially with oral administration, as 7-hydroxymitragynine did. Interestingly, both MGM-9 and 7-hydroxymitragynine have a favorable bioavailability. The ratios of oral/subcutaneous ED_{50} values of MGM-9 and 7-hydroxymitragynine were much smaller than those of morphine in the tail-flick, hot-plate, and writhing tests. In comparisons of the time courses of the antinociceptive effects of MGM-9 and morphine after subcutaneous or oral administration, the duration of the effect of MGM-9 was shorter than that of morphine. Especially with oral administration, MGM-9 induced a much more potent and rapid effect than morphine. This difference may be due to the difference in the pharmacokinetics of these drugs. It is known that the oral dose of morphine required to elicit an antinociceptive effect is much higher than the parenteral dose because of its high first-pass effect. This pharmacokinetic property of morphine is a factor in its lower potency compared to that of MGM-9.

We have studied the opioid receptor binding affinities of mitragynine and related compounds. These compounds, including 7-hydroxymitragynine and mitragynine, have a relatively higher affinity for μ-opioid receptors than for δ- and κ-receptors (Takayama et al. 2002). MGM-9 has a high affinity for μ- and κ-opioid receptors (Matsumoto et al. 2008). Naloxone completely blocked the antinociceptive effect of MGM-9 in the tail-flick, hot-plate, and writhing tests, confirming the involvement of opioid receptor systems in its effect. The μ-antagonist β-FNA and the κ-antagonist norBNI partially and significantly inhibited MGM-9-induced antinociception in the three tests. Furthermore, a different κ-antagonist GNTI significantly inhibited the effect of MGM-9, as did norBNI. Co-administration of β-FNA and norBNI completely reversed the effect of MGM-9. The selective δ-antagonist naltrindole was ineffective. Isolated guinea pig ileum was used to examine μ- and κ-opioid receptors, and mouse vas deferens was used to examine δ-opioid receptors. The data obtained from functional bioassays using electrical stimulation in isolated guinea pig ileum and mouse vas deferens supported the hypothesis that the potent opioid agonistic actions of MGM-9 were mediated by μ- and κ-opioid receptors because MGM-9 showed potent agonistic effects at μ- and κ-opioid receptors in guinea pig ileum but weak activity in mouse vas deferens. Taken together, the data obtained from in vitro and in vivo assays clarified that the potent antinociceptive effects of MGM-9 resulted from its combined action at both μ- and κ-opioid receptors.

In addition, MGM-9 induced less hyperlocomotion and fewer rewarding effects than morphine. The rewarding effect of MGM-9 was blocked by a μ-antagonist and enhanced by a κ-antagonist. The 7-hydroxymitragynine derivative MGM-9 is a promising novel analgesic that has a stronger antinociceptive effect and weaker adverse effects than morphine.

13.4.2 ANTI-ALLODYNIC EFFECT ON NEUROPATHIC PAIN

Morphine has been proposed for the treatment of neuropathic pain, but it has incomplete efficacy and dose-limiting adverse effects (Gilron et al. 2005). To develop additional analgesics, derivatives have been synthesized by simplification and introduction of substituents into morphine's chemical structure (Corbett et al. 2006). Thus most morphine-derived opioid analgesics used clinically have μ-receptor agonist profiles. Although all three major types of opioid receptors (μ, δ, and κ) are able to mediate analgesia and antinociception, they have different pharmacological activities. Recently, a prominent role of δ-opioid receptors in chronic pain such as neuropathic pain and inflammatory pain was reported using mutant animals and selective agonists (Nadal et al. 2006; Nozaki et al. 2012). In a neuropathic pain rat model, δ-opioid receptor protein expression was increased compared to a sham operation in the dorsal root ganglion (Kabli and Cahill 2007). δ-Opioid receptor activation leads to decreased chronic pain but weakly influences acute pain, in contrast to μ-opioid receptor activation (Gaveriaux-Ruff et al. 2011). A novel strategy for pain management is to use dual acting and/or mixed opioid agonists (Matsumoto et al. 2008; Cremeans et al. 2012). Therefore opioid agonists, which act on not only δ- but also μ-opioid receptor subtypes, might be broad-spectrum analgesics useful in treating a variety of painful conditions.

7-Hydroxymitragynine-related indole alkaloids have interesting pharmacological profiles, such as high oral potency and few side effects. Therefore, further investigation of the development of novel analgesics against acute and chronic pain is warranted. We hypothesized that a dual-acting μ- and δ-opioid agonist derived from 7-hydroxymitragynine could induce not only potent antinociceptive effects against acute pain but also an anti-allodynic effect against neuropathic pain. We synthesized novel μ-/δ-opioid dual agonists MGM-15 and MGM-16 (Figure 13.4) and clarified their pharmacological profiles using in vitro and in vivo experiments

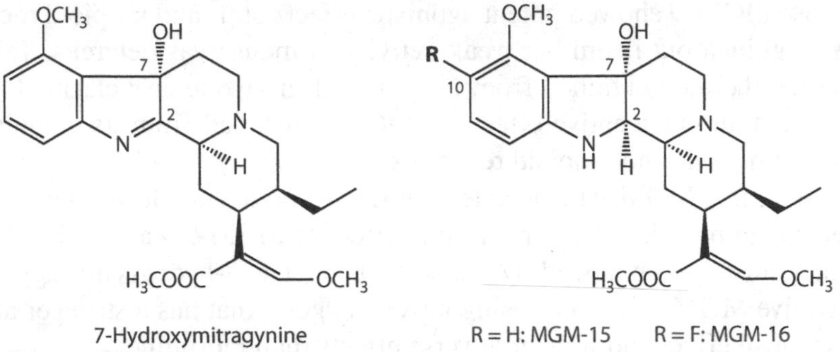

7-Hydroxymitragynine   R = H: MGM-15   R = F: MGM-16

FIGURE 13.4 Chemical structure and conformation of 7-hydroxymitragynine, MGM-15, and MGM-16.

TABLE 13.4

ED_{50} Values of Antinociceptive Effects Produced by Subcutaneous and Oral Administration of Morphine, 7-Hydroxymitragynine (7-OHMG), MGM-15, and MGM-16 in Mice

	Morphine	7-OHMG	MGM-15	MGM-16
s.c.	4.57 (3.12–6.69)	0.80 (0.48–1.33)	0.30 (0.18–0.50)	0.064 (0.040–0.10)
p.o.	63.0 (37.2–106.8)	4.43 (1.57–6.93)	1.26 (0.84–1.88)	0.263 (0.165–0.420)

ED_{50} represents the median effective dose (mg/kg) (95% confidence limits).

under physiological conditions. Furthermore, anti-allodynic effects of MGM-16 were investigated in a mouse sciatic nerve–ligation model (Matsumoto et al. 2013).

Antinociceptive effects of MGM-15 and MGM-16 were investigated in acute thermal pain tests in mice. MGM-15 and MGM-16 induced dose-related antinociceptive responses in the tail-flick tests after subcutaneous and oral administration (Matsumoto et al. 2013). The effect peaked at 15 to 30 minutes after injection. The antinociceptive effect of MGM-15 was about 15 and 50 times more potent than morphine after subcutaneous and oral administration, respectively. The antinociceptive effect of MGM-16 was about 71 and 240 times more potent than morphine after subcutaneous and oral administration, respectively (Table 13.4).

To determine the opioid receptor-type selectivity of MGM-15 and MGM-16 antinociception, mice were pretreated with selective opioid receptor antagonists in the tail-flick tests (Matsumoto et al. 2013). The antinociceptive effect of MGM-15 and MGM-16 was completely blocked by the irreversible μ-opioid receptor selective antagonist β-FNA. These effects were partially and significantly blocked by the δ-opioid receptor selective antagonist naltrindole. The selective κ-opioid antagonist norBNI was ineffective against MGM-15– and MGM-16–induced antinociception.

We next investigated the ability of MGM-15 and MGM-16 to activate G-proteins in CHO-K1 cells expressing recombinant μ- and δ-opioid receptors (Matsumoto et al. 2013). μ-Opioid agonist DAMGO and δ-opioid agonist [Met]-enkephalin each produced a concentration-dependent increase in [^{35}S] GTPγS binding to the CHO-K1 cell membrane. MGM-16 also showed a concentration-dependent increase in [^{35}S]GTPγS binding to CHO-K1 cell membranes expressing recombinant μ- and δ-opioid receptors. In contrast, MGM-15 showed a smaller increase in the binding of [^{35}S]GTPγS than that of MGM-16.

MGM-16 showed potent μ-/δ-dual agonistic activities, and demonstrated further higher potency than MGM-15 in vitro and in an in vivo acute pain model. Thus, we investigated the anti-allodynic effect of MGM-16 in a neuropathic pain model. Mice with partial sciatic nerve ligation exhibited marked neuropathic pain-like behavior on the ipsilateral side 7 days after the nerve ligation. We evaluated the anti-allodynic effects induced by subcutaneous and oral administration of MGM-16 in sciatic nerve-ligated mice using von Frey filaments (Matsumoto et al. 2013). MGM-16 (0.1, 0.2, and 0.4 mg/kg) dose-dependently increased the ipsilateral paw withdrawal threshold in sciatic nerve–ligated mice, and maximal anti-hyperalgesic responses were seen at 15 or 30 minutes after subcutaneous administration of MGM-16 (Matsumoto et al. 2013). MGM-16 (0.2 mg/kg) reversed the threshold to control level in sciatic nerve–ligated mice. In oral administration, MGM-16 (0.5, 1, and 2 mg/kg) dose-dependently increased the ipsilateral paw withdrawal threshold in sciatic nerve–ligated mice, and maximal anti-hyperalgesic responses were seen 30 minutes after administration (Matsumoto et al. 2013). MGM-16 at 1 mg/kg and gabapentin at 100 mg/kg reversed the threshold to control level in sciatic nerve–ligated mice.

To investigate the contribution of the opioid receptor subtypes in the anti-allodynic effect of MGM-16, sciatic nerve–ligated mice were pretreated with selective opioid receptor antagonists (Matsumoto et al. 2013). The anti-allodynic effect of MGM-16 was completely blocked by β-FNA and naltrindole in the chronic pain model. The selective κ-opioid antagonist norBNI was ineffective against MGM-16–mediated antinociception.

We have previously synthesized opioid analgesics from the indole alkaloid mitragynine and surveyed these compounds for their opioid agonistic activities in vitro to elucidate the specific structure necessary for its pharmacophore to bind to opioid receptors. A nitrogen atom, a benzene residue, and an oxygen function play a significant role in producing opioid agonistic activity (Matsumoto et al. 2005). The conversion of an indolenine moiety in 7-hydroxymitragynine (MGM-15) and 10-fluoro-7-hydroxymitragynine into an indoline derivative (MGM-16) led to an increase in agonistic potency. This indicates that the sp^3 carbon at the C2 position (Figure 13.3), which spatially configured the benzene ring in the ligands, was more efficient at exerting the opioid activity.

Takayama et al. (2006) reported that the dimension or electronegativity of the functional group at the C10 position of mitragynine derivatives is important in eliciting opioid agonistic effects. Among those derivatives, the C10-fluorinated derivative, MGM-9, showed the highest potency (Takayama et al. 2006). MGM-16, the C10- fluorinated derivative of MGM-15, showed the most potent opioid agonistic effects among the derivatives tested previously (Takayama et al. 2002; 2006; Matsumoto et al. 2008). MGM-16 showed full agonistic properties on μ- and δ-opioid receptors. MGM-16 also showed higher affinities and intrinsic efficacies than MGM-15 on μ- and δ-opioid receptors in vitro. A benzene residue of 7-hydroxymitragynine-related indole alkaloids plays an essential role in producing opioid agonistic activity for the specific structure necessary for its pharmacophore to bind to opioid receptors. It is speculated that the fluorine group at the C10 position on benzene residue strengthens pharmacophore binding to opioid receptors because lone pair of fluorine easily forms hydrogen bonding with the receptor molecule (Takayama et al. 2006). Based on receptor binding studies and tail-flick tests using selective antagonists, MGM-15 and MGM-16 showed similar opioid receptor-type selectivity. Taken together, fluorination at the C10 position increased the affinity to μ- and δ-opioid receptors but did not change the selectivity to the opioid receptor subtypes.

Subcutaneously and orally administered MGM-16 showed dose-dependent and anti-allodynic effects in sciatic nerve–ligated mice. The antinociceptive effect of MGM-16 was completely and partially blocked by the μ-selective antagonist β-FNA and by the δ-selective antagonist naltrindole, respectively, in an acute pain model. The anti-allodynic effect of MGM-16 was completely blocked by β-FNA and naltrindole in a chronic pain model. The contribution of δ-opioid receptors to the anti-allodynic effect of MGM-16 in the neuropathic pain model was higher than in the acute pain test. As previously reported, up-regulation of δ-opioid receptors is induced in dorsal root ganglia in a rat peripherally nerve injured model (Kabli and Cahill 2007). δ-opioid receptor knockout mice showed an increase in mechanical allodynia under neuropathic pain induced by partial sciatic nerve ligation (Nadal et al. 2006; Gaveriaux-Ruff et al. 2011). However, δ-opioid receptor knockout mice showed normal pain responses to acute pain. Furthermore, δ-opioid selective agonists induce clear anti-allodynic effects in various neuropathic pain animal models but do not show clear antinociceptive effects on acute nociception (Gaveriaux-Ruff et al. 2011). Thus, δ-opioid receptor activation leads to controlled mechanical allodynia. This indicates that the δ-opioid receptor agonistic properties of MGM-16 are involved in its anti-allodynic effects in sciatic nerve–ligated mice.

Several studies showed the effectiveness of μ-opioid receptor agonists and demonstrated that μ-opioid receptors contribute to the control of mechanical allodynia in neuropathic pain (Mansikka et al. 2004; Finnerup et al. 2010). Indeed, μ-opioid-receptor knockout mice showed an increase in mechanical allodynia under neuropathic pain induced by partial sciatic nerve ligation (Nadal et al. 2006; Gaveriaux-Ruff et al. 2011). Furthermore, μ-opioid receptor agonists attenuate mechanical allodynia in sciatic nerve–ligated mice (Mansikka et al. 2004). This suggests that the μ-opioid receptor contributes to alleviating mechanical allodynia. In this study, we found that the anti-allodynic effect of MGM-16 was significantly blocked by a μ-opioid receptor antagonist. Therefore the μ-opioid receptor agonistic activities of MGM-16 also contribute to its anti-allodynic effect in partial sciatic nerve–ligated mice.

Previously, there were no reports on G protein action following the binding of mitragynine-related indole alkaloids to opioid receptors (Raffa et al. 2013). In this study, we show that MGM-16 dose-dependently increases GTPγS binding to CHO-K1 cells expressing recombinant μ- and δ-opioid receptors. We investigated the antinociceptive effects of MGM-16 ingested orally, based on the traditional usage of *M. speciosa* and the relevance of this route for clinical administration. When administered orally, the antinociceptive effect of MGM-16 was about 240 times more potent than that of morphine in the mouse tail-flick test. Gabapentin, a synthetic analog of GABA, was initially developed as an anticonvulsant and later licensed for the management of postherpetic neuralgia (Gilron et al. 2005). In our results, gabapentin is effective for mechanical allodynia, but very high doses (100 mg/kg, p.o.) were needed to alleviate it, as reported previously. Under this experimental condition, the anti-allodynic effect of MGM-16 is about 100 times more potent than that of gabapentin in partial sciatic nerve–ligation mice. MGM-16 exhibits a potential ability to alleviate chronic pain more effectively than existing drugs.

7-Hydroxymitragynine-related indole alkaloids have interesting pharmacological characteristics such as high oral potency and few side effects. Therefore, further investigation for the development of novel analgesics against acute and chronic pain was warranted. We synthesized the new dual-acting μ- and δ-opioid agonist MGM-16, which showed higher opioid agonist potency than that of 7-hydroxymitragynine. When administered orally, MGM-16 showed 100-fold higher analgesic potency than gabapentin in sciatic nerve–ligated mice. Therefore MGM-16 could be useful for the treatment of chronic pain such as mechanical allodynia. This derivative may become a new class of seed compound with potential therapeutic utility for treating neuropathic pain.

13.5 CONCLUDING REMARKS

Physical dependence and analgesic tolerance are a major concern during morphine administration. 7-Hydroxymitragynine is structurally different from clinically used opioid analgesics such as morphine, fentanyl, and buprenorphine. We speculated that the pharmacophore binding of 7-hydroxymitragynine to opioid receptors is different from that of morphine. This may lead to a potential difference between the opioid effects of 7-hydroxymitragynine and morphine. Therefore, the derivatives of 7-hydroxymitragynine, such as MGM-9 and MGM-16, have the potential to induce novel and beneficial pharmacological effects, different from those of the existing opioid analgesics. In addition, the analgesic effect of 7-hydroxymitragynine is much more potent than that of morphine when orally administered. The studies on pharmacological effects of 7-hydroxymitragynine and its derivatives are therefore a quick route in developing novel opioid analgesics. Unknown ancient treasures are still hidden in the Thai traditional medicine *M. speciosa*, and we should hunt for them.

REFERENCES

Adkins, J.E., Boyer, E.W., and McCurdy, C.R. 2011. *Mitragyna speciosa*, a psychoactive tree from Southeast Asia with opioid activity. *Curr Top Med Chem* 11: 1165–75

Bodnar, R.J., Williams, C.L., Lee, S.J., and Pasternak, G.W. 1988. Role of µ1-opiate receptors in supraspinal opiate analgesia: a microinjection study. *Brain Res* 447: 25–34.

Burkill, I.H. 1935. *A Dictionary of the Economic Products of the Malay Peninsula.* Vol. II., Crown Agents for the Colonies, London, pp. 1480–83.

Corbett, A.D., Henderson, G., McKnight, A.T., and Paterson, S.J. 2006. 75 years of opioid research: The exciting but vain quest for the Holy Grail. *Br J Pharmacol* 147: S153–62.

Cremeans, C.M., Gruley, E., Kyle, D.J., and Ko, M.C. 2012. Roles of µ-opioid receptors and nociceptin/orphanin FQ peptide receptors in buprenorphine-induced physiological responses in primates. *J Pharmacol Exp Ther* 343: 72–81.

Dhawan, B.N., Cesselin, F., Raghubir, R. et al. 1996. International Union of Pharmacology. XII. Classification of opioid receptors. *Pharmacol Rev* 48: 567–92.

Duwiejua, M., Woode, E., and Obiri, D.D. 2002. Pseudo-akuammigine, an alkaloid from *Picralima nitida* seeds, has anti-inflammatory and analgesic actions in rats. *J Ethnopharmacol* 81: 73–79.

Finnerup, N.B., Sindrup, S.H., and Jensen, T.S. 2010. The evidence for pharmacological treatment of neuropathic pain. *Pain* 150: 573–81.

Gaveri aux-Ruff, C., Nozaki, C., Nadal, X. et al. 2011. Genetic ablation of delta opioid receptors in nociceptive sensory neurons increases chronic pain and abolishes opioid analgesia. *Pain* 152: 1238–48.

Gilron, I., Bailey, J.M., Tu, D., Holden, R.R., Weaver, D.F., and Houlden, R.L. 2005. Morphine, gabapentin, or their combination for neuropathic pain. *N Engl J Med* 352: 1324–34.

Grewal, K.S. 1932. Observations on the pharmacology of mitragynine. *J Pharmacol Exp Ther* 46: 251–71.

Horie, S., Koyama, F., Takayama, H. et al. 2005. Indole alkaloids of a Thai medicinal herb, *Mitragyna speciosa*, that has opioid agonistic effect in guinea-pig ileum. *Planta Med* 71: 231–36.

Kabli, N., and Cahill, C.M. 2007. Anti-allodynic effects of peripheral delta opioid receptors in neuropathic pain. *Pain* 127: 84–93.

Ling, G.S., and Pasternak, G.W. 1983. Spinal and supraspinal opioid analgesia in the mouse: The role of subpopulations of opioid binding sites. *Brain Res* 271: 152–56.

Macko, E., Weisbach, J.A., and Douglas, B. 1972. Some observations on the pharmacology of mitragynine. *Arch Int Pharmacodyn Ther* 198: 145–61.

Mansikka, H., Zhao, C., Sheth, R.N., Sora, I., Uhl, G., and Raja, S.N. 2004. Nerve injury induces a tonic bilateral mu-opioid receptor-mediated inhibitory effect on mechanical allodynia in mice. *Anesthesiology* 100: 912–21.

Matsumoto, K., Mizowaki, M., Thongpradichote, S. et al.1996a. Central antinociceptive effects of mitragynine in mice: Contribution of descending noradrenergic and serotonergic systems. *Eur J Pharmacol* 317: 75–81.

Matsumoto, K., Mizowaki, M., Thongpradichote, S. et al. 1996b. Antinociceptive action of mitragynine in mice: Evidence for the involvement of supraspinal opioid receptors. *Life Sci* 59: 1149–55.

Matsumoto, K., Horie, S., Ishikawa, H. et al. 2004. Antinociceptive effect of 7-hydroxymitragynine in mice: Discovery of an orally active opioid analgesic from the Thai medicinal herb *Mitragyna speciosa*. *Life Sci* 2143–55.

Matsumoto, K., Horie, S., Takayama, H. et al. 2005. Antinociception, tolerance and withdrawal symptoms induced by 7-hydroxymitragynine, an alkaloid from the Thai medicinal herb *Mitragyna speciosa*. *Life Sci* 78: 2–7.

Matsumoto, K., Takayama, H., Ishikawa, H. et al. 2006. Partial agonistic effect of 9-hydroxycorynantheidine on mu-opioid receptor in the guinea-pig ileum. *Life Sci* 78: 2265–71.

Matsumoto, K., Takayama, H., Narita, M. et al. 2008. MGM-9 [(E)-methyl 2-(3-ethyl-7a,12a-(epoxyethanoxy)-9-fluoro-1,2,3,4,6,7,12,12b-octahydro-8-methoxyindolo[2,3-a]quinolizin-2-yl)-3-methoxyacrylate], a derivative of the indole alkaloid mitragynine: A novel dual-acting μ- and κ-opioid agonist with potent antinociceptive and weak rewarding effects in mice. *Neuropharmacology* 55: 154–65.

Matsumoto, K., Narita, M., Muramatsu, N. et al. 2013 The orally active opioid μ/δ dual agonist MGM-16, a derivative of indole alkaloid mitragynine, exhibits potent analgesic effect on neuropathic pain in mice. *J Pharmacol Exp Ther,* in press.

McCurdy, C.R., and Scully, S.S. 2005. Analgesic substances derived from natural products (natureceuticals). *Life Sci* 78: 476–84.

Nadal, X., Baños, J.E., Kieffer, B.L., and Maldonado, R. 2006. Neuropathic pain is enhanced in delta-opioid receptor knockout mice. *Eur J Neurosci* 23: 830–34.

Narita, M., Funada, M., and Suzuki, T., 2001. Regulations of opioid dependence by opioid receptor types. *Pharmacol Ther* 89: 1–15.

Narita, M., Imai, S., Itou, Y., Yajima, Y., and Suzuki, T. 2002. Possible involvement of mu1-opioid receptors in the fentanyl- or morphine-induced antinociception at supraspinal and spinal sites. *Life Sci* 70: 2341–54.

Nozaki, C., Le Bourdonnec, B., Reiss, D. et al. 2012. δ-Opioid mechanisms for ADL5747 and ADL5859 effects in mice: Analgesia, locomotion, and receptor internalization. *J Pharmacol Exp Ther* 342: 799–807.

Paul, D., Bodnar, R.J., Gistrak, M.A., and Pasternak, G.W. 1989. Different mu receptor subtypes mediate spinal and supraspinal analgesia in mice. *Eur J Pharmacol* 168: 307–14.

Raffa, R.B., Beckett, J.R., Brahmbhatt, V.N. et al. 2013. Orally active opioid compounds from a non-poppy source. *J Med Chem* 56: 4840–8.

Sakurada, S., Zadina, J.E., Kastin, A.J. et al. 1999. Differential involvement of mu-opioid receptor subtypes in endomorphin-1- and -2-induced antinociception. *Eur J Pharmacol* 372: 25–30.

Suwanlert, S. 1975. A study of kratom eaters in Thailand. *Bull Narc* 27: 21–27.

Takayama, H., Ishikawa, H., Kurihara, M. et al. 2002. Studies on the synthesis and opioid agonistic activities of mitragynine-related indole alkaloids: Discovery of opioid agonists structurally different from other opioid ligands. *J Med Chem* 45: 1949–56.

Takayama, H. 2004. Chemistry and pharmacology of analgesic indole alkaloids from the rubiaceous plant, *Mitragyna speciosa*. *Chem Pharm Bull* 52: 916–28.

Takayama, H., Misawa, K., Okada, N. et al. 2006. New procedure to mask the 2,3-π bond of the indole nucleus and its application to the preparation of potent opioid receptor agonists with a Corynanthe skeleton. *Org Lett* 8: 5705–8.

Thongpraditchote, S., Matsumoto, K., Tohda, M. et al. 1998. Identification of opioid receptor subtypes in antinociceptive actions of supraspinally administered mitragynine in mice. *Life Sci* 62: 1371–78.

Tohda, M., Thongpraditchote, S., Matsumoto K. et al. 1997. Effects of mitragynine on cAMP formation mediated by δ-opiate receptors in NG108-15 cells. *Biol Pharm Bull* 20: 338–40.

Watanabe, K., Yano, S., Horie, S., and Yamamoto, L.T. 1997. Inhibitory effect of mitragynine, an alkaloid with analgesic effect from Thai medicinal plant *Mitragyna speciosa*, on electrically stimulated contraction of isolated guinea-pig ileum through the opioid receptor. *Life Sci* 60: 933–42.

Watanabe, K., Yano, S., Horie, S. et al. 1999. Pharmacological properties of some structurally related indole alkaloids contained in the Asian herbal medicines, hirsutine and mitragynine, with special reference to their Ca^{2+} antagonistic and opioid-like effects. In: *Pharmacological Research on Traditional Herbal Medicines*, Watanabe, H., and Shibuya, T. (eds.), 163–177. Amsterdam: Harwood Academic Publishers.

Yamamoto, L.T., Horie, S., Takayama, H. et al. 1999. Opioid receptor agonistic characteristics of mitragynine pseudoindoxyl in comparison with mitragynine derived from Thai medicinal plant *Mitragyna speciosa*. *Gen Pharmacol* 33: 73–81.

14 Non-Analgesic CNS Effects

Jaclyn R. Beckett, Justin R. Nixon, and Ali H. Tejani

CONTENTS

14.1 INTRODUCTION

A major clinical utility of poppy-derived opioids derives from their ability to ameliorate pain. The analgesic (antinociceptive) properties of *Mitragyna speciosa*, extracts, and analogs of these extracts are covered in other chapters. This chapter focuses on some non-analgesic central nervous system (CNS) effects of these compounds.

M. speciosa, extracts, and analogs of extracts produce a variety of CNS effects. These (side) effects can be broadly broken down into typical opioid-like effects (those commonly resulting from classic opioids such as morphine and codeine) and other effects (those not commonly resulting from classic opioids and more specific to mitragynine and related *M. speciosa* compounds). The atypical effects can be further divided into antidepressant, stimulant, and motor/locomotor activities that are mediated through several anatomical and neurotransmitter systems, including the hypothalamic–pituitary–adrenal (HPA) axis, and serotonergic (5-HT), noradrenergic, and dopaminergic pathways. Other pathways, including the endogenous cannabinoid system, are still being investigated (Shamima et al. 2012).

14.2 CLASSIC OPIOID-LIKE EFFECTS

14.2.1 INHIBITION OF COUGH

Inhibition of cough (antitussive action) is a classic opioid effect. The only study to date that has reported on the antitussive activity of a kratom analog was conducted using unanaesthetized dogs (Macko et al. 1972). A cough reflex was evoked by activating an electromagnet that had previously been surgically implanted in the wall of the trachea. The dogs were administered either codeine or mitragynine, and the inhibition of elicited coughs was recorded for the subsequent 30-second period. The oral ED_{50} for mitragynine was 2.3 mg/kg and for codeine 3.5 mg/kg. The authors concluded that "mitragynine and codeine were approximately equipotent in suppressing the cough reflex."

14.2.2 RESPIRATORY DEPRESSION

Agonists at mu-opioid receptor (MOR) are known to cause respiratory depression. And respiratory depression is the usual cause of death by most pure opioids. In a study by Macko et al. (1972), oral administration of codeine (176 mg/kg) to mice produced gasping, convulsions, and death; oral administration of mitragynine (46–920 mg/kg), however, produced some depressed respiratory rate but no evidence of lethal toxicity. In the same study, nalorphine (10 mg/kg, i.p.) was administered to cats one hour after i.p. administration of morphine (1.9–3.8 mg/kg), codeine (1.7–14.1 mg/kg), or mitragynine (46 mg/kg). They observed an increased respiratory rate after administration of nalorphine in the cats pretreated with morphine or codeine, but not with mitragynine. However, intravenous administration of mitragynine at a dose of 4.6 mg/kg led to a cat's death from respiratory failure. Oral administration of 8–80 mg/kg of mitragynine to anesthetized dogs did not produce any observable side effects. Intravenous mitragynine at 4.6 mg/kg did not result in significant side effects, but 9.2 mg/kg i.v. produced some respiratory slowing, and severe respiratory depression was noted at 31.8 mg/kg i.v. Based on these results, mitragynine can produce respiratory depression, but not as potently as morphine or codeine. To our knowledge, there have not been any clinical studies of the respiratory effect of mitragynine (or of its analogs) in humans.

14.2.3 EUPHORIA

Euphoria (a feeling of general well being) is a principal reason that abusers initially seek out opioids (and other drugs that induce the effect). Opioid-induced euphoria is believed to be mediated through activation of the G-protein–coupled MOR type following receptor activation by an agonist. Mitragynine has been shown in several studies to have affinity for MOR and intrinsic activity at the receptor (agonist action), so it has the potential of producing opioid-like euphoric effect. However, as described in reviews (e.g., Raffa et al. 2013) and other chapters in this book, there appears to be some other component (possibly non-opioid) in the overall pharmacology of mitragynine. The effects of this putative non-opioid component appear to predominate at lower doses of mitragynine, and euphoria is not usually described with the use of low doses. The opioid component is more prevalent at higher doses, and users report more opioid-like or even mixed opioid/non-opioid effects at higher dose. The higher doses can be experienced as euphoric or dysphoric, depending on the individual user and amount of prior experience.

14.2.4 CONDITIONED PLACE PREFERENCE

Conditioned place preference (CPP) is a test that is used to help assess the "likeability" of drugs (Suzuki et al. 1991). In this test, animals learn to associate one discernible location ("place") with the administration of a drug (established during a conditioning phase) and an identifiable different place where another drug, no drug, or a control is administered. At the end of the conditioning period, the animals are tested to determine their preferred place, i.e., to display a place preference. The strength of the preference (or aversion) of a drug is measured by the difference between the amount of test time that they spend in the drug-administered place compared with the amount of test time that they spend in the drug-neutral place (R0).

7-Hydroxymitragynine   ((αE,2S,3S,7aS,12bS)-3-ethyl-1,2,3,4,6,7,7a,12b-octahydro-7a-hydroxy-8-methoxy-α-(methoxymethylene)indolo[2,3-a]quinolizine-2-acetic acid methyl ester) (7-OH-MG) and analog ((E)-methyl 2-(3-ethyl-7a,12a-(epoxyethanoxy)-9-fluoro-1,2,3,4,6,7,12,12b-octahydro-8-methoxyindolo[2,3-a]quinolizin-2-yl)-3-methoxyacrylate)   (MGM-9)   have been studied in a CPP test with mice (20). The test was conducted using a shuttle box divided into two initially equally preferable compartments. The experiment was divided into three phases (a pre-conditioning phase, a conditioning phase, and a test phase). During the pre-conditioning phase, mice were placed in either compartment, and the time spent in each compartment during a 15-minute observation session was recorded automatically using an infrared beam sensor. The conditioning phase (3 days for drugs and 3 days for control) was initiated after the pre-conditioning phase and conducted once daily for 6 days. Morphine (5 mg/kg), 7-OH-MG (0.5, 1, or 2 mg/kg), MGM-9 (0.5, 1, or 2 mg/kg), or saline were administered subcutaneously. The mice were then placed for 1 hour into the compartment opposite to that in which they had spent most of their time during the preconditioning phase. On alternate days, the mice were placed in the other compartment for 1 hour. In the test phase, the mice were allowed free access to both compartments, and the amount of time that they spent in each was recorded. The mice displayed CPP for the compartment the where they had received morphine or 7-OH-MG. The magnitude of the CPP effect was dose-related. That is, at higher doses, the mice spent more time in the compartment where the drug had been administered (20). At similar antinociceptive doses, 7-OH-MG (ED_{50} = 0.8) and morphine (ED_{50} = 4.57) exhibited almost equal CPP.

The MG analog MGM-9 did not induce a significant CPP. While CPP is not exclusive for an opioid mechanism of action, these findings are consistent with other results with 7-OH-MG (R0). Nevertheless, no CPP study on these compounds to date has demonstrated the opioid component of the effect by showing reversal by an opioid receptor antagonist or showing absence of effect in genetically modified MOR-deficient knockout (KO) mice.

In a very recent study by Sufka et al. (2014) a significant CPP of mice for MG (5 and 30 mg/kg, i.p.) was again demonstrated. In this study, S(+)-amphetamine (1 mg/kg) served as a positive control and haloperidol (1 mg/kg, i.p.) served as negative (aversive) control. The M. speciosa extract and alkaloid-enriched fraction increased preference scores, but to a lesser extent than did mitragynine. This could suggest that the extract did not contain sufficiently high concentration of active compounds or that the extract contains opioid antagonist compounds in addition to the agonist ones.

14.2.5 Withdrawal

14.2.5.1 Zebrafish

Zebrafish have dopaminergic projections in forebrain regions analogous to the mammalian mesolimbic system, and they display behavioral signs and biochemical changes during withdrawal from psychotropic drugs, including morphine (Khor et al. 2011; Raffa et al. 2013). They are increasingly being used as an invertebrate model to study drug withdrawal (Khor et al. 2011). In a recent study, zebrafish were exposed to morphine (1.5 mg/L) for 2 weeks, then were tested for signs of abstinence-induced withdrawal. Withdrawal from morphine produced strong anxiogenic behaviors (similar to opioid withdrawal in humans). However, mitragynine (2 mg/L), given 1 hour before discontinuation of morphine, attenuated the abstinence-induced withdrawal from morphine. Cortisol levels were also reduced. There are several possible interpretations of this study. One is that mitragynine attenuates the anxiety and stress associated with opioid withdrawal and might explain the practice of using kratom to ameliorate withdrawal symptoms during weaning from opioid use.

14.2.5.2 Mice and Cats

Morphine-dependent mice and 7-OH-MG–dependent mice display withdrawal signs when treated with naloxone (naloxone-precipitated withdrawal) (Matsumoto et al. 2005). Mice were injected with morphine or 7-OH-MG twice daily (starting at 8 mg/kg and incrementally increasing to 45 mg/kg), and naloxone (3 mg/kg) was injected 2 hours after the final dose. Similar withdrawal signs were observed in the morphine- and 7-OH-MG–treated mice, including jumping, rearing, urination, and forepaw tremor. Withdrawal from 7-OH-MG seemed to produce somewhat less diarrhea (possibly relevant to the finding that 7-OH-MG seems to produce less constipation than does morphine at equi-antinociceptive doses) than did morphine. However, these results were observational and further study is warranted regarding the gastrointestinal endpoint.

Very mild nalorphine-precipitated withdrawal in cats has been reported. In the study (Macko et al. 1972), mitragynine (18 mg/kg, i.p.) produced very mild stimulation, mydriasis, and restlessness in cats. One hour after mitragynine administration, the opioid receptor antagonist nalorphine (10 mg/kg, i.p.) was given to determine whether it would precipitate withdrawal. The authors reported difficulty in seeing any changes in behavior following nalorphine administration. On the contrary, cats treated with equi-antinociceptive doses of codeine or morphine showed marked stimulation following nalorphine administration (Macko et al. 1972).

14.2.5.3 Humans

The classic signs of opioid withdrawal in humans are well known and generally include effects opposite to those produced by opioids. Signs include diarrhea, anxiety, agitation, chills ("cold turkey"), muscle aches, rhinorrhea, and aggressive behavior. It has been reported that chronic consumption of kratom (in Thailand) followed by abstinence elicits withdrawal symptoms (Suwanlert et al. 1975).

14.3 NON-OPIOID-LIKE EFFECTS

14.3.1 ANTIDEPRESSANT (SEROTONERGIC PATHWAYS)

Clinical depression is multifactorial and is thought to involve neurohormonal imbalances in the HPA axis (e.g., cortisol levels) and in pathways of the neurotransmitters serotonin, (nor)epinephrine, and dopamine (Idayu et al. 2010). In vitro, mitragynine displays some affinity for $5\text{-}HT_{2C}$ and $5\text{-}HT_{7}$ receptors (Boyer et al. 2008). It has also been suggested that 7-OH-MG resembles serotonin or 4-hydroxytryptophan and acts as a $5\text{-}HT_{2A}$ ligand. In vivo, compounds from *M. speciosa* inhibit serotonin-induced head-twitch in mice, an action possibly at $5\text{-}HT_{2A}$ (and/or alpha-2 adrenergic receptors) (Matsumoto et al. 1996).

Mitragynine also displays antidepressant-like effects in behavioral despair tests in mice (Kumarnsit et al. 2007). Behavioral despair tests, which include forced swim tests (FST), tail suspension tests (TST), and open-field tests (OST), are animal models generally considered relevant to the study of depression and the development of new antidepressant compounds. OST is also used to rule out non-specific drug effects, for example hyperkinesia, which can result from stimulant, anticonvulsant, or anticholinergic activity. In a study by Idayu et al. (2010), mice were randomly assigned to a control or one of six experimental groups: MG at 5 mg/kg, MG at 10 mg/kg, MG at 30 mg/kg, fluoxetine at 20 mg/kg (a positive control), amitriptyline at 10 mg/kg (another positive control), and amphetamine at 1 mg/kg. Each drug was administered i.p. once, with appropriate consideration for times to peak level. The results of FST, TST, and OST revealed a presumptive antidepressant activity of mitragynine. The duration of immobility was significantly reduced when mitragynine was administered, and a 30-mg/kg dose of mitragynine reduced immobility almost as much as did fluoxetine or amitriptyline. Negative results in the OST test appear to rule out other mechanisms, suggesting that the reduction in immobility was the result of an antidepressant mechanism of mitragynine. Corticosterone levels were also measured. With no treatment, serum corticosterone levels significantly rose during the stress test. Mitragynine significantly attenuated the stress-induced increase in corticosterone levels to about the same extent as did fluoxetine or amitriptyline. These results are consistent with a reduction in HPA axis hyperactivity and an antidepressant activity of mitragynine (Idayu et al. 2010).

14.3.2 STIMULANT ACTIVITY (ADRENERGIC PATHWAYS)

Mitragyna speciosa is a stimulant at low doses, used traditionally to combat work fatigue. Classified under the coffee family Rubiaceae, *M. speciosa* also has stimulant properties. This action might be related to mitragynine's affinity for postsynaptic alpha$_2$ adrenergic receptors (Boyer et al. 2008). Descending noradrenergic and serotonergic pathways have both been found to play a role in the antinociceptive effects of supraspinally administered mitragynine against physical noxious stimuli; only descending noradrenergic pathways appear to be involved against thermal noxious stimuli (Matsumoto et al. 1996).

The majority of information regarding stimulant activity of kratom in humans has been inferred from its traditional usage and from adverse effect and abuse reporting. From this information, it is clear that the stimulant activity of mitragynine is dose-dependent. Stimulant activity of *M. speciosa* is apparent mainly at low to moderate doses of about 1–5 g of raw leaf material; at higher doses, MG produces sedation (Prozialeck et al. 2012). Experience-reporting sites, such as Erowid (http://www.erowid.org/) and Sage Wisdom (http://sagewisdombotanicals.com/), report a variety of stimulant-like effects that vary among individuals. Some express a feeling of anxiety and agitation, whereas others describe the stimulant effects as pleasant, mild, and not as strong as amphetamine-induced effects. Patients using mitragynine for pain management report that the stimulant effects are more favorable than are the sedative effects that often accompany the use of classic opioids. For further details, see Chapter 16.

Much of the abuse and acute adverse effects that have been reported for mitragynine seems to be related to its stimulant, in addition to its opioid, properties (Prozialeck et al. 2012). Excess stimulant effects are manifested as anxiety, irritability, and increased aggressiveness. In individuals with long-term addiction to kratom, other more unusual effects occur (such as tremor, anorexia, weight loss, and psychosis), which may be manifestations of long-term excess stimulant activity (Prozialeck et al. 2012).

14.3.3 LOCOMOTOR ACTIVITY (DOPAMINERGIC PATHWAYS)

Although not well characterized, mitragynine seems to have some influence on dopamine pathways. Data are conflicting as to whether the D1 (Stolt et al. 2013) or D2 (Boyer et al. 2008) receptors (or both) are involved. Stolt et al. (2013) also conducted in vivo studies in mice to determine dopaminergic effects of kratom. Animals were injected with one of the following: saline subcutaneously, apomorphine 0.3 mg/kg subcutaneously (APO), kratom (containing mitragynine 2 mg/kg and paynantheine 0.1 mg/kg) intraperitoneally, or kratom + APO. Mice were placed in a Moti-Test box, and locomotor activity was measured. The results suggested that MG has presynaptic dopaminergic activity, essentially acting as a functional antagonist. However, this mechanism alone would suggest that kratom extracts would not likely have abuse potential (Stolt et al. 2013), but this contrasts with various case reports.

14.4 CONCLUSION

M. speciosa, extracts, and mitragynine and analogs produce various effects on the CNS other than analgesia. They have stimulant, antidepressant, and locomotor depression activity, apparently by affecting various CNS subsystems, including the HPA axis and the neurotransmitters serotonin, norepinephrine, and dopamine pathways.

REFERENCES

Boyer, E.W., Babu, K.M., Adkins, J.E., McCurdy, C.R., and Halpem, J.H. 2008. Self-treatment of opioid withdrawal using kratom (*Mitragyna speciosa* Korth.). *Addiction* 103: 1048–50.

Idayu, N.F., Hidayat, M.T., Moklas, M.A., Sharida, F., Raudzah, A R., Shamima, A.R., and Apryani, E. 2010. Antidepressant-like effect of mitragynine isolated from *Mitragyna speciosa* Korth in mice model of depression. *Phytomedicine* 18: 402–407.

Jansen, K.L., and Prast, C.J. 1988. Ethnopharmacology of kratom and the *Mitragyna* alkaloids. *J Ethnopharmacol* 23: 115–19.

Khor, B.S., Jamil, M.F., Adenan, M.I., and Shu-Chien, A.C. 2011. Mitragynine attenuates withdrawal syndrome in morphine-withdrawn zebrafish. *PLoS One* 6: e28340.

Kumarnsit, E., Keawpradub, N., and Nuankaew, W. 2007. Effect of *Mitragyna speciosa* aqueous extract on ethanol withdrawal symptoms in mice. *Fitoterapia* 78: 182–85.

Macko, E., Weisbach, J.A., and Douglas, B. 1972. Some observations on the pharmacology of mitragynine. *Arch Int Pharmacodyn Ther* 198: 145–61.

Matsumoto, K., Mizowaki, M., Suchitra, T. et al. 1996. Central antinociceptive effects of mitragynine in mice, contribution of descending noradrenergic and serotonergic systems. *Eu. J Pharmacol* 317: 75–81.

Matsumoto, K., Horie, S., Takayama, H. et al. 2005. Antinociception, tolerance and withdrawal symptoms induced by 7-hydroxymitragynine, an alkaloid from the Thai medicinal herb *Mitragyna speciosa*. *Life Sci* 78: 2–7.

Prozialeck, W.C., Jivan, J.K., and Andurkar, S.V. 2012. Pharmacology of kratom: An emerging botanical agent with stimulant, analgesic, and opioid-like effects. *J Amer Osteopath Assoc* 112: 792–99.

Raffa, R.B., Beckett, J. R., Brahmbhatt, V.N. et al. 2013. Orally active opioid compounds from a non-poppy source. *J Med Chem* 56: 4840–48.

Shamima, A.R., Fakurazi, S., Hidayat, M.T., Hairuszah, I., Moklas, M.A., and Arulselvan, P. 2012. Antinociceptive action of isolated mitragynine from *Mitragyna speciosa* through activation of opioid receptor system. *Int J Mol Sci* 13: 11427–42.

Stolt, A.C., Schroder, H., Neurath, H. et al. 2013. Behavioral and neuro-chemical characterization of kratom (*Mitragyna speciosa*) extract. *Psychopharmacology* 231(1): 13–25.

Sufka, K.J., Loria, M.J., Lewellyn, K. et al. 2014. The effect of *Salvia divinorum* and *Mitragyna speciosa* extracts, fraction and major constituents on place aversion and place preference in rats. *J Ethnopharmacol* 151: 361–64.

Suwanlert, S. 1975. A study of kratom eaters in Thailand. *Bull Narc* 27(3): 21–27.

15 Other Opioid-Associated Endpoints

Steven T. Orlando and Vivek N. Brahmbhatt

CONTENTS

15.1 INTRODUCTION

A classic therapeutic effect of opioids is the treatment of diarrhea, while a classic adverse effect of opioid drugs used for pain relief is constipation. This usually occurs at or only slightly above analgesic doses, and it is almost an inevitable outcome during long-term opioid use (or abuse). The effects are mediated by both the central (brain and spinal cord) and peripheral (directly on the gastrointestinal tract) actions of opioids.

Nausea and vomiting are other frequently encountered opioid adverse effects, particularly in opioid-naïve individuals. This chapter summarizes studies that have examined these and some other opioid-associated endpoints for mitragynine (MG) and related compounds.

Another classic effect of certain opioids—those that bind to the kappa-opioid receptor (KOR)—is diuresis. Because MG and related compounds have been reported to have some affinity for KOR, we also looked for evidence that they produce diuresis.

15.2 CONSTIPATION

Opioid use is commonly associated with antidiarrheal actions (a therapeutic effect) or constipation (an adverse effect). Opium has been used as an antidiarrheal agent for centuries. In the 1600s, Thomas Sydenham's recipe of 1 lb sherry wine, 2 oz opium, 1 oz saffron, 1 oz cinnamon powder, and 1 oz clove powder was reported to be effective for the treatment of diarrhea (Holzer 2009). Kratom has been used to treat diarrhea in Southeast Asia, and constipation has been reported as a side effect by recreational users (EMCDDA 2013). While tolerance develops to some other side effects of opioids with chronic use (such as nausea and vomiting, and sedation), constipation often persists throughout treatment (Holzer 2009).

Opioids' constipating effect is classified as a Type A adverse drug reaction (ADR-A), meaning that it is related to the drugs' pharmacological action and is therefore predictable. Other ADRs may be considered Type B, that is, unpredictable based on a drug's pharmacological action. Approximately 80% of all ADRs fall into the Type A class (Ritter 2008). Other common examples of an ADR-A are bleeding produced by the anticoagulant warfarin and decreased heart rate produced by a β-adrenergic blocker such as metoprolol. Warfarin is considered a blood thinner, and too much warfarin can make one's blood too "thin" (overly anticoagulated), leading to easy bleeding and bruising. As would be expected, ADR-As are usually dose-related (Ritter 2008). A patient who requires a higher dose of an opioid to maintain or achieve better pain relief will likely experience more constipation.

15.2.1 OPIOIDS

Opioid receptors are widely distributed in the brain and in the gastrointestinal (GI) tract. The constipation-inducing effect of opioids involves the contribution of both sites and derives from the activation of central and peripheral opioid receptors by opioid receptor agonists (Holzer 2009; Matsumoto et al. 2006). However, the concentration of opioid in the gut rather than in the brain correlates better with constipation, and opioid antagonists that do not cross the blood–brain barrier can completely antagonize morphine-induced constipation in canine and rat models (Holzer 2009).

The body's endogenous ligands for opioid receptors (such as met- and leu-enkephalin, β-endorphin, and dynorphin) bind to and activate μ-, κ-, and δ-opioid receptors located in the GI tract. The relative amount of each opioid receptor subtype (MOR, KOR, DOR) varies with species and GI layer and region. In humans, MORs are the most prevalent opioid receptor in the GI tract and are located on myenteric and submucosal neurons in the enteric nervous system and on immune cells in the lamina propria.

The myenteric and submucosal plexuses are found in all layers of the alimentary canal. The widespread distribution of opioid receptors within the GI tract facilitates an important role in digestion (Holzer 2009).

Opioid receptors modulate GI motility and secretion. Agonist binding at these receptors results in constipation through inhibition of enteric nerve activity, inhibition of propulsive motor activity, and inhibition of ion and fluid secretion into the tract. It is a reduction of enteric nerve excitability and pre- and postsynaptic inhibition of excitatory and inhibitory pathways that lead to inhibition of the enteric nervous system. Propulsive motor activity is inhibited by an elevation of muscle tone, induction of non-propulsive motility patterns, and an inhibition of distention-induced peristalsis (Holzer 2009). The strong peripheral influence of MOR on the GI tract is revealed by loperamide, an opioid agonist used in the treatment of diarrhea rather than pain. Loperamide has high affinity for MOR, but because it does not cross the blood–brain barrier in sufficient amounts it produces no analgesia. It does, however, act on MOR located in the GI tract, where it produces its opioid antisecretory and antitransit (antidiarrheal) effects (Holzer 2009).

15.2.2 MG AND 7-OH-MG

Matsumoto et al. (2006) compared 7-OH-MG–induced antinociception and inhibition of GI transit in mice with morphine's effects in the same tests. Standard tail-flick and hot-plate tests were used to quantify antinociception. To quantify GI transit inhibition, mice were fasted for 18 hours, and then were injected subcutaneously (s.c.) with 7-OH-MG, morphine, vehicle, or saline. Charcoal in an aqueous suspension was administered 15 minutes later. The small intestine was removed 30 minutes following charcoal administration and both the length of the small intestine from the pylorus to cecum and the farthest distance to which the charcoal suspension had traveled were measured. Gastrointestinal transit (GIT, %) and inhibition of GIT (%) were calculated as follows.

Both morphine and 7-OH-MG inhibited GIT in a dose-dependent manner (Figure 15.1). The two dose-response curves essentially overlap. The ED_{50} value (dose calculated from the dose-response curve to produce 50% effect) for GIT inhibition was 1.07 mg/kg for morphine and 1.19 mg/kg for 7-OH-MG. Hence the two are essentially equally potent in this endpoint. But when the ratios of ED_{50} values for antinociception (therapeutic effect) and the inhibition of GIT (ADR) are compared, the ratio for 7-OH-MG is larger than the ratio for morphine (Table 15.1). This indicates a greater separation between pain relief and constipation for 7-OH-MG compared to morphine (about 5- to 6.5-fold) (Matsumoto et al. 2006).

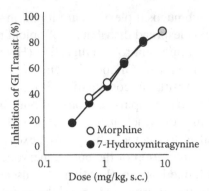

FIGURE 15.1 Dose-response curve for inhibition of gastrointestinal transit by morphine and 7-hydroxymitragynine following subcutaneous administration to mice. (Redrawn from Matsumoto et al. 2006.)

TABLE 15.1

Antinociception, Inhibition of GI Transit (GIT), and Relative Potency of 7-OH-MG Compared with Morphine in Mice

	TF (ED_{50})	HP (ED_{50})	IGIT (ED_{50})	Ratio (TF/IGIT)	Ratio (HP/IGIT)
Morphine	4.57	4.08	1.07	4.27	3.81
7-OH-MG	0.80	0.93	1.19	0.67	0.78
Ratio	5.7	4.4	0.9	—	—

Abbreviations: TF, tail-flick test; HP, hot-plate test; IGIT, inhibition of gastrointestinal transit; ED_{50}, median effective dose (mg/kg); 7-OH-MG, 7-Hydroxymitragynine.
Source: Adapted from Table 5 in Matsumoto et al. (2006).

In this same study, the GI inhibitory effects of 7-OH-MG and morphine were antagonized by pretreatment with β-funaltrexamine (β-FNA), a selective MOR antagonist, indicating that the GIT inhibitory effects were due to an agonist effect at MOR. Naloxonazine, a centrally acting μ_1-selective antagonist, only slightly attenuated the effects of 7-OH-MG and morphine (no statistical difference in antagonism between 7-OH-MG and morphine). Interestingly, whereas β-FNA and naloxonazine produced equal effect on 7-OH-MG- and morphine-induced inhibition of GI transit, the non-selective opioid receptor antagonist naloxone only antagonized morphine, while only slightly antagonizing 7-OH-MG (Matsumoto et al. 2006). These results suggest differences between 7-OH-MG and morphine yet to be elucidated.

A similar study was conducted in rats (Macko et al. 1972). In this study, mitragynine was compared with codeine and morphine. The rats were fasted for 20–24 hours, and then at a predetermined time prior to charcoal administration (based on time of peak effect), the rats were administered test drug. The GI tracts were removed 30 minutes following the charcoal meal and the distance the charcoal traveled was measured. Both morphine and codeine produced dose-related inhibition of transit by oral or intraperitoneal routes (ED_{50} = 3.53 mg/kg and 25.2 mg/kg, respectively, by the i.p. route). In contrast, mitragynine produced <50% inhibition by both oral and i.p. routes (55.2 mg/kg orally produced an average of 16% inhibition; 36.8 mg/kg i.p. produced only 9% inhibition).

15.3 EMESIS

Emesis was not observed in cats administered mitragynine i.p. (up to 46 mg/kg) or dogs administered mitragynine orally (up to 24 mg/kg) (Macko et al. 1972). In the same study, equi-analgesic doses of codeine produced pronounced emesis.

15.4 FOOD AND WATER INTAKE, WEIGHT BALANCE

The pharmacologic effects of mitragynine appear to involve opioid and also adrenergic and serotonergic neurotransmitter systems. Each of these systems are involved in the modulation of food and water intake and in the maintenance of proper weight. Whether MG also produced such effects was investigated by Kumarnsit et al. (2006). Mitragynine was extracted from young leaves of *M. speciosa* from Thailand. Male Wistar rats (180–200 g) were used to study the effects. Each rat was individually placed in a wire mesh cage, fed, and observed for a week before beginning the experiment. To measure acute effects, either intraperitoneal saline or mitragynine (15, 30, 45, 50 mg/kg) was given at 09:00; imipramine (40 mg/kg) was used as a positive control. Food and water intake were recorded 24 hours after injection. Chronic effects were studied by dividing the rats into two groups. Each group received either saline or extract (40 mg/kg, i.p.) once daily for 60 days. Food, water intake, and body weight were measured at 24 hours after injection. Saline-treated rats consumed more food compared with the imipramine treated rats, 88.4 vs. 44.7 g/kg/d, respectively. A single injection of extract at 45 and 50 mg/kg also resulted in reduced food intake (47.8 and 45 g/kg/d, respectively), but the lower doses (15 and 30 mg/kg) did not have significant effects. A similar pattern was noted for water consumption. Saline-treated rats consumed on average 117.7 g/kg/d water, while imipramine-treated rats consumed only 32.4 g/kg/d during the first 24 hours of administration. Significant reduction was also noted in 45 and 50 mg/kg extract-treated rats (81.4 and 57.3 g/kg/d respectively), but not in the rats treated at the lower doses (15 and 30 mg/kg). Chronic administration of extract (40 mg/kg, i.p.) resulted in significant reduction in food intake compared with the saline-treated rats. For the duration of 60 days, the extract-treated group ate an average 1133 ± 44 g/rat, whereas the saline-treated group ate an average 1382 ± 30 g/rat. Cumulative water intake was also assessed during the 60 days and showed that extract-treated rats consumed 25.6% less than the saline-treated rats. The weight gain of the extract-treated group was 17.6% less than the saline-treated group.

Thus, acute and chronic intraperitoneal administration of *M. speciosa* extract strongly suppresses voluntary food consumption in rats. This confirms anorectic effects of extract similar to those observed with imipramine. Long-term suppression on food consumption during the 60-day period suggests that tolerance does not develop to this effect. Since previous studies have shown that, in addition to the opioid pathway, central adrenergic and serotonergic pathways are involved in the central effects of mitragynine, a possible mechanism for the anorectic effects of the extract could involve non-opioid serotonergic and noradrenergic systems that promote central serotonergic transmission. Drugs such as fenfluramine or desmethylimipramine are well known to reduce food intake and body weight. Activation of serotonin receptors might be the factor that contributes to early satiety and loss of body weight. Though inhibitory patterns of the extract on food and water intakes were similar, the mechanisms were shown to be distinct. Water intake is inhibited by activation of serotonin receptors. 5-HT $2_A/2_C$ receptor subtypes were found to mediate inhibitory effects on water intake. Additionally, injection of clonidine, an α_2-adrenergic agonist, also resulted in inhibition of water intake. However, activation of the α_2-adrenergic and serotonergic mechanisms seemed to act independently to inhibit water intake. These findings suggest that the extract may act on the central serotonergic or adrenergic systems to suppress water intake.

15.5   DIURESIS

We could not find a study of emesis related to *M. speciosa* extract, mitragynine, 7-OH-MG, or analogs. Diuresis was not reported in toxicity summaries (Macko et al. 1972).

15.6 CONCLUSION

Kratom has been used as an antidiarrheal therapy in several countries, and kratom users have noted constipation. These properties are consistent with opioids. Mitragynine and 7-hydroxymitragynine have been shown to inhibit GI transit in mice and rats. However, the inhibition by 7-hydroxymitragynine and mitragynine is less than that of equi-analgesic doses of codeine or morphine. This presents a possible clinical benefit, as opioids are commonly associated with constipation. The greater separation between constipation and analgesia displayed by 7-hydroxymitragynine compared with morphine suggests the possibility of additional (non-opioid) pharmacology (Raffa et al. 2013).

REFERENCES

EMCDDA. 2013. Kratom (*Mitragyna speciosa*). http://www.emcdda.europa.eu /publications/drug-profiles/kratom (accessed December 29, 2013).

Holzer, P. 2009. Opioid receptors in the gastrointestinal tract. *Regulatory Peptides* 155: 11–17.

Kumarnsit, E., Keawpradub, N., and Nuankaew, W. 2006. Acute and long-term effects of alkaloid extract of *Mitragyna speciosa* on food and water intake and body weight in rats. *Fitoterapia* 77(5): 339–45.

Macko, E., Weisbach, J.A., and Douglas, B. 1972. Some observations on the pharmacology of mitragynine. *Arch Int Pharmacodyn Ther* 198: 145–61.

Matsumoto, K., Hatori, Y., Murayama, T. et al. 2006. Involvement of mu-opioid receptors in antinociception and inhibition of gastrointestinal transit induced by 7-hydroxymitragynine, isolated from Thai herbal medicine *Mitragyna speciosa*. *Eur J Pharmacol* 549: 63–70.

Raffa, R., Beckett, J., Brahmbhatt, V. et al. 2013. Orally active opioid compounds from a non-poppy source. *J Med Chem* 56(12): 4840–48.

Ritter, J.M. 2008. *A Textbook of Clinical Pharmacology and Therapeutics*. London: Hodder Arnold.

16 The Kratom Experience from Firsthand Reports

Earth Erowid and Fire Erowid

CONTENTS

"There was a noticeable psychoactive effect, which can be summed up as feeling simultaneously stimulated and relaxed. Kratom tea makes me feel talkative, social, and energized, but at the same time mellow and chilled out." (Hypersphere 2014)

16.1 INTRODUCTION

This chapter summarizes the experience of using kratom based on self reports from users of dried leaves, commercially produced kratom powders, and extr   acts. We include a discussion of firsthand reports as a data source, the review process Erowid uses to collect these reports, an overview of dosage, duration, and experiential effects, a sample experience report, and a series of excerpts from experience reports that document specific effect types.

16.1.1 Peer-Reviewed Self-Case Reports

First-person experience reports offer a direct way to document the use*
and effects of psychoactive plants and drugs in humans. Users describe in their own words the substances they take, how they take them, the experiences they have, and the impact a substance such as *Mitragyna speciosa* (kratom) has on their lives. Individual reports can be compared to medical case reports in peer-reviewed literature or to the FDA's MedWatch Adverse Event Reporting Program for the general public (Craigle 2007).

Sometimes dismissed as anecdotal, self-reports highlight the fundamentally subjective nature of having one's thoughts and feelings influenced by taking a psychoactive substance. As with medical case reports, a single experience report should not be assumed to be representative of the wider population; it is an individual data point about what happened to one person who used a particular psychoactive on a particular day at a particular dose. In isolation, any single report is just one person's opinion, but those opinions can be discussed objectively and mined for useful, even quantifiable data. Firsthand reports of psychoactive substance use are not new, but the number and detail available through Internet communications has changed how they can be approached as data.

---

* Throughout this chapter, we use the term "use" rather than "abuse" to describe the ingestion of kratom. Although many medical and research texts use "abuse" as a technical term of art, Erowid's longstanding policy is to reserve "abuse" to describe clearly problematic, self-destructive use of a substance rather than all recreational or unapproved use.

In some ways, first-person experience reports are weaker than medical case reports, which are generally written by an attending physician. Most self-reports are anonymous, for legal and privacy reasons, so it's often not possible to contact the author for follow-up questions. It is also rare for experience reports to include toxicology information validating the identity or amount of the substance(s) taken.

But in other ways, experience reports are more valuable than medical case reports, which are usually restricted to events that resulted in a medical emergency. Self-reports are not limited to individuals making dangerous choices that require clinical intervention, and thus they represent a broader spectrum of the population. Self-reports provide insight into the thoughts of a large number of users, the way they make choices, and the types of experiences they have.

Despite the inherent problems of subjectivity, bias, and memory errors, firsthand reports are *primary* sources of information that are increasingly available to medical professionals, educators, and the public. There are many sources of firsthand reports; resource sites such as Erowid, web forums such as Bluelight and Drugs-Forum, and a variety of digital and print publications offer an unprecedented opportunity for people to learn about the effects of psychoactive drugs from those who use them.

16.1.2 EROWID'S EXPERIENCE VAULTS

Erowid Center is an educational nonprofit organization with a mission of providing and facilitating access to objective, accurate, and nonjudgmental information about psychoactive plants, chemicals, technologies, and related issues. One long-term project of the organization is collecting, reviewing, categorizing, and publishing first-person reports about the use of psychoactive substances. Over the past 18 years, we have compiled on Erowid.org a collection of more than 100,000 firsthand descriptions of the recreational, medical, and spiritual use of psychoactive drugs, including more than 700 related to kratom.

With hundreds of kratom reports, we can document and distill out the reasons why people use kratom, the routes and methods of self-administration, the range of doses used, how dose relates to subjective effects, the range of effects, and what doses or substance combinations are most likely to lead to health problems.

16.1.3 REVIEW PROCESS

Reports in Erowid's collection are not "posts" to a forum. Each report undergoes a multi-step review process. When a report is submitted, it enters a triaging system where it is read, graded from A (exceptional) to F (unpublishable), and commented on by at least two trained volunteers. It is then vetted and edited by an experienced reviewer—someone well read about a wide variety of psychoactive substances—who categorizes and gives a final rating to the report before publishing. The goal of this process is to evaluate each report for interest, quality, accuracy, and general believability prior to publication. This also allows Erowid to ensure that no one's privacy is compromised. We prioritize publishing high-quality reports, and those that include valuable data such as dangerous interactions, health benefits, and new drug descriptions.

16.1.4 KRATOM REPORTS

Of the 714 experience reports about *Mitragyna speciosa* and kratom products that have been submitted to Erowid as of January 2014, 218 have been published. Among submitted reports, when author gender is specified, 91% were written by men and 9% by women. The earliest reports in our collection describe experiences from 2000, which marks the beginning of commercial mail-order availability of kratom in North America, Australia, and Europe.

In 2000, kratom products were generally dried leaf, whole or crushed. Based on experience reports and Erowid's vendor monitoring, leaf extracts and extract-enhanced leaf first became available in 2002 and were widely available by 2004. Interest in kratom information grew slowly but steadily during the subsequent decade. In 2004, the kratom index page was the 50th most visited substance index on Erowid.org. By early 2014, it had risen to 8th place.

16.2   KRATOM USE OVERVIEW

Generally, in North America, the dried leaves of *M. speciosa* are taken as a bitter tea or in capsules. Kratom users commonly describe paradoxically stimulating and sedating effects that are most often likened to the effects of opioids such as codeine or oxycodone. It is used recreationally, as an analgesic, as a self-treatment for opioid withdrawal symptoms, and less commonly as a method for interrupting another addiction (e.g. smoking cessation). Reports from Southeast Asia include significantly more descriptions of chewed (buccal) whole leaves used as a daytime work drug, taken for the purpose of relieving boredom and the discomfort of manual labor. Kratom is considered addictive; reports of physical and psychological dependence, tolerance, and physical withdrawal symptoms are prevalent, though not universal, in experience reports.

16.2.1 Dosage

The amount of kratom leaf or leaf powder used per dose varies from one gram to fifty grams. Because kratom is a natural plant product and the commercial products are unregulated, potencies can vary dramatically from one capsule, Mylar package, or baggie of leaf powder to another. Many commercial kratom products in 2014 include kratom leaf extract redeposited on crushed or powdered leaf, making them more potent than untreated kratom leaves. Tinctures, resins, and straight extracts (not deposited on leaf) are also sold. Dosage of these products can vary depending on the concentration of *M. speciosa* alkaloids present in the source material.

Retail kratom products available to the general public range from the low potency "commercial grade" crushed leaf, to "premium," "super premium," "enhanced," and "super enhanced" leaf. Extracts are usually labeled as "5x," "15x," or similar, putatively indicating relative strengths. These, plus additional terms such as "Maeng Da" (aka "Pimp Grade"), plant provenance such as "Indonesian," and other vendor-supplied details are used in wine-tasting-style comparisons of the quality and potency of products. Although some ethnobotanical vendors reliably distinguish weaker from more potent kratom preparations using these terms, other vendors incorporate them into marketing jargon with no connection to the potency of the product.

Most kratom used in North America is ingested orally, with a few reports of buccal, smoked, or (more rarely) rectal administration. Many kratom users take repeated doses in a single session/day, often redosing between one to three hours after the previous dose. A day or session of kratom use might involve up to eight administrations, though typical usage involves between one and three doses in a day. The dosage estimates for single oral administration in non-habituated users are given in Table 16.1.

TABLE 16.1

Dosage Estimates for Single Oral Administration of Kratom in Non-Habituated Users

	Commercial Grade Dried Leaf	Premium Dried Leaf	Super Premium Dried Leaf	Extract-Enhanced Dried Leaf
Threshold	3–7 g	2–4 g	1–2 g	1 g
Light	5–10 g	3–5 g	2–4 g	1–2 g
Common	10–20 g	4–10 g	3–5 g	2–3 g
Strong	20–50 g	8–15 g	4–8 g	3–6 g

16.2.2 Duration

Kratom teas and extracts, when taken orally on an empty stomach, normally begin to take effect within 15 minutes. On the other end of the spectrum, the effects of capsules swallowed on a full stomach can be delayed 1–3 hours. Higher doses result in longer durations. The primary stimulating phase lasts 1–2 hours, with the more sedating effects lasting another 1–3 hours, for a total duration of 2–5 hours for a single oral dose.

Users report lingering aftereffects lasting several hours before they feel "down" from the experience. Some kratom users report alcohol-like hangovers following moderate or strong doses. Taking kratom daily can result in physical dependence. In such instances, symptoms described as similar to those of opioid withdrawal usually begin 18–24 hours after the last dose and continue for 1–14 days.

16.2.3 Effects

The most commonly reported kratom effects range from euphoric stimulation to sleepy dreamlike reverie and from pedestrian caffeine- or alcohol-like intoxication to overwhelming nausea and vertigo. The character of the experience changes as dose increases, beginning with a light relaxing stimulation at low doses and becoming a strong, nodding sedation at high doses. Euphoria and moderate intoxication are characteristic of the mid-dose range.

Positive/Desired Effects
- Simultaneous stimulation and relaxation
- Euphoria
- Increased sociability, talkativeness, and empathy
- Analgesia
- Vivid "nodding" dreams
- Reduced discomfort and boredom during manual labor
- A lasting "glow" the following day (at lower doses)
- Reduction or cessation of opioid withdrawal symptoms
- Pro-sexual, aphrodisiac qualities
- Sleepy sedation (at higher doses)

Neutral Effects
- Pupil pinning/miosis (reduced pupil size)
- Change in ability to focus eyes

Negative/Unwanted Effects

- Very bitter taste, difficulty consuming as tea
- Itchiness/pruritus
- Dizziness, nausea, and vomiting at higher doses or when used in combination with alcohol, benzodiazepines, or opioids
- Mild depression during and/or after
- Perceived increase in body temperature (feeling hot and sweaty)
- Hangover similar to alcohol
- Reduced sexual desire or performance
- Desire to repeat experience more frequently than intended
- Tolerance building quickly after consecutive days of use
- Addiction, difficulty controlling use, and withdrawal symptoms
- Liver issues, hepatitis, and symptoms of hepatotoxicity (unknown whether caused by *M. speciosa* or product contamination)

16.3 A BASIC KRATOM EXPERIENCE

The following is a firsthand report written in 2012 and submitted to the Erowid Experience Vaults. It was chosen as representative of a typical North American college-educated user without an existing opioid or kratom habit.

I am lounging by the pool in a mansion in Alaska, which used to belong to a crooked, gangster-affiliated strip mall developer. It is now a youth hostel that caters to backpackers in summer.

I am a 28-year-old male with significant drug experience who works about 60 hours a week. This morning, I took a short swim and ate a small bowl of Raisin Bran. I'm waiting about two hours after this mini breakfast before ingestion of Bali Kratom, in order to ensure that I am riding on an empty stomach. My current mind-state is clear, content and optimistic. It is Christmas Day.

My experience with kratom so far has been in the 4–6 gram range, producing the expected floaty-yet-focused stimulation. The source I use is excellent and places a lot of emphasis on customer satisfaction.

T+0:00 (6:48 am Alaska time): I pour myself a teacup of warm water and swallow the first ten capsules. These caps contain a little over half-a-gram each. My total dose will be just below eleven grams.

T+0:15: I head to the kitchen for another cup of water and sit back down by the pool to continue slowly taking the pills. This area of the hostel is serene. There are groups of people in the living room watching movies on the big-screen TV and hanging out in the kitchen. I love this little poolside area because it is quiet. There are some foreigners here at the hostel and I intermittently hear conversations in Arabic, Korean, and Mandarin Chinese coming from other rooms.

T+0:20: I flip on YouTube and listen to the studio version of "4th of July, Asbury Park (Sandy)" by Bruce Springsteen. "Sandy" is my favorite song, and it reminds me how important it is to hold on to one's beliefs. When I first heard it at twelve years old, I realized it basically explained everything I've ever wanted out of life—love, freedom, and transcendence.

T+0:30: First-alerts abound. Here comes the skin flushing, body-warmth, and very slight numbness. I decide to splash some cool pool-water on my face.

T+0:45: I'm grooving to tracks off the first two Springsteen albums.

I belch and taste the Kratom. I've read reports of people saying that Kratom tea is vile. I don't know. The kratom-burps don't taste as good as some fine Alaskan marijuana, but they don't taste nearly as bad as a San Pedro slushy, or the rug-burn-on-your-tongue taste of pure 2C-I, or that small chunk of pretzel you pick out of the carpet and smoke when you're coming down from crack at 6 am. These are experiences we've all had, right? :-)

T+1:00: A Middle-Eastern dude and an African dude sit down at a table on the far side of the pool and their Arabic conversation bounces off the walls of this cavernous room. The language sounds a bit harsh, but they are giggling and obviously having a good talk. The opiate-esque flushing continues, now accompanied by a lovely tingle/itch on my scalp.

T+1:30: I go outside to smoke a half a cigarette, and chat a little bit with my friend. We reflect on the year and our struggles.

Once back inside, I take a nice healthy piss and check my look in the mirror. My eyes are pinned. As I sit back down in my chill-spot, I feel that familiar sense of stimulation mixed with a relaxing body buzz. The warmth and itching are increasing and I feel no pain. This is most certainly akin to a moderate dose of prescription opiates. My head is getting slightly fuzzy and my thoughts are coming in shorter, tighter sentences.

T+2:00: At this point, I must have hit the peak. I doubt it is going to get any stronger than this. But I don't need it to. I feel great. My motor skills are a little bit off, but not badly. I make a cup of green tea and continue relaxing. I think about the "reason for the season." I work as a merchandiser, contracted by several big-box retailers and I always dread the Christmas season because of the crowds and the general anxiety which prevails in the stores while I am trying to concentrate and work. But as I lie here, listening to the Arabic intonations of my fellow hostel mates, I remind myself that anger is simply wasted energy, no matter who or what it is directed towards.

I feel that little head-bob coming on, and it's like my body wants to catch a nod, but knows it won't actually get there. I walk around the pool area, stretching and exercising my arms a little bit. I wade in the shallow end for a few minutes. This is good.

T+5:00: Effects plateau and then gently subside. I spend the rest of the morning reading, listening to music, and reflecting and feeling good. (Fleming 2014, ExpID: 98719)

16.4 EFFECTS CHARACTERIZATION EXCERPTS

Each of the key effects of kratom is listed below with a brief summary and excerpts from experience reports that document and detail the effect.

16.4.1 STIMULATION AND SEDATION

Users report that stimulant effects occur earlier and at lower doses. Sedative effects kick in slightly later and at higher doses. Perhaps the quality of the kratom experience most often remarked on in user reports is the seemingly paradoxical sense of being stimulated and sedated at the same time.

> For an hour I just thought "pure stimulant." Then a mild stoning phase. Lying down eyes closed with good sensual unfolding. Afterwards talkative. (DesT 2001, ExpID: 9657)
>
> ***
>
> My typical kratom experience begins about 15 to 20 minutes after consuming the plant material. It begins with a little excitation, and I begin to feel like I'm on a large dose of caffeine, only less jittery. Stimulant-like euphoria increases from there, and at around the 45-minute mark, I am completely absorbed in wave after wave of euphoria. The stimulant properties tend to wear off and I begin to feel... not quite sleepy, nor drowsy—sedated or calm is probably the best word. If I take as much as 9 grams, I tend to nod off and remain only tenuously connected to consensus reality, unless I choose to come out of this state (e.g., to talk to someone). (MIM 2005, ExpID: 47037)
>
> ***
>
> I live in Thailand where this stuff grows naturally, and eat it from time to time, mostly when my Thai friends offer it. I've read of other experiences with this leaf on the Internet and the dosages people are taking

are way beyond what anyone here takes. I do quite well with just one half or one leaf. They split down the center vein quite easily, so it's easy to take it in half-leaf doses.

For me, krathom [an alternate spelling for kratom] is primarily stimulating in its effect. I often eat a half leaf or more when embarking on a long drive, to keep me alert and make the trip a lot more interesting. Basically it feels like a mild dose of E[cstasy], with less jitters—in fact no jitters, just mild elation and alertness. It wears off in about two hours and there is virtually no crash (just a tad less than for caffeine I'd say). (Pike 2004, ExpID: 22387)

***

I feel edgy and lazy at the same time. There is a tiny bit of sedation that makes sitting in a comfy chair extra nice. Thinking and visualizing come very easy, ambient music is a perfect match for my condition. (Cautious Psychonaut 2014, ExpID: 46384)

16.4.2 Euphoria

Mood improvements ranging from mild to gushing euphoria are reported by some kratom users.

I began to feel a slight tingling and calm spread through me, at the same time, ever stronger waves of intense euphoria hit me again and again. I noticed quite a strong itch, which strangely enough wasn't unpleasant. Although I have zero tolerance, and zero experience with opiates/oids it is everything I imagine them to be. After an hour or so, I felt sort of like in a pleasant sleepy trance, with my eyes half open, just utterly numbed out bliss, with amazingly pleasant, relaxed waking dreams. (Looking back, after some opioid experience, this was my first nod.) (Limpet Chicken 2014, ExpID: 43574)

***

This was a delightfully euphoric high—I was just very happy to be alive. I was trembling a bit as if I were on MDMA or another amphetamine-like substance, but my body felt both an Ecstasy-type electrical energy and an opioid-type warmth and comfort, and my mind was lucid but definitely very lit up. (Guttersnipe 2004, ExpID: 33101)

***

Five minutes in, a noticeable mild euphoria (which I wasn't expecting) creeps in. I check for signs of placebo effect but my face keeps smiling sorta without me. Bit of a "third eye" buzz. Still ⅓ of the cup to drink, but it's easier to handle the bitterness once you like the effect. (Darklight 2002, ExpID: 18480)

***

At about 20 minutes I started to feel euphoric and chatty although I was noticing a slight slur in my voice and a laziness in my desire to form words clearly. (aquapan 2003, ExpID: 29517)

16.4.3 ANALGESIA

Many descriptions of kratom use mention analgesia and "pain killer" effects, often comparing kratom to pharmaceutical opioids. Some users say they prefer the pain relief offered by kratom to that of opioid pharmaceuticals.

First thing I noticed is a feeling of great euphoria and energy. Not really a messed up feeling, but something was starting to happen. After an hour I noticed that I was feeling like I took a painkiller or two, and I was very happy @ this point. (Sirk 2004, ExpID: 28819)

***

Smoke one small leaf off a kratom plant. Decrease in body sensations/pain killer effect like Vicodin. (Renwick 2001, ExpID: 9843)

***

An old weightlifting injury causes me to have migraines occasionally and strong headaches at least a few times a week. So a substance that has analgesic properties piqued my interest.

T+0:00: 10 grams on an empty stomach. [...]

T+0:30: Starting to have the familiar feeling of an opioid drug, limbs are getting heavy and feelings of euphoria start to creep in. Pleasurable warmth comes over the body and increases up until about an hour in. [...]

T+1:00: I seemed to have reached the zenith of the high and really do not feel any pain or at least notice it. It is easy to speak. Words flow and even the most stressful thing is manageable. [...] My thoughts on the plant: Kratom so far has been very helpful to me. In a sense many opiates are more like pain management instead of painkillers as people would have you believe. They get rid of some of the pain, while helping you to deal with the rest. As a migraine sufferer, this provides a good alternative to using the aforementioned medications as they have many more side effects and can have more serious long-term health problems. (Engrgamer 2009, ExpID: 71290)

***

The additional stamina kratom seems to provide allows me to push my workouts harder than with a vitamin/caffeine stack alone. Quite possibly the mild euphoria helps me push myself. The painkilling effect may also contribute to this, allowing me to work through some of my joint pain. [...] If I had to make a dose to dose comparison I'd say 1 Tbsp. of kratom = 10 mg hydrocodone, for pain relief. It's very close to the same in terms of sedation also. (Jonas 2007, ExpID: 62727)

***

I currently suffer from hemorrhoids. It's the second time I've had this symptom, the first time was 16 years ago, at a time when I was a heavy tobacco smoker and relatively heavy drinker. This spring, for professional reasons I had to take many long rides by train and by car with some moderate social drinking and smoking. Suddenly, the same vein of my rectum that had swollen 16 years ago started to swell again. [...] I usually don't think of kratom as an herbal medicine, more as a mild recreational/work psychoactive, with potentially unpleasant effects when overdone and a small but real addictive potential. I had the usual good effects of kratom: empathy, pot enhancement, moderate stimulation mixed with relaxation, nice warmth and so on. Best of all: 30 minutes into the effects, I noticed that I felt no more discomfort in my anal area. I thought that it might have been an analgesic effect. Kratom is supposed to have some and a friend of mine successfully manages back pain with this plant, but I've never noticed them myself until then. Something cool was that this effect lasted after I came down: I felt less pain the next day as well. (bloodfreak 2014, ExpID: 100242)

***

I was recently diagnosed with a form of tendonitis that gave me considerable pain in my right wrist at the thumb ("de Quervain's tenosynovitis") and my options were surgery to correct the problem (followed by physical therapy for several months coupled with opiates for pain for a time until the surgery healed) OR baby the wrist and allow the body to heal itself which takes approximately 12 months when it is as advanced as mine was.

I was prescribed Ultram ER (100 mg) and when that did nothing for the pain I was prescribed Ultram 50 mg. This also hardly dented the pain I experienced. Two aspirin tablets a few times a day DID help the pain somewhat but only enough to make it bearable so I stopped taking Ultram. [...]

I take 3 grams of private reserve kratom (about three slightly rounded teaspoons of the crushed leaves). I drink this mixture along with two aspirin tablets and in 20 minutes begin to feel relaxed and slightly euphoric, within 45 minutes to an hour all my pain disappears. Gone. As if I had taken Oxycontin, yet without the flying high feeling.

[...] I am thrilled that I have found a legal herb that can alleviate pain for me as effectively as opiates. I am careful to not take it recreationally, though it would be very easy to do so because the euphoria is very nice, since I'd rather be able to treat pain as needed without having to take large doses. [...]

Kratom is a very positive experience for me, for pain relief all I need is 3 grams and two aspirin. I hope others who may be frustrated in their community trying to deal with chronic pain without being able to get relief try kratom and find it as effective. (NaturalHealing 2010, ExpID: 79133)

16.4.4 Vivid "Nodding" Dreams

At higher doses, users report an opioid-like "nodding" experience that often includes dream imagery. Some users report the night's sleep after taking kratom has increased dream vividness and activity.

T+1:10: Drink last of first extraction.

T+1:15: Poured 1 of 3 new cups in the second extraction. Take a small sip and then fairly suddenly overcome and slip into a twi-light dreamy stage, on the couch. I notice it's a beautiful day and the sounds seem to travel further. Very pleasant, begin to fall into a light sleep. Deeply relaxing to the muscles and mind. Dreams shifting and full of people, objects and places—seems no particular reason, just flowing dreams. (Reville 2002, ExpID: 18482)

***

I lay down in bed, closed my eyes, and something wonderful began to occur. As I lay there, I began to experience two or three second fragments of what I can only describe as dreams. It was like my con-sciousness would be plopped down into the middle of a dream and a moment later be extracted just as suddenly. I would return to reality only [to] be plopped down into the middle of another completely dif-ferent dream a few moments later.

I have very little recollection of the actual content of any of these dreams—I must have experienced well over 100 of them, each one totally random and unique. The most interesting part is that I was totally conscious throughout all of this, I wasn't sleepy at the time and I am positive that I didn't drift off. I spent an extremely pleasant 45 minutes in this state. (Incarnadine 2004, ExpID: 30055)

16.4.5 Easing Manual Labor

One of the more common traditional types of kratom use described in Southeast Asia is consuming *M. speciosa* to ease daytime fieldwork. A number of reports from North American, Australian, and European users reflect the boredom relief, mental stimulation, and physical ease that kratom can bring.

My job as a full-time produce stocker is very physical and boring. I'm usually drained and grumpy all day at work and for a few hours after. I tried 0.75 grams of kratom 10x extract, mixed in warm water and downed in a gulp on a half-empty stomach during work.

The initial effects of stimulation and good cheer came on in about 10 minutes and lasted for 2 hours. It felt like a very strong but clean coffee buzz combined with a "polite" and scaled-back MDMA euphoria. When customers wasted my time with obvious or rude questions, I answered cheerily, and volunteered to stock watermelons. It made a day at work about as fun as a day off without drugs. I felt a lot more outgoing than usual. [...] At the end of the shift, I felt groggy, but quickly rebounded after it totally wore off (4.5 hours after drinking). (seaborg 2007, ExpID: 63393)

***

I have a pick-and-pack job at a local warehouse. I average sixty hours of work per week, pulling ten- and twelve-hour days full of heavy lifting and constant motion. I would return home from these outings completely exhausted and washed-out, unable to muster the energy to do much of anything except eat and go to bed. [...]

About a month into the job, I decided to experiment with kratom as a work aid. I bring a spoonful (about 5 grams) of kratom with me to work in a small resealable plastic bag. Instead of making it into tea, I take it orally, as one would with a powdered dietary supplement, washing it down with soda or sweet tea. I usually break the five grams into three doses which I take during a ten-hour day: the first about half an hour after arriving, the second about three hours into my day, and the third about an hour after lunch (six hours into my day). I am careful to avoid taking doses just before or just after eating, as kratom deadens my hunger and tends to cause mild nausea if taken in conjunction with food.

The effects set in gradually over ten or fifteen minutes; increased energy, a sense of reverie or calm even during the most strenuous labor, and (most notably) a peculiar effect on time-sense. This is possibly the most odd of all the effects, since I've experienced no loss of lucidity, and can account for everything I've done, but time seems to sneak by much faster than normal. [...] I find it clears my head, keeps me awake, and gives a sense of buoyant relaxation, so that even the most boring task is endurable. (Maturin 2008, ExpID: 68954)

16.4.6 REDUCTION OR CESSATION OF OPIOID WITHDRAWAL SYMPTOMS

Kratom is widely used by opioid addicts to stop or reduce opioid withdrawal symptoms. In some cases, kratom use helps individuals stop taking opioids entirely, but in other cases kratom simply replaces the other opioid addiction with daily kratom use.

I had a serious injury that kept me in the hospital and on narcotics including fentanyl, morphine, oxycodone, and hydrocodone for a period of about six weeks. One exacerbating circumstance is that it required HIGH doses of opiates for any analgesic properties to be recognized. Needless to say withdrawal was a nightmare.

I found references to the use of kratom as a way to ease withdrawal pains and discomfort. It doesn't just ease the problems, it totally wipes them out. At lower doses my energy level is increased to an extent, I feel great, a little opiation is present. I had no trouble with nausea at low to moderate doses. High doses can turn my stomach a bit.

For people struggling with opiate withdrawal I would say this is a miracle, and I'm not easily impressed. (AnonymousPatient 2004, ExpID: 38806)

***

My story with kratom begins about 3 years ago, while I was in a detox and rehabilitation program for a morphine habit. I ordered several ounces of kratom leaf prior to entering detox, and it arrived shortly afterward. After completing treatment with buprenorphine, the physical and emotional symptoms of morphine withdrawal were still nearly unbearable. Every fiber of my body ached for what seemed like months, though the most painful portion lasted only about 10 days. I could sleep for no longer than 4 hours at a time, but even this sleep was accompanied by horrendous nightmares. I lost weight, and when I saw myself in the mirror for that first month, I saw an emaciated, drawn, and wide-eyed version of myself that I barely recognized.

During the course of the detox, around day 3 or 4 post-buprenorphine treatment, my stoicism broke and I made a glass of kratom tea. I crushed about 5 tablespoons of the dried leaf in my hands and added boiling water and lemon. The "tea" was very bitter and astringent. Within 30 minutes the tea alleviated all of my symptoms. I felt like me again. Normal, in no particular pain, but for the pain of the guilt that I had relapsed. I flushed the rest of the leaf down the toilet in fear of ruining my detox. (BB 2007, ExpID: 56768)

***

This report tracks the successful and largely withdrawal-free metha-done detox of a 46-year-old, 167 lb. woman with hepatitis C who had been addicted to methadone (concurrently with Xanax, cigarettes, and coffee). The detox broadly went through three stages: a metha-done taper, a substitution of kratom for methadone, and an ibogaine experience to detox from the kratom.

The detox began with the subject attempting to taper off of methadone. Some false starts were observed, such as when she thought she could taper one milligram per day. Also, bureaucratic problems slowed down the process as the clinic made it easy to increase her dose, but requests to decrease her dosage required the staff to conference, and this process could take weeks if they did not outright lose the request. The taper that worked once it finally got started took her from 125 mg to 55 mg [of methadone] on a schedule of 5 mg every two weeks. Her plan was to con-tinue tapering until she could discontinue usage altogether; however, at 55 mg, she was no longer able to taper and had her clinic hold. [...]

[She wanted to try ibogaine therapy, but was told there would be a long wait.] She decided to try kratom in the interim.

After taking the kratom tea, the first day off of methadone left the subject completely amazed. After 11 years of methadone use, it seemed like a "miracle" that she could leave it behind for a legally available herb. The subject started taking kratom roughly a cup at a time just before noon, and by 23:00 in the evening, she had drunk tea made from about 50 grams of kratom. Early in the day, she was in good spirits. By the evening, she was showing some anxiety, but it was manageable and nowhere near what she would be experiencing with methadone withdrawals. [...]

Kratom withdrawal differs from methadone withdrawal. The most marked symptom of kratom withdrawal is an overwhelming anxiety. The subject became hooked on kratom and was unable to taper down because of this anxiety and also because of intense spasms. Instead, she began to slowly increase her tolerance and dose. Methadone addicts are terrified of withdrawal, so she insisted on staying on kratom for far too long. She feels like when she started to actually enjoy the kratom to get high, addiction was the obvious outcome. All told, the length of her kratom addiction was about six months, and she acknowledges that becoming hooked on kratom to avoid the possibility of feeling any methadone withdrawals was not useful; however, even though she was unable to quit kratom on her own, there was the silver lining that her kratom addiction was much easier to dispose of with ibogaine than a full-blown methadone addiction. (Transmigraine 2010, ExpID: 86136)

***

I have used kratom off and on for a couple of years now. I was addicted to tramadol as a painkiller and found that taking a tablespoon of powdered kratom leaf (not extract) followed by a glass of juice would stave off any withdrawal symptoms. I would buy plain leaf and grind it in my coffee grinder. The taste was bad, but with a strong juice I could quickly gag it down. I thought it was a miracle drug that could cure any type of opiate/opioid addiction.

I began taking larger and larger amounts of kratom both in the morning and at night. If I didn't dose, I would get achy, sweaty, and have tons of anxiety. The most troubling was this anxiety, which made me feel very, very bad. The only thing that would relieve these symptoms was more kratom. (Anonymous 2007, ExpID 62381)

***

I'm a 29-year-old female living in the US. I have an off and on active opiate addiction of about 15 years (and more on the "on" side than "off"). By opiates, I mean any and all of them, started off with Vicodin, then went to Percocet, and on to Oxycontin, heroin, and eventually the methadone monster: I was taking 100 mg a day in my methadone clinic.

I recently moved from NYC to upstate New York, thinking that the clinic in this area could at the very least give me a guest dose until they could take me on as a full-time patient. To my dismay (and horror) the local clinic here had a waiting list of six to twelve months, which meant that my choices were to a) move back to the city, b) go to a rehab and detox, or c) come off of methadone cold turkey. Being that I didn't really need to go to rehab, and that moving back to the city JUST to stay on or taper off slowly of methadone was a very pathetic prospect, my choice was limited to coming off cold turkey.

Four days later and ridiculously ill, with symptoms such as nausea, vomiting, diarrhea, and muscle aches, I received my order of powdered Bali kratom leaf and began my journey into the Land of No Opiate Withdrawals.

T+0:00: I mix 28 grams of kratom into a pot of 4 cups of H_2O using a wire whisk, continuously stirring, and bring the sludge to a gentle boil, then turn it down to simmer for 15 minutes (stirring every few minutes). I add about ⅔ cup of sugar to the hot mix, let it cool for a few minutes, and divvy the water with the plant material still in it (stirring as I pour) into 4 glasses (to measure 4 doses of 7 grams).

T+1:00: Nausea and churning of stomach definitely gone, almost 100%. [...] My limbs feel a little heavier, in a good way. My mood has gone from despair to contentment. I wouldn't say I am experiencing euphoria or a "high" but I liken the feeling to getting my methadone after missing a dose of it one day.

T+1:30: I have energy to do things for the first time since I stopped my methadone. I get up and begin to clean the house, doing dishes, and listening to music. The music is enjoyable, and I find myself wishing I had company around.

T+2:00: The initial energy burst I got seems to have tapered off, but I now feel relaxed, content, and have almost NO symptoms of withdrawal from the methadone at this point.

T+2:25: I feel like I can actually fall asleep for the first time in days, so I take another 7 gram dose of kratom so that it lasts me through the night (I have read that mytragynine has a very short half-life) and proceed to lie down.

T+2:40: I feel myself fade off to sleep—no issue whatsoever.

The next morning, I wake at approximately 8:30am (T+10:30) with some very mild anxiety and fatigue, but after dosing again with the kratom, get energy and anti-anxiety effects almost immediately. For me, kratom is extremely effective as an opiate withdrawal remedy when used in the right doses and frequently (2–3 times a day) as its effects do not seem to last as long as methadone or buprenorphine. I plan to use this as a long-term taper (several months) and possibly in place of methadone maintenance. (msFancyPants 2011, ExpID: 92099)

16.4.7 PRO-SEXUAL AND ANTI-SEXUAL QUALITIES

Some people report mildly increased sexuality and prolonged intercourse, while others describe decreased sexual desire and ability to perform.

T+1:30: Attempted sex with partner (primarily for experimental purposes only, as desire was not there due to drug action). Sexual responses significantly depressed, difficulty maintaining an erection. No sexual interest but enjoying touch and intimacy from the perspective of comfort and warmth. (Agean Moss 2005, ExpID: 46142)

***

I have noticed that kratom has wonderful anti-anxiety, anti-depressant, painkilling, diuretic, prolonged sexual intercourse, and even nasal decongestant qualities. (AnonymousK 2008, ExpID: 62926)

***

A subtle aphrodisiac effect, increasing sexual desire as well as enhancing performance. In some people this effect is very significant while others find it relatively minor. (Yolk 2014, ExpID: 98929)

16.4.8   PUPIL PINNING/MIOSIS

As with most opioid agonist drugs, miosis is common. Pupils decrease in size and responsiveness as a function of dose.

Kratom's euphoriant effects lose their power after 1.5–2 hours, but still remain in pinpoint pupils, general relaxation, and pleasant thoughts. (Seifuru 2004, ExpID: 36932)

***

To anyone who may think this is placebo the mild dry mouth and constricted pupils suggest otherwise. (420_Psych 2014, ExpID: 46806)

***

My pupils were pinpoints and would not dilate even in pitch blackness. (Lahey 2014, ExpID: 56786)

***

After an hour or two, we contentedly walked around my neighborhood, holding hands, enjoying the sun. I noticed around this time that my pupils were practically non existent, and I looked to myself like I was on morphine. (BB 2006, ExpID: 44536)

16.4.9   ITCHINESS/PRURITUS

Users often report itchiness after taking kratom, similar to the itching people experience when using opioids. A few reports describe itching lasting for days after last use.

T+0:45: By this time I am feeling pretty euphoric. I begin to get a little bit itchy. Nothing different than a typical opiate itch. [...] The itching involved seemed less severe than it would have been if I had taken an equivalent amount of Vicodin or Percocet. (Investigator 2007, ExpID: 57960)

***

About 1.5 hours after kratom consumption, skin was itchy, a symptom common to opioid drugs. This was bearable. (Psychopharm 2004, ExpID: 35770)

***

Now, approximately 2 hours after ingestion, I've got the itchies that are a common side effect from prescription narcotics. (ExisT 2014, ExpID: 41939)

***

Sleep is very comfortable, long and deep at times, with occasional but pleasant waking during the night. Sporadic, mild bodily itching (although not intense or overly bothersome), in all warm spots of body, crotch, behind knees and elbows, neck and head. (Agean Moss 2005, ExpID: 46142)

***

The Third Day after use: I woke up feeling a great deal better, no more cold symptoms, no more back ache, no more fever. However, my body was quite itchy. When I had nothing to do, I found myself scratching my hands, arms and legs a great deal. I went to the mall that day to see some of my friends, still itching on and off. Other than the itching, I felt baseline, and was able to interact with everyone normally. By the fourth day the itch was completely gone. (Baron Von Bad Guy 2005, ExpID: 47211)

***

The only obnoxious effect I get nowadays in very high doses (like when I've been chewing leaves all day at work) is itchiness. This itch is almost always concentrated on my belly and/or chest but sometimes on the top of my head and behind my knees. It's nothing terrible, and it feels really good to scratch it. Not like a PCP feeling of creepy crawlies. It kind of adds to the euphoria. (Tomtom 2007, ExpID: 53894)

16.4.10 Dizziness, Nausea, and Vomiting

Many people report nausea and or vomiting associated with higher doses of kratom or when mixing it with other drugs.

I did actually vomit on two occasions that I've used kratom tea, but vomiting on an opiate, or opiate-like drug, is really not all that unpleasant. I mean, you know, it's somewhat unpleasant, but not as bad as when one is sober, and certainly not anywhere near as bad as vomiting from alcohol overindulgence. For this reason, it is highly advisable to use kratom on an empty stomach. That will also increase the effects, as it will be absorbed better. (rev. cosmo 2004, ExpID: 32178)

***

T+1:00: The walking is no longer helping but only making things worse. I decide that vomiting is going to be inevitable so I go inside and wait by the toilet. I gag once but nothing comes up. I am trying to just remain calm, not necessarily to resist vomiting. As long as I just sit here next to my good friend, the toilet, staring mindlessly at the floor, my nausea is manageable. Closing my eyes helps mellow the dizziness too and I keep drifting off to sleep for minutes at a time.

T+1:20: Opening my eyes or standing up brings on the queasiness full-force, but somehow I am able to retrieve the phone and sober up long enough to make a due phone call. My porcelain companion remains beside me throughout the correspondence. Talking is very difficult because of the nausea and my comprehension is slightly deteriorated. The phone call lasts ten minutes and then I'm free again to wallow in my sickness.

T+1:45: I manage to abandon my post and make it to the den to smoke a little cannabis. It certainly does help, but not as much as I wish. I lie down and enter a twilight sleep state for the next 7 hours, waking twice to smoke more cannabis and drink some water. The water almost brings the onslaught of vomiting, but I am able to keep it down since I know I'm dehydrated. (Jesper 2004, ExpID: 30154)

***

Preparation: 1 oz. dried crushed kratom leaves. Leaves placed into pan, squirted with ~½ cup lemon juice. 2 cups water added. Solution left to boil for 20 minutes. Repeated two more times with a total of 6 cups of water, leaves discarded.

Note: 1st decanted liquid was a   dark brown color, 2nd was much lighter, and 3rd was nearly clear. The resulting liquid was divided evenly between two of us. It was one of the foulest drinks I have experienced, but I choked it down.

T+0:15: I begin to feel a pressure on my face and body similar to what I feel during the onset of a cannabis high. My body temperature seems elevated, but I have no thermometer to confirm this. [...]

T+0:25: I am feeling more nausea, so I get up to drink some orange juice to see if it will help. It doesn't, I vomit. Afterwards I feel instantly better. I am still sweating.

T+0:35: My nausea returns and I step outside for some fresh air, it doesn't help. I vomit again. I can definitely still feel an overall numbness and body heaviness even after vomiting.

T+0:45: My hands are shaking, but not constantly. I still feel nausea and I am still sweating. Typing is difficult. If I lay back and close my eyes I feel much better.

T+0:55: I feel as though I could go to sleep with ease, but I do not attempt to.

T+1:00: Most of my nausea has subsided. I still feel very relaxed but I can tell that all effects are decreasing in strength (I am assuming from losing the tea when I vomited).

T+1:25: I experience a sudden wave of nausea and vomit.

T+1:40: I am almost completely back to baseline but with the very slightest headache possible.

T+2:00: I am definitely baseline.

Conclusion: Overall I would say that this was a mostly negative experience. The euphoria and positive body feelings were quite enjoyable but the dizziness and nausea just totally ruined it. (J-to-the 2004, ExpID: 36154)

16.4.11   HANGOVER

Reports of hangover effects range from feeling better than normal the day following kratom use, to mild alcohol-type hangovers, to more serious nausea and dysphoria. For the purposes of this hangover grouping, withdrawal effects following discontinuation of frequent use are not included.

[After taking eight grams of crushed leaf at 10 pm with no kratom or opioid tolerance.] The next morning I awaken at 8 am. I have a very slight headache. Would I describe it as a hangover? No. I am still mentally content and relaxed, but the light pain in my head is noticeable enough to be annoying. A heavy cloud lies over my eyes, and colors seem a little more vivid. I am still very tired, and physical activity is simply out of the question. Foggy, is the word I am looking for. Foggy, relaxed, warm, content, and very, very sleepy. (Wallcrawler 2014, ExpID: 100269)

***

[After six grams of kratom leaf powder consumed in capsules taken along with smoked cannabis.] This combo definitely produces an after-effect the following day. I have just returned from a lunch meeting and I must say that I felt dulled-out and ever-so-slightly detached the entire time. It is like the most intense weed hangover I have ever experienced. I am reconfirmed in my conviction that, if I'm ever going to try this combo again, I MUST have the next day completely free. (CosmicCharlie 2012, ExpID: 94222)

***

I got into kratom as an alternative to popping pills. For a little background, I love getting high. But I'm grown now with responsibility that can't be ignored, even for half a day, so really getting high is a thing of the past. Not only can't I afford the time to really get wrecked, there's also no time for hangovers or detox. [...] I've found the magic to be in the 2–3 gram range, taken all at once. Four grams and up leads to more sedation, more hangover, more constipation, and more opportunity for dependence. (TooOldForDrugs 2014, ExpID: 97078)

**

I have been using kratom for a few months, about once a week on average, with both the resin and powder extracts forms. Before a couple of weeks ago, the only negative after effects for me were a hangover the following day consisting of a mild to moderate headache and lethargy, although sometimes I felt quite energized the next day. There is also a tendency to feel depressed for a couple of days, with a slight feeling of unreality, and quite severe constipation for a couple of days. I haven't used on a constant basis, so this is just from single doses. (PB 2007, ExpID: 51161)

**

[After 8 grams of "kratom super powder" resulting in a "strong experience."] Sleep came easily, and other than a little grogginess there were no hangover effects the next morning. (Dark Matter 2014, ExpID: 47357)

**

After taking high doses of kratom, and not falling asleep right afterwards, I feel a "hangover" that can be characterized as being irritable, and in just a bad mood. I did not experience this the first few times I drank kratom, however the more I drink it on a regular basis, the more I notice these ill-effects. (apolytonjs 2014, ExpID: 42959)

16.4.12   KRATOM ADDICTION AND WITHDRAWAL

Perhaps the most problematic characteristic of kratom is that many users report wanting to take it daily and have difficulty controlling their intake. As with other opioid-agonist drugs, kratom use leads to tolerance, desire to redose, and withdrawal effects after daily use.

It's been almost a year since I began using kratom, and I feel it necessary to say some things about it that I was blind to or didn't yet know about previously.

During the beginning of my kratom usage I loved it. However, soon after, my already shaky willpower gave way and I began using it almost every day, sometimes more than once. This turned into a three and a half month long binge where my dosages increased from 10 grams, gradually up to a minimum of 15 grams, and to get really "kratomified" I would require a dose of 18–20 grams. This was using standard-strength *Mitragyna speciosa* leaves, rather than the alleged super-kratom. Anyway, I didn't feel that it was detrimental to my health at all, and I was quite happy nearly all the time. However, it was starting to get me into some serious credit card debt, at $130 a pound with shipping, so I decided to quit.

After the first 20 hours of having no kratom, I began to seriously crave it, but I figured, hey, this stuff doesn't seem physically addictive, and I've beaten my caffeine addictions (at least temporarily) on numerous occasions, so I'll just ignore the cravings, keep on smoking my sacred herbs, and move on with my life.

Then I reached the 24th hour, and all hell broke loose. I was a mess. What began as a psychological craving quickly became physical. By the second night, I was so restless that I couldn't sleep more than a few winks the whole night. It was as if my body was filled with a restless electrical energy that made me just HAVE to move my arms and legs. I'd get frustrated and flail about violently, to try to make the feeling go away, but it just got worse. By the next morning, I had disregarded my troubled financial situation and spent $130 for another pound. I figured I'd only have one more night to spend in misery until my blessed package arrived.

It's kind of funny, I've come to associate the USPS truck with kratom, and even now, every time I see one, I get a little rush in my gut like the initial rush of kratom. Anyway, after another miserable night of little sleep, even with the help of excessive amounts of valerian root, I eagerly awoke, thinking my package would come by noon.

I was wrong. To make what could be a long story short, the normally fast company I order from had slipped up and my package didn't arrive until the fifth day. During this time, I almost completely broke down. Up until then, the only thing I'd been physically addicted to was caffeine, and I had always considered that an easy one to break away from. What's a little headache compared to this psychological torture? I just can't stress enough how uncomfortable I felt in my own skin, like my skeleton was going to crawl out of my body. I'd wake up after barely sleeping, completely sore from my twisted muscles.

I had more knots in my back than I've ever had before. And something inside me changed then, I fear forever. The iron control I've always felt that I've had over my emotions slipped, faltered, and was gone. I felt like I couldn't control the feeling of utter desolation and depression that would overcome me. From days one to four of my forced detox, I spent most of my time curled up in a ball, grinding my teeth, sweating, and trying to ward off the illogical feelings of depression and anxiety that were consuming me. Ironically, on the fifth day, I woke up feeling mostly fine, as if the addiction had left my body (physically, anyway).

Then the package arrived. After living through this hell of a detox, I was afraid to start again, but here was this pound of kratom, beckoning. So of course I brewed some up, and drank it. Ah, bliss! And now it only took me 12 grams to get the same effect that 18 used to give me! I went through the pound quickly, in about two weeks, and to my surprise, I was able to remember the horrors of my previous detox and I successfully quit before I was that hooked again. This time detox was very mild and quite bearable. (Xorkoth 2005, ExpID: 46082)

***

I've used kratom tea for two years in an attempt to stop a drinking habit that had gotten out of hand. I found kratom to completely mute alcohol cravings—something that had always beaten my prior attempts to moderate or stop drinking. I have to give kratom total credit for my not drinking, or even really wanting to drink, since then.

However, on the down side, I broke the rules of kratom use and began to use it every day. So, of course, I became addicted. I've decided to give it up entirely since I can't seem to moderate well. I'm now going through the withdrawal, which so far has been only slightly miserable. At the end of week one I've had mild craving (not for alcohol, just kratom), extreme fatigue, and some depression. (Susan 2007, ExpID: 49475)

***

In the last two years, my kratom consumption varied, 25–50 grams daily, but peaked at a kilogram per month. Crushed leaf was always my preferred brew, occasionally powdered leaf, always the "standard" variety. A short time ago I began to experiment with encapsulated super extracts (15x, 50x, 250x). Bad mistake. The extracts have a sharp and potent onset much closer to real opium than the leafy teas. I began to feel the euphoria I had sought for so many years, and once again my kratom intake surged. [...] I was strung out for months, riding a kratom rollercoaster, up, down, up ... (Jojoba 2006, ExpID: 55648)

***

I heeded the warnings about addiction, but didn't think it could happen to me, a recovering alcoholic with 8 years sober. Within a week I noticed that around halfway through my work day I would start to feel achy and sick, all the problems kratom took care of so beautifully would return... sore legs, bad circulation, etc. So I'd get home and make some tea and feel better, oh so much better! Gradually my afternoon work performance suffered so much I started bringing the tea to work in a Snapple bottle and drinking a bunch after lunchtime. This was about 9 months ago. Since then I've been a kratom junky. Who would have thought? (Dr. Robert 2006, ExpID: 49447)

***

[After 10 grams of kratom per day] I am three days into my "turkey" now and it is really bad. No sleep at all! My arms and legs are tingling and I thrash about all night, cold sweats, goose bumps. I can't sleep in bed with my wife, too twitchy. Concentration span is next to nothing, no appetite etc. Probably a lot like coming off heroin.

Three days in and I am starting again. This time a controlled weaning I think starting with 10 grams and reducing by 2 g per day over a week or so. Never 10 grams per day to zero in a day. (the good doctor 2014, ExpID: 85769)

***

I'm writing this to warn people. I have been a regular user of opiates for about 7 years. I have been through withdrawals from morphine, Oxycontin and hydrocodone so I feel like I have a pretty good grasp of what opiate detox is like.

That being said, kratom really snuck up on me. I decided to try the kratom tincture, which was advertised as the strongest most potent formulation available on the vendor's website. It came in 20 milliliter bottles and I initially took about 3 ml a day via medicine dropper under the tongue. It really was powerful; I can describe it by comparing it to 15 mg of Oxycontin insufflated, for someone without a tolerance. [...] My tolerance increased and I was going through a 20 ml bottle every 3 days when I ran out of money, roughly 4 months into my kratom experience.

I thought I would be fine, but after going a day without, I woke up to familiar aches and pain and the beginnings of anxious feelings. This got progressively worse and I did not sleep for about 3 days until I went to my doctor and got some temazepam (Restoril) 30 mg for sleep. [...] After about 14 solid days I stopped feeling really bad and just felt pretty bad for another 14 days. All in all, the worst and most unexpected detox I've ever gone through. (danyv1782 2014, ExpID: 82797)

***

I was a heavy opiate user for about 6 years, taking regular doses as high as 500 mg of morphine per day. I went into withdrawals on a regular basis. I'd feel a little sick 8 hours after the last dose. By the second day, I was a total mess. "The most terrible flu ever" doesn't quite capture it. Buckets of cold sweat. My guts churned and squeezed painfully. The diarrhea was terrible and constant. All senses were amplified: sound was oppressive, light hurt. I lay in bed with my fingers splayed because I couldn't stand the feeling of them touching together. There was a near-complete loss of energy. I was incapable of doing the simplest thing.

Years after my final kick and now I'm sipping kratom tea. My regular dose is about a heaping half-cup full, about 20 grams worth, steeped in boiling water for 45 minutes—one in the morning, and sometimes a second in the evening. After a year of this dosage, I decided to stop. I was a little scared, remembering my opiate kicking experiences, so I set aside a whole weekend to sweat it out. Last dose came Friday morning.

By Saturday morning, I felt a little under the weather. Not sniffles or anything, just mild disquiet, almost malaise. I was able to go downstairs and make coffee. Then breakfast. Then make phone calls. I watched some TV, went out for a walk, worked in the woodshop. Sunday comes, and things are pretty much the same. I feel a little slow, but not much different than yesterday. I do a crossword. I do laundry. I sweep out the garage, taking little breaks now and again to sit down.

By Monday, I was back at work. The malaise was nearly gone. No sniffles, no diarrhea, and no outward signs that I'd just stopped taking heavy doses of a drug that some governments put in the same class as heroin. By Tuesday, there are no symptoms at all. A week passes, then a month, nothing lingers.

I've now done this 2–3 times, with similar experience each time. In my experience, kratom withdrawal really doesn't compare to opiate withdrawal. The effects were so subtle as to almost go unnoticed. I have more troubles when I catch a slight cold. (Efraim Ivener 2014, ExpID: 89029)

16.4.13 HEPATITIS AND HEPATOTOXIC SYMPTOMS

A small number of kratom reports describe hepatitis and/or hepatotoxic symptoms. This effect has also been reported in the medical literature (Kapp et al. 2011). Most of the reports of liver-health issues involve kratom extracts rather than unenhanced dried leaves. It is unknown whether these are attributable to kratom alkaloids, extract-production byproducts, or other contaminants.

A few months ago I purchased 15x extract after reading of the opium-like quality of this legal (in the US) plant. I depleted the whole baggie within two weeks time, taking a strong dosage usually every other or third night. The effects were incredible at first: euphoria, warmth, bliss, everything you read in forums. By the fourth dosage, everything went wrong. Within hours, I felt an intense, steadily increasing pain in my abdomen. The pain became so great that I was eventually curled up in a ball on my couch, vomiting helplessly on the floor. At the time, I figured I simply ate a dinner contaminated with some microbe. The next day however, I felt chills, my urine was the color of black tea, and I experienced an intense nausea. This condition would not go away. By the fifth day, I figured this was not a regular food contamination problem. By the fifth day, my whole body and eyes were a dark yellow color … jaundice had set in.

After several blood tests, and doctor's visits I was diagnosed with cholestatic hepatitis, a non-infectious liver disorder where your gall bladder essentially shuts down for some time. My liver-panel blood test showed elevated ALT, AST, alkaline phosphate (a marker for gall-bladder health), bilirubin (the chemical that causes the yellow color of jaundice), and serum albumin levels.

My condition lasted two weeks. […] It very well could have been that the extract was tainted with lab chemicals. (Sly 2008, ExpID: 71949)

***

In mid-October I began using kratom. For the next two weeks, I used it almost daily, taking 3 grams of crushed premium leaf in the mornings.

On Halloween weekend I suddenly became very ill and immediately stopped use. I experienced fatigue and an extreme loss of appetite, along with a sharp pain in my abdomen. I could not keep food down for the first week or so. I became extremely jaundiced and, as is common with jaundice, my entire body was itchy.

The diagnosis of hepatitis was made when my ALT (liver enzyme) was measured at 500, which is around ten times normal. Two weeks later after onset this number peaked at around 1,400. For completeness, my AST (liver enzyme, 40 is normal) peaked at around 300, and my bilirubin topped out at six (one is normal, this causes jaundice).

I have seen specialists about this illness and a cause (other than kratom) cannot be found. (Nlogn 2011, ExpID: 88678)

16.5   CONCLUDING COMMENTS

Weighing the benefits and risks of a traditional plant used as a medicine, work aid, and recreational drug is remarkably complex. Modern commercialization technologies transform plants into products; potency increases, and the user base shifts and expands. Naturally occurring as a bitter, relatively mild herbal stimulating sedative, *Mitragyna speciosa* and its chemical constituents are now available in highly potent, colorfully packaged products.

It is a compelling yet equally concerning fact that kratom offers a plant-based option for managing opioid withdrawal symptoms. Some savvy, Internet-age heroin users have stepped outside the cycle of black market street opiates and methadone clinics by substituting a relatively affordable tea. The dilemma is that kratom, like many pharmacological opioid addiction treatments, can itself be addictive.

Hundreds of first-person kratom experience reports document everything from enjoyable evenings to concerns of liver toxicity. It is our hope that the collection of these stories can enhance society's understanding of the benefits and problems associated with the use of psychoactive substances such as kratom and will provide information that is directly useful to students, researchers, medical professionals, and the general public.

ACKNOWLEDGEMENTS

Thanks to Sylvia Thyssen, Michael Pinchera, Agathos, Dendron, the experience report authors, and the Erowid crew for contributing to this chapter and the public record about kratom use.

REFERENCES

Craigle V. 2007. MedWatch: The FDA Safety Information and Adverse Event Reporting Program. *J Med Lib Assoc* 95(2): 224–25. http://www.ncbi.nlm.nih.gov/pmc/articles/PMC1852611/ (accessed Jan 30, 2014).

Kapp F.G., Maurer H.H., Auwärter V., Winkelmann M., and Hermanns-Clausen M. 2011. Intrahepatic cholestasis following abuse of powdered kratom (*Mitragyna speciosa*). *J Med Toxicol* 7(3): 227–31.

EROWID EXPERIENCE REPORTS (ACCESSED FEB 4, 2014)

Agean Moss. "Left Me Feeling Indifferently: An Experience with Kratom Powder." Erowid.org. Oct 21, 2005. Erowid.org/exp/46142

Anonymous. "New Dependence: An Experience with Kratom Leaves." Erowid.org. Jun 25, 2007. Erowid.org/exp/62381

AnonymousK. "My New Love: An Experience with Kratom." Erowid.org. Jun 10, 2008. Erowid.org/exp/62926

AnonymousPatient. "Great for Easing RX Withdrawal: An Experience with Kratom." Erowid.org. Dec 30, 2004. Erowid.org/exp/38806 apolytonjs. "In Between Dreams: An Experience with Kratom." Erowid.org. Jan 23, 2014. Erowid.org/exp/42959

aquapan. "Follow Recommended Doses: An Experience with Kratom." Erowid.org. Dec 27, 2003. Erowid.org/exp/29517

Baron Von Bad Guy. "The Experience, the Hangover, the Itch: An Experience with Kratom, Cannabis, & Alcohol." Erowid.org. Dec 22, 2005. Erowid.org /exp/47211

BB. "Helped with Opiate Withdrawal: An Experience with Kratom." Erowid.org. Aug 20, 2006. Erowid.org/exp/44536

BB. "The Nightmare of Detox: An Experience with Kratom Leaves." Erowid.org. Mar 12, 2007. Erowid.org/exp/56768

bloodfreak. "Unexpected Hemorrhoid Remedy: An Experience with Kratom." Erowid.org. Jan 23, 2014. Erowid.org/exp/100242

Cautious Psychonaut. "Intense Euphoria: An Experience with Kratom Extract." Erowid.org. Feb 4, 2014. Erowid.org/exp/46384

CosmicCharlie. "Dynamic Duo: An Experience with Kratom & Cannabis." Erowid.org. Jan 27, 2012. Erowid.org/exp/94222

Danyv1782. "Severe Withdrawals from Tincture: An Experience with Kratom." Erowid.org. Jan 23, 2014. Erowid.org/exp/82797

Dark Matter. "A Rewarding Experience: An Experience with Kratom." Erowid.org. Feb 3, 2014. Erowid.org/exp/47357

Darklight. "Mitragyna Bioassay: An Experience with Mitragyna speciosa (Kratom)." Erowid.org. Oct 27, 2002. Erowid.org/exp/18480

DesT. "Stimulant then Stoning: An Experience with Kratom Leaves." Erowid.org. Sep 25, 2001. Erowid.org/exp/9657

Dr. Robert. "I Was a Junkie: An Experience with Kratom." Erowid.org. Mar 15, 2006. Erowid.org/exp/49447

Efraim Ivener. "Light Withdrawals from Tea: An Experience with Kratom." Erowid.org. Jan 23, 2014. Erowid.org/exp/89029

Engrgamer. "Legitimate Alternative: An Experience with Kratom Powder." Erowid.org. Aug 28, 2009. Erowid.org/exp/71290

ExisT. "Legally Excellent: An Experience with Kratom." Erowid.org. Feb 3, 2014. Erowid.org/exp/41939

Fleming. "A Klondike Kratom Kristmas: An Experience with Kratom Capsules." Erowid.org. Feb 4, 2014. Erowid.org/exp/98719

Guttersnipe. "Unique, Rewarding Range of Experiences: An Experience with Kratom, Cannabis & Benzodiazepines." Erowid.org. Aug 23, 2004. Erowid.org/exp/33101

Hypersphere. "A Medicinal and Recreational Ally: An Experience with Kratom." Erowid.org. Feb 4, 2014. Erowid.org/exp/81158

Incarnadine. "Dream Fragments: An Experience with Kratom Leaves." Erowid.org. Mar 18, 2004. Erowid.org/exp/30055

Investigator. "The Secret Sedative: An Experience with Kratom Extract (15x)." Erowid.org. Feb 14, 2007. Erowid.org/exp/57960

J-to-the. "The Wicked Devil Tea: An Experience with Kratom." Erowid.org. Aug 23, 2004. Erowid.org/exp/36154

Jesper. "Instant Hangover Tea: An Experience with Kratom." Erowid.org. Jan 19, 2004. Erowid.org/exp/30154

Jojoba. "It's Good... Too Good: An Experience with Kratom." Erowid.org. Sep 28, 2006. Erowid.org/exp/55648

Jonas. "Easy and Consistent: An Experience with Kratom." Erowid.org. Jun 6, 2007. Erowid.org/exp/62727

Lahey. "Overdid It: An Experience with Kratom Extract (15x)." Erowid.org. Jan 23, 2014. Erowid.org/exp/56786

Limpet Chicken. "Blissful Numb Euphoria: An Experience with Kratom & Hash." Erowid.org. Feb 3, 2014. Erowid.org/exp/43574

Maturin. "Moderation and Manual Labor: An Experience with Kratom." Erowid.org. Jun 2, 2008. Erowid.org/exp/68954

MIM. "Kratom Preparation and Combinations: An Experience with Kratom." Erowid.org. Oct 20, 2005. Erowid.org/exp/47037

msFancyPants. "Zapped Away Methadone Withdrawals: An Experience with Kratom." Erowid.org. Aug 22, 2011. Erowid.org/exp/92099

NaturalHealing. "An Amazing Analgesic: An Experience with Kratom Leaves." Erowid.org. Jan 28, 2010. Erowid.org/exp/79133

nlogn. "Kratom-Induced Hepatitis?: An Experience with Kratom Leaves." Erowid.org. Mar 7, 2011. Erowid.org/exp/88678

PB. "Extreme Abdominal Pain: An Experience with Kratom Resin." Erowid.org. Mar 17, 2007. Erowid.org/exp/51161

Pike. "Getting Off on Native Krathom: An Experience with Kratom Leaves." Erowid.org. Mar 19, 2004. Erowid.org/exp/22387

Psychopharm. "Easy & Euphoric: An Experience with Kratom." Erowid.org. Aug 9, 2004. Erowid.org/exp/35770

Renwick. "Notes from a Kratom Experience: An Experience with Smoked Kratom (Mitragyna speciosa), Nitrous & GBL." Erowid.org. Mar 24, 2004. Erowid.org/exp/9843

rev. cosmo. "Very Pleasant, Much Like Opium: An Experience with Kratom & Cannabis." Erowid.org. Mar 24, 2004. Erowid.org/exp/32178

Reville. "Second Trial: An Experience with Mitragyna speciosa (Kratom)." Erowid.org. Oct 27, 2002. Erowid.org/exp/18482

seaborg. "Ideal for Manual Labor: An Experience with Kratom Extract (10x)." Erowid.org. Aug 9, 2007. Erowid.org/exp/63393

Seifuru. "Not Bad If You're into Opiates: An Experience with Kratom Powder." Erowid.org. Oct 5, 2004. Erowid.org/exp/36932

Sirk. "My New Favorite: An Experience with Kratom." Erowid.org. Mar 18, 2004. Erowid.org/exp/28819

Sly. "Kratom-Induced Hepatotoxicity: An Experience with Kratom Extract (15x)." Erowid.org. Aug 15, 2008. Erowid.org/exp/71949

Susan. "Quitting and Withdrawal: An Experience with Kratom." Erowid.org. Jun 27, 2007. Erowid.org/exp/49475

the good doctor. "Next Time a Controlled Weaning: An Experience with Kratom." Erowid.org. Jan 23, 2014. Erowid.org/exp/85769

Tomtom. "The Collector: An Experience with Fresh Kratom Leaves." Erowid.org. Jun 26, 2007. Erowid.org/exp/53894

TooOldForDrugs. "Really Nice at Low Doses: An Experience with Kratom." Erowid.org. Jan 23, 2014. Erowid.org/exp/97078

Transmigraine. "Detox in Three Stages: An Experience with Methadone, Kratom & Ibogaine." Erowid.org. Aug 18, 2010. Erowid.org/exp/86136

Wallcrawler. "Powerful Relaxation Profound Happiness: An Experience with Kratom Leaves." Erowid.org. Feb 3, 2014. Erowid.org/exp/100269

Xorkoth. "Subversive Herb—A Follow-up Report: An Experience with Kratom." Erowid.org. Oct 14, 2005. Erowid.org/exp/46082

Yolk. "An Unexpected Love Affair: An Experience with Kratom." Erowid.org. Jan 23, 2014. Erowid.org/exp/98929

420_Psych. "Warm Euphoric Goodness: An Experience with Kratom Powder." Erowid.org. Jan 23, 2014. Erowid.org/exp/46806

17 The Toxicology of Indole Alkaloids

Jennifer L. Ingram-Ross

CONTENTS

17.1 INTRODUCTION

Alkaloids are a structurally diverse group of substances that derive from amino acid precursors and contain a heterocyclic nitrogen atom. Well over 10,000 distinct alkaloids have been identified from various living organisms including plants, bacteria, fungi, and animals (Evans 2009); of these, approximately half originate from plants (Makkar et al. 2007). Alkaloids are often pharmacologically active and thus have been used as medications (e.g., morphine, codeine) or recreational drugs (e.g., cocaine, psilocybin). Additionally, alkaloids such as mescaline have been used in religious or spiritual rituals.

Alkaloids are classified based on the similarity of their structures. Among the numerous classes or derivatives of alkaloids are the indole alkaloids, which derive from tryptophan and contain a six-membered benzene ring fused to a five-membered nitrogen-containing pyrrole ring. More than 4,000 indole alkaloids have been identified, making it one of the largest classes of alkaloids (Seigler 1998).

Indole alkaloids have been used in traditional medicine for the treatment of numerous diseases and continue to be studied for medicinal applications; however, they also have been associated with various toxicities in both animals and humans. This chapter will provide a brief overview of the non-clinical and clinical toxicities related to several plant-derived indole alkaloids.

17.2 BRUCINE

Strychnos nux-vomica, commonly known as poison nut or Quaker buttons, is a tree that is native to India and Southeast Asia. It has been used extensively in Chinese medicine for the improvement of blood circulation and treatment of rheumatic pain and cancer (Yin et al. 2003; Chen et al. 2011). As many as sixteen alkaloids have been identified in *S. nux-vomica*, with strychnine and brucine comprising approximately 70% (Chen et al. 2013). Strychnine has long been known for its dramatic toxicity in humans (Makarovsky et al. 2008) and thus does not have any clinical applications. Brucine, however, has been studied for its analgesic and anti-inflammatory properties (Yin et al. 2003) as well as for the potential treatment of various types of cancer (Li et al. 2012; Ma et al. 2012; Qin et al. 2012; Luo et al. 2013).

Strychnine is well known for its neurotoxicity via antagonism of the glycine receptor (reviewed by Philippe et al. 2004). Conversely, there have been few reports of brucine-associated toxicity in animals or humans. A single oral, intravenous, or intraperitoneal administration of brucine to Swiss-Webster mice resulted in central nervous system depression followed by convulsions, seizures, and death (Malone et al. 1992). Brucine also was tested for its genotoxic potential and found to be non-mutagenic in the Ames assay (Cosmetic Ingredient Review Expert Panel 2008).

Naik and Chakrapania (2009) reported convulsions and muscle spasms in a male patient who, during a religious ritual, consumed a drink made from the bark of the *S. nux-vomica* tree. Five days after admission to the hospital, the patient experienced weakness of the lower limbs along with abdominal pain and vomiting; he was diagnosed with rhabdomyolysis and acute renal failure. Following dialysis, the patient fully recovered. A case of suicide by brucine ingestion was reported by Teske et al. (2011); brucine was detected in the blood, urine, bile, liver, brain, and gastric contents. Achappa et al. (2012) reported vomiting and tonic clonic convulsions in a male patient who consumed a drink made from *S. nux-vomica* tree bark. Upon hospitalization, the patient went into cardiac arrest and subsequently died. In this case, the presence of brucine and absence of strychnine in the drink was confirmed by HPTLC analysis.

17.3 HARMALA ALKALOIDS

17.3.1 Sources and Uses

Peganum harmala is a plant native to Mediterranean countries and introduced to the southwestern United States in 1928 (Davison and Wargo 2001). It is known by several names, among them Syrian rue, African rue, and harmal. *P. harmala* contains several β-carboline alkaloids, including harmaline, harmalol, harmol, harmine, and harmane. The harmala alkaloids have been measured in seed and root extracts at concentrations ranging from 1 to 10% (Kartal et al. 2003; Hemmateenejad et al. 2006; Herraiz et al. 2010), with no one compound consistently reported as the major alkaloid in the plant.

Historically, *P. harmala* has been used in the preparation of "Turkey red" dye (Davison and Wargo 2001), and extracts of *P. harmala* have been used in traditional African and Middle Eastern medicine for the treatment of numerous conditions, including fever, diarrhea, and pain (Farouk 2008). Harmala alkaloids and *P. harmala* extracts have been tested for the potential treatment of dermatological conditions (El-Rifaie 1980) and studied for their antitumor potential (Ishida et al. 1999; Chen et al. 2004; Song et al. 2004), antileishmanial activity (Di Giorgio et al. 2004), and antibacterial properties (Darabpour et al. 2011).

17.3.2 Nervous System Effects

Neurotoxicity is a common finding associated with harmala alkaloids. In C57/BL6 mice, a single intravenous administration of harmine and its derivatives resulted in acute toxicity manifesting as trembling, twitching, jumping, tetanus, supination, and death (Chen et al. 2004). A single intraperitoneal (IP) administration of harmaline to Swiss albino mice resulted in tremors within 3–4 minutes of dose administration (Mehta et al. 2003). Single or repeated IP administration of harmaline to Wistar or Sprague Dawley rats (Biary et al. 2000; Shourmasti et al. 2012; Stanford and Fowler 1998; Wilms et al. 1999) or a single subcutaneous administration of harmaline to New Zealand White rabbits (Du and Harvey 1997) also resulted in tremors within minutes. These findings were attributed to the modulation of neurotransmitters in the olivo-cerebellar system.

Effects on learning and memory in male Wistar rats were observed following a single IP administration of harmane (Celikyurt et al. 2013). Injections were administered 30 minutes prior to testing; harmane impaired working memory in a three-panel runway test and learning in a passive avoidance test. Possible mechanisms for these findings included effects on the cholinergic, dopaminergic, and/or serotonergic systems. Harmaline also was shown to affect learning when administered subcutaneously to New Zealand White rabbits (Du and Harvey 1997). This was demonstrated as a delay in the acquisition of a conditioned response (extension of the nictitating membrane in response to a tone conditioned stimulus), an effect that was ameliorated by administration of the NMDA channel blocker dizocilpine. This study suggested a relationship between harmaline and NMDA receptors in the inferior olive.

Hypothermia also was observed in Wistar rats following a single IP administration of harmaline, harmine, or total alkaloid extract (Abdel-Fattah et al. 1995); this finding was attributed to indirect stimulation of the serotonin 1A receptor via elevation of endogenous serotonin levels.

17.3.3 OTHER EFFECTS

Repeated (twice weekly for one month) subcutaneous injection of the alcoholic extract of *P. harmala* seeds to Balb/c mice resulted in histologic damage to the kidneys and liver (Mohamed et al. 2013). In the kidneys, observations included hemorrhage in the connective tissue along with degeneration and necrosis in the tubules. In the liver, findings included widening of sinusoids, congestion in central veins, hemorrhage, cellular degeneration, hypertrophy, necrosis, and fibrosis. Conversely, repeated (once daily for six weeks) intramuscular administration of the aqueous extract of *P. harmala* seeds to Wistar rats showed no adverse effects on the liver or kidney (Muhi-eldeen et al. 2008), although a severe local inflammatory response was observed at the injection sites.

The genotoxic potential of harmala alkaloids has been assessed using both in vitro and in vivo techniques (Picada et al. 1997). Of the five alkaloids tested, only harmalol did not demonstrate mutagenicity in the Ames assay. Harman and harmol were mutagenic in the TA97 strain of *Salmonella typhimurium*, while harmine was mutagenic in strains TA97 and TA98 (only in the presence of S9 microsomal enzyme mix), and harman, harmine, and harmaline were mutagenic in strain TA102. Both harman and harmine were negative in the in vivo micronucleus assay in Swiss Webster mice. However, harman and harmine were found to induce chromosomal aberrations and DNA damage in V79 Chinese hamster lung fibroblasts (Boeira et al. 2001).

There have been few reports of *P. harmala*–associated toxicity in humans. Mahmoudian et al. (2002) reported vomiting, abdominal pain, low blood pressure, convulsions, tremors of the limbs and facial muscles, and hallucinations in a male patient who ingested 150 g of *P. harmala* seeds while undergoing treatment for opium addiction; the patient recovered within hours of hospitalization. Yuruktumen et al. (2008) reported nausea, vomiting, hallucinations, diaphoresis, hypertension, tachycardia, tachypnea, and coma in a female patient who consumed 100 g of *P. harmala* seeds in a hot tea preparation to decrease anxiety; while hospitalized, she also exhibited abnormal kidney and liver function, as evidenced by elevated levels of urea, creatinine, aspartate aminotransferase (AST), and alanine aminotransferase (ALT). In this case, the patient recovered following 10 days of hospitalization, but her AST and ALT levels remained elevated upon discharge. These effects in humans were attributed to the modulation of neurotransmitters and reversible inhibition of monoamine oxidase by the harmala alkaloids.

17.4   RESERPINE

Rauvolfia serpentina is a plant native to Southeast Asia and commonly known as snakeroot or Indian snakeroot. In traditional Indian medicine, *R. serpentina* was used for the treatment of various ailments, including snake bites and mental illness (Dey and De 2011). Extracts of *R. serpentina* leaves also have been studied for the possible treatment of diarrhea (Ezeigbo et al. 2012).

   R. serpentina contains the indole alkaloid reserpine, which has long been used in the treatment of hypertension (Doyle et al. 1955). Reserpine is marketed under numerous brand names for the treatment of hypertension, schizophrenia, and psychosis (*Mosby's Drug Consult* 2006). Preclinical studies conducted in support of marketing authorization for reserpine demonstrated an increased incidence of mammary fibroadenomas in female mice, seminal vesicle tumors in male mice, and adrenal medullary tumors in male rats; in addition, reserpine was teratogenic in rats and embryocidal in guinea pigs (*Mosby's Drug Consult* 2006). Side effects in humans include depression, drowsiness, tiredness, and confusion (Aronson 2006).

   When administered as repeated subcutaneous injections to rats, reserpine caused a syndrome of oral dyskinesia that was marked by tongue protrusion, chewing movements, and facial twitching (Neisewander et al. 1991; Neisewander et al. 1994). These findings were attributed to a decrease in dopamine neurotransmission. The severity of the findings increased with age (Bergamo et al. 1997; Burger et al. 2004) and could be at least partially ameliorated by the administration of antioxidants (Burger et al. 2003, 2004; Pereira et al. 2011), suggesting that these effects also may have been due to oxidative stress caused by reserpine.

   Repeated (once daily for 48 days) intraperitoneal (IP) administration of reserpine to Sprague-Dawley rats resulted in liver damage that was characterized by elevations in serum transaminases and damage to cell membranes, organelles, and cytoplasm as noted by electron microscopy (Al-Bloushi et al. 2009). These findings were reversible following simultaneous or subsequent treatment with antioxidant green tea extract. In addition, Ma et al. (2010) demonstrated that a single oral or IP administration of reserpine to Sprague-Dawley rats resulted in lesions of the gastric mucosa characterized by edema, ulcers, and/or petechial points. These findings were partially inhibited by vagotomy and reversible following cessation of reserpine treatment. The effects of reserpine on gastric mucosa were attributed to the activation of vagal nerves and regulation of gastric acid secretion via inhibition of the sympathetic nervous system.

Reserpine was found to be non-genotoxic in the SOS chromotest assay and non-mutagenic in a diploid yeast assay (von Poser et al. 1990).

17.5   VINCA ALKALOIDS

Catharanthus roseus is a plant native to Madagascar but now widely cultivated; it is known by several names, including Madagascar periwinkle, rosy periwinkle, and vinca. Extracts of *C. roseus* have been used in traditional medicine for the treatment of diabetes (Singh et al. 2001) and more recently have been tested in animal models of the disease (Singh et al. 2001; Nammi et al. 2003; Chattopadhyay 1999). *C. roseus* contains several alkaloids, including vinblastine and vincristine, which are used for the treatment of various types of cancer (*Mosby's Drug Consult* 2006). Side effects in humans include nervous system disorders, cardiovascular and respiratory effects, hematological effects, gastrointestinal discomfort, and localized skin irritation (Aronson 2006).

When administered as a single intravenous (IV) injection to mice, vinblastine and vincristine were toxic to developing spermatogonia (Lu et al. 1979). Topical application of vinblastine or vincristine to rabbit skin resulted in erythema, scabs, and ulcers (Lobe et al. 1998). Repeated (2-week) IV administration of vincristine to Sprague-Dawley rats resulted in hyperalgesia (Aley et al. 1996) or allodynia (Nozaki-Taguchi et al. 2001), while 7-day repeat intraperitoneal administration to ICR mice also resulted in allodynia (Saika et al. 2009). In addition, IV administration of vincristine to Wistar rats resulted in decreased rates of gastric emptying, gastrointestinal motility, and fecal output (Peixoto Júnior et al. 2009). Preclinical studies conducted in support of marketing authorization for vincristine demonstrated clinical neurotoxicity (uncoordinated movements, weakness, reduced muscle tone, limited use of the limbs), peripheral nerve fiber degeneration, and skeletal muscle atrophy in rats administered repeated (once per week for 6 weeks) IV injections (Talon Therapeutics Inc. 2012).

The genotoxic potential of vinca alkaloids has been assessed using both in vitro and in vivo techniques. Vinblastine was positive in the in-vitro micronucleus test in mouse lymphoma cells (Collins et al. 2007) and human TK6 cells (Bryce et al. 2007), and both vinblastine and vincristine were positive in the in vivo micronucleus test in mouse peripheral blood and rat bone marrow (Cammerer et al. 2007).

17.6   CONCLUSIONS

Trees and shrubs rich in indole alkaloids can be found growing in many parts of the world. In some cultures, particularly those in countries to which these plants are native, the physiologically active substances that can be extracted from these plants have found their way into traditional medicine or religious rituals; however, these experiences sometimes come at a cost, as many indole alkaloids are known to be toxic. Cases of accidental overdose have been reported for compounds such as brucine and the harmala alkaloids. These practices have led researchers to investigate the effects of various alkaloids in both in vitro and in vivo systems, resulting in an understanding of the mechanism behind some of the observed toxicities. One finding common to the alkaloids discussed herein is neurotoxicity, which manifests in animals and humans as symptoms ranging from drowsiness to seizures and death. Compounds such as reserpine and the harmala alkaloids reportedly have adverse effects on the liver, while both harmala and vinca alkaloids are reported to have adverse effects on the skin as well as demonstrated genotoxicity. Despite this predilection for toxicity, indole alkaloids continue to be refined by researchers for use in modern medicine as the potential benefits of these compounds are studied in animal and cell-based models of disease.

REFERENCES

Abdel-Fattah, A.-F.M., Matsumoto, K., Gammaz, H.A.-K., and Watanabe, H. 1995. Hypothermic effect of harmala alkaloid in rats: Involvement of serotonergic mechanism. *Pharmacol Biochem Behav* 52: 421–26.

Achappa, B., Madi, D., Babu, Y.P.R., and Mahalingam, S. 2012. Rituals can kill: A fatal case of brucine poisoning. *Australasian Med J* 5: 421–23.

Al-Bloushi, S., Safer, A.-M., Afzal, M., and Mousa, S.A. 2009. Green tea modulates reserpine toxicity in animal models. *J Toxicol Sci* 34: 77–87.

Aley, K.O., Reichling, D.B., and Levine, J.D. 1996. Vincristine hyperalgesia in the rat: A model of painful vincristine neuropathy in humans. *Neuroscience* 73: 259–65.

Aronson, J. K., ed. 2006. *Meyler's Side Effects of Drugs.* 15th ed. Oxford: Elsevier.

Bergamo, M., Abílio, V. C., Queiroz, C.M.T., Barbosa-Júnior, H.N., Abdanur, L.R.A., and Frussa-Filho, R. 1997. Effects of age on a new animal model of tardive dyskinesia. *Neurobiol Aging* 18: 623–29.

Biary, N., Arshaduddin, M., Al Deeb, S., Al Moutaery, K., and Tariq, M. 2000. Effect of lidocaine on harmaline-induced tremors in the rat. *Pharmacol Biochem Behav* 65: 117–21.

Boeira, J.M., da Silva, J., Erdtmann, B., and Henriques, J.A.P. 2001. Genotoxic effects of the alkaloids harman and harmine assessed by Comet assay and chromosome aberration test in mammanlian cells in vitro. *Pharm Tox* 89: 287–94.

Bryce, S.M., Bemis, J.C., Avlasevich, S.L., and Dertinger, S.D. 2007. In vitro micronucleus assay scored by flow cytometry provides a comprehensive evaluation of cytogenetic damage and cytotoxicity. *Mutat Res* 630: 78–91.

Burger, M.E., Alves, A., Callegari, L., Athayde, F.R., Nogueira, C.W., Zeni, G., and Rocha, J.B.T. 2003. Ebselen attenuates reserpine-induced orofacial dyskinesia and oxidative stress in rat striatum. *Prog Neuro-Psychopharmacol Biol Psych* 27: 135–40.

Burger, M., Fachinetto, R., Calegari, L., Paixao, M. W., Braga, A. L., and Rocha, J.B.T. 2004. Effects of age on reserpine-induced orofacial dyskinesia and possible protection of diphenyl diselenide. *Brain Res Bull* 64: 339–45.

Cammerer, Z., Elhajouji, A., Kirsch-Volders, M., and Suter, W. 2007. Comparison of the peripheral blood micronucleus test using flow cytometry in the rat and mouse exposed to aneugens after single-dose applications. *Mutagenesis* 22: 129–34.

Celikyurt, I.K., Utkan, T., Gocmez, S.S., Hudson, A., and Aricioglu, F. 2013. Effect of harmane, an endogenous β-carboline, on learning and memory in rats. *Pharmacol Biochem Behav* 103: 666–71.

Chattopadhyay, R.R. 1999. A comparative evaluation of some blood sugar lowering agents of plant origin. *J Ethnopharmacol* 67: 367–72.

Chen, J., Xiao, H.-L., Hu, R.-R., Hu, W., Chen, Z.-P., Cai, H., Liu, X., Lu, T.-L., Fang, Y., and Cai, B.-C. 2011. Pharmacokinetics of brucine after intravenous and oral administration to rats. *Fitoterapia* 82: 1302–08.

Chen, J., Hu, W., Qu, Y.-Q., Dong, J., Gu, W., Gao, Y., Fang, Y., Fang, F., Chen, Z.-P., and Cai, B.-C. 2013. Evaluation of the pharmacodynamics and pharmacokinetics of brucine following transdermal administration. *Fitoterapia* 86: 193–201.

Chen, Q., Chao, R., Chen, H., Hou, X., Yan, H., Zhou, S., Peng, W., and Xu, A. 2004. Antitumor and neurotoxic effects of novel harmine derivatives and structure-activity relationship analysis. *Int J Cancer* 114: 675–82.

Collins, J.E., Ellis, P.C., White, A.T., Booth, A.E.G., Moore, C.E., Burman, M., Rees, R.W., and Lynch, A.M. 2008. Evaluation of the Litron in vitro MicroFlow® Kit for the flow cytometric enumeration of micronuclei (MN) in mammalian cells. *Mutat Res* 654: 76–81.

Cosmetic Ingredient Review Expert Panel. 2008. Final report of the safety assessment of alcohol denat., including SD alcohol 3-A, SD alcohol 30, SD alcohol 39, SD alcohol 39-B, SD alcohol 39-C, SD alcohol 40, SD alcohol 40-B, and SD alcohol 40-C, and the denaturants, quassin, brucine sulfate/brucine, and denatonium benzoate. *Int J Toxicol* 27: 1–43.

Darabpour, E., Bavi, A.P., Motamedi, H., and Nejad, S.M.S. 2011. Antibacterial activity of different parts of *Peganum harmala* L. growing in Iran against multi-drug resistant bacteria. *EXCLI J* 10: 252–63.

Davison, J. and Wargo, M. 2001. Recognition and control of African rue in Nevada. Fact Sheet FS-01-45. University of Nevada Cooperative Extension, Reno, Nevada.

Dey, A. and De, J.N. 2011. Ethnobotanical aspects of *Rauvolfia serpentina* (L). Benth. ex Kurz. in India, Nepal and Bangladesh. *J Med Plants Res* 5: 144–50.

Di Giorgio, C., Delmas, F., Ollivier, E., Elias, R., Balansard, G., and Timon-David, P. 2004. In vitro activity of the β-carboline alkaloids harmane, harmine, and harmaline toward parasites of the species *Leishmania infantum*. *Exp Parasitol* 106: 67–74.

Doyle, A.E., McQueen, E.G., and Smirk, F.H. 1955. Treatment of hypertension with reserpine, with reserpine in combination with pentapyrrolidinium, and with reserpine in combination with veratrum alkaloids. *Circulation* 11: 170–81.

Du, W. and Harvey, J.A. 1997. Harmaline-induced tremor and impairment of learning are both blocked by dizocilpine in the rabbit. *Brain Res* 745: 183–88.

El-Rifaie, M.E.-S. 1980. *Peganum harmala*: Its use in certain dermatoses. *Int J Dermatol* 19: 221–22.

Evans, W.C. 2009. *Trease and Evans Pharmacognosy*. 16th ed. Edinburgh: Saunders/Elsevier.

Ezeigbo, I.I., Ezeja, M.I., Madubuike, K.G., Ifenkwe, D.C., Ukweni, I.A., Udeh, N.E., and Akomas, S.C. 2012. Antidiarrhoeal activity of leaf methanolic extract of *Rauwolfia serpentina*. *Asian Pac J Trop Biomed* 2: 430–32.

Farouk, L., Laroubi, A., Aboufatima, R., Benharref, A., and Chait, A. 2008. Evaluation of the analgesic effect of alkaloid extract of *Peganum harmala* L.: Possible mechanisms involved. *J Ethnopharmacol* 115: 449–54.

Hemmateenejad, B., Abbaspour, A., Maghami, H., Miri, R., and Panjehshahin, M.R. 2006. Partial least squares-based multivariate spectral calibration method for simultaneous determination of beta-carboline derivatives in *Peganum harmala* seed extracts. *Anal Chim Acta* 575: 290–99.

Herraiz, T., González, D., Ancín-Azpilicueta, C., Arán, V.J., and Guillén, H. 2010. β-Carboline alkaloids in *Peganum harmala* and inhibition of human monoamine oxidase (MAO). *Food Chem Toxicol* 48: 839–45.

Ishida, J., Wang, H.-K., Bastow, K. F., Hu, C.-Q., and Lee, K.-H. 1999. Antitumor agents 201: Cytotoxicity of harmine and β-carboline analogs. *Bioorg Med Chem Lett* 9: 3319–24.

Kartal, M., Altun, M.L., and Kurucu, S. 2003. HPLC method for the analysis of harmol, harmalol, harmine and harmaline in the seeds of *Peganum harmala* L. *J Pharm Biomed Anal* 31: 263–69.

Li, P., Zhang, M., Ma, W.-J., Sun, X., and Jin, F.-P. 2012. Effects of brucine on vascular endothelial growth factor expression and microvessel density in a nude mouse model of bone metastasis due to breast cancer. *Chin J Integr Med* 18: 605–09.

Lobe, D.C., Kreider, J.W., and Phelps, W.C. 1998. Therapeutic evaluation of compounds in the SCID-RA papillomavirus model. *Antiviral Res* 40: 57–71.

Lu, C.C. and Meistrich, M.L. 1979. Cytotoxic effects of chemotherapeutic drugs on mouse testis cells. *Cancer Res* 39: 3575–82.

Luo, W., Wang, X., Zheng, L., Zhan, Y., Zhang, D., Zhang, J., and Zhang, Y. 2013. Brucine suppresses colon cancer cells growth via mediating KDR signalling pathway. *J Cell Mol Med* 17: 1316–24.

Ma, X.-J., Lu, G.-C., Song, S.-W., Liu, W., Wen, Z.-P., Zheng, X., Lü, Q.-Z., and Su, D.-F. 2010. The features of reserpine-induced gastric mucosal lesions. *Acta Pharmacol Sin* 31: 938–43.

Ma, Y., Zhao, J., Wang, Y., Li., Z., Feng, J., and Ren, H. 2012. Effects of brucine on bone metabolism in multiple myeloma. *Mol Med Rep* 6: 367–70.

Mahmoudian, M., Jalilpour, H., and Salehian, P. 2002. Toxicity of *Peganum harmala*: Review and a case report. *Iran J Pharmacol Ther* 1: 1–4.

Makarovsky, I., Markel, G., Hoffman, A., Schein, O., Brosh-Nissimov, T., Tashma, Z., Dushnitsky, T., and Eisenkraft, A. 2008. Strychnine: A killer from the past. *Israel Med Assoc J* 10: 142–45.

Makkar, H. P. S., Siddhuraju, P., and Becker, K. 2007. Plant secondary metabolites. In *Methods in Molecular Biology*, ed. J.M. Walker. Totowa, New Jersey: Humana Press Inc.

Malone, M.H., St. John-Allan, K.M., and Bejar, E. 1992. Brucine lethality in mice. *J Ethnopharmacol* 35: 295–97.

Mehta, H., Saravanan, K.S., and Mohanakumar, K.P. 2003. Serotonin synthesis inhibition in olivo-cerebellar system attenuates harmaline-induced tremor in Swiss albino mice. *Behav Brain Res* 145: 31–36.

Mohamed, A.H.S., Al-Jammali, S.M.J., and Naki, Z.J. 2013. Effect of repeated administration of *Peganum harmala* alcoholic extract on the liver and kidney in albino mice: A histo-pathological study. *J Sci Innov Res* 2: 585–97.

Mosby. 2005. *Mosby's Drug Consult 2006*. 16th ed. Mosby/Elsevier.

Muhi-eldeen, Z., Al-Shamma, K. J., Al-Hussainy, T. M., Al-Kaissi, E. N., Al-Daraji, A. M., and Ibrahim, H. 2008. Acute toxicological studies on the extract of Iraqi *Peganum harmala* in rats. *Eur J Sci Res* 22: 494–500.

Naik, B. S. and Chakrapani, M. 2009. A rare case of brucine poisoning complicated by rhabdomyolysis and acute renal failure. *Malays J Pathol* 31: 67–69.

Nammi, S., Boini, M.K., Lodagala, S.D., and Behara, R.B.S. 2003. The juice of fresh leaves of *Catharanthus roseus* Linn. reduces blood glucose in normal and alloxan diabetic rabbits. *BMC Complement Altern Med* 3: 4.

Neisewander, J.L., Castaneda, E., and Davis, D.A. 1994. Dose-dependent differences in the development of reserpine-induced oral dyskinesia in rats: Support for a model of tardive dyskinesia. *Psychopharmacology* 116: 79–84.

Neisewander, J.L., Lucki, I., and McGonigle, P. 1991. Behavioral and neurochemical effects of chronic administration of reserpine and SKF-38393 in rats. *J Pharmacol Exp Ther* 257: 850–60.

Nozaki-Taguchi, N., Chaplan, S.R., Higuera, E.S., Ajakwe, R.C., and Yaksh, T.L. 2001. Vincristine-induced allodynia in the rat. *Pain* 93: 69–76.

Pereira, R.P., Fachinetto, R., Prestes, A. de S., Wagner, C., Sudati, J.H., Boligon, A.A., Athayde, M.L., Morsch, V.M., and Rocha, J.B.T. 2011. *Valeriana officinalis* ameliorates vacuous chewing movements induced by reserpine in rats. *J Neural Transm* 118: 1547–57.

Peixoto Júnior, A.A., Teles, B.C.V., Castro, E.F.B., Santos, A.A., de Oliveira, G.R., Ribeiro, R.A., Rola, F.H., and Gondim, F.A.A. 2009. Vincristine delays gastric emptying and gastrointestinal transit of liquid in awake rats. *Brazilian J Med Biol Res* 42: 567–73.

Philippe, G., Angenot, L., Tits, M., and Frédérich, M. 2004. About the toxicity of some *Strychnos* species and their alkaloids. *Toxicon* 44: 405–16.

Picada, J.N., da Silva, K.V.C.L., Erdtmann, B., Henriques, A.T., and Henriques, J.A.P. 1997. Genotoxic effects of structurally related β-carboline alkaloids. *Mutat Res* 379: 135–49.

Qin, J.-M., Yin, P.-H., Li, Q., Sa, Z.-Q., Sheng, X., Yang, L., Huang, T., Zhang, M., Gao, K.-P., Chen, Q.-H., Ma, J.-W., and Shen, H.-B. 2012. Anti-tumor effects of brucine immuno-nanoparticles on hepatocellular carcinoma. *Int J Nanomedicine* 7: 369–79.

Saika, F., Kiguchi, N., Kobayashi, Y., Fukazawa, Y., Maeda, T., Ozaki, M., and Kishioka, S. 2009. Suppressive effect of imipramine on vincristine-induced mechanical allodynia in mice. *Biol Pharm Bull* 32: 1231–34.

Seigler, D.S. 1998. *Plant Secondary Metabolism*. Norwell, MA: Kluwer Academic.

Shourmasti, F.R., Goudarzi, I., Lashkarbolouki, T., Abrari, K., Salmani, M. E., and Goudarzi, A. 2012. Effects of riluzole on harmaline induced tremor and ataxia in rats: Biochemical, histological and behavioral studies. *Eur J Pharmacol* 695: 40–47.

Singh, S.N., Vats, P., Suri, S., Shyam, R., Kumira, M.M.L., Ranganathan, S., and Sridharan, K. 2001. Effect of an antidiabetic extract of *Catharanthus roseus* on enzymic activities in streptozotocin induced diabetic rats. *J Ethnopharmacol* 76: 269–77.

Song, Y., Kesuma, D., Wang, J., Deng, Y., Duan, J., Wang, J.H., and Qi, R.Z. 2004. Specific inhibition of cyclin-dependent kinases and cell proliferation by harmine. *Biochem Biophys Res Commun* 317: 128–32.

Stanford, J.A. and Fowler, S.C. 1998. At low doses, harmaline increases forelimb tremor in the rat. *Neurosci Lett* 241: 41–44.

Talon Therapeutics Inc. 2012. Marqibo [package insert]. South San Francisco, CA: Talon Therapeutics Inc.

Teske, J., Weller, J.-P., Albrecht, U.-V., and Fieguth, A. 2011. Fatal intoxication due to brucine. *J Anal Toxicol* 35: 248–53.

von Poser, G., Andrade, H.H.R., da Silva, K.V.C.L., Henriques, A.T., and Henriques, J.A.P. 1990. Genotoxic, mutagenic and recombinogenic effects of rauwolfia alkaloids. *Mut Res* 232: 37–43.

Wilms, H., Sievers, J., and Deuschl, G. 1999. Animal models of tremor. *Mov Disord* 14: 557–71.

Yin, W., Wang, T.-S., Yin, F.-Z., and Cai, B.-C. 2003. Analgesic and anti-inflammatory properties of brucine and brucine N-oxide extracted from seeds of *Strychnos nux-vomica*. *J Ethnopharmacol* 88: 205–14.

Yuruktumen, A., Karaduman, S., Bengi, F., and Fowler, J. 2008. Syrian rue tea: A recipe for disaster. *Clin Toxicol* 46: 749–52.

18 Opioid-Induced Adverse Effects and Toxicity

Joseph V. Pergolizzi, Jr.

CONTENTS

18.1 INTRODUCTION

Opioids include a class of drugs in multiple formulations and products that interact with the mu, kappa, and/or delta receptors. These receptors are located in the brain and spinal cord (central nervous system) and in the peripheral nervous system, allowing them to have concurrent central and peripheral effects. Opioid toxicities can be complex, in that the mix of receptors and sub-type receptors activated by a particular opioid is unique to that molecule; the opioid-receptor profiles and genetic expression of their metabolizing enzymes in the patient is unique; and, finally, their interplay is unique. While broad statements can be made about specific opioids, routes of administration, or patient characteristics, clinicians must be mindful that two similar patients may react quite differently to the same opioid analgesic agent.

Pure mu-opioid-receptor agonists do not have ceiling doses in the conventional sense; their "ceiling" is defined mainly by their toxicity (Vella-Brincat and Macleod 2007). As a general rule, opioid toxicity is dose related (Ripamonti et al. 1997). Partial mu-opioid-receptor agonists and kappa-opioid-receptor agonists do possess a dose ceiling (Vella-Brincat and Macleod 2007) and may also have treatment-limiting toxicities.

Few studies or guidelines aid the clinician in determining the degree of opioid exposure in terms of dose, duration, and specific opioid product associated with toxicity (Rhodin et al. 2010). There is a paucity of head-to-head comparisons of specific opioid products in the settings of chronic cancer and non-cancer pain. Thus, we know that opioid-associated adverse events are prevalent, but have few reliable tools for risk stratification to guide prophylaxis.

Although opioid-associated adverse events are prevalent, they are by and large manageable. These side effects can be uncomfortable, unpleasant, and even distressing to patients, but they can be controlled with the right care. Although neuronal apoptosis has been observed with opioid use, opioid therapy is not associated with irreversible organ damage—even over the long term and at high doses (Vella-Brincat and Macleod 2007).

This chapter groups opioid-associated side effects under broad headings and recognizes that this is an imperfect system. Some topics, such as sleep, may fit under more than one heading or could even be a heading unto itself.

18.2 BOWEL DYSFUNCTION

Opioid-induced bowel dysfunction (OIBD) can manifest as a wide variety of effects, including nausea, vomiting, gastro-esophageal reflux, dysphagia, dry mouth, bloating, abdominal pain, and constipation. There is great inter-patient variability in OIBD and other opioid-associated side effects, but there are specific risk factors for OIBD, namely advanced age, female gender, inadequate hydration or nutrition, and limited mobility. Symptoms associated with OIBD may also be caused, in whole or part, by the patient's underlying disease, comorbidities, or drug therapies. For example, in cancer patients, the stage of cancer or chemotherapies may confer added risk for nausea, vomiting, or bowel disorders (Wirz et al. 2009).

Much like the central nervous system, the gut's enteric nervous system (ENS) is composed of a vast complex of billions of intercommunicating neurons and interneurons. The ENS relays information from sensory neurons to motor and effector systems in a network regulated by two plexi: the myenteric plexus and the submucosal plexus. Motor neurons coordinate their functions mainly from input from local sensory neurons, but also receive inputs from the central nervous system (both sympathetic and parasympathetic pathways) (Brock et al. 2012). Redundant visceral nerve structures create convoluted pathways within this system that are not clearly understood. Broadly, the myenteric plexus regulates motor activity of the gut, while the submucosal plexus directs secretion and absorption. Specialized electrical interstitial cells generate neuromuscular "slow wave" activity, regulating smooth muscle action potentials. Gut motility is regulated by the myenteric plexus mainly through the neurotransmitters acetylcholine (AC), serotonin (5-HT), vasoactive intestinal peptide (VIP), and nitric oxide (NO). Since opioids inhibit the release of these neurotransmitters, opioids can interfere with gut motility (Brock et al. 2012). Thus, although both sites are involved, it is the effects of opioids on the peripheral nervous system, not the central nervous system, that usually predominate in OIBD (Wang and Yuan 2013).

The ENS has six sphincters, and opioids may adversely affect sphincter function and tone (Brock et al. 2012). For instance, an opioid-induced sphincter of Oddi disorder can result in pancreato-biliary pain (Sharma 2002).

Following is a brief discussion of some of the main manifestations of OIBD.

18.2.1 CONSTIPATION

The true prevalence of opioid-induced constipation is difficult to establish with any certainty because constipation may occur as a combination of any number of underlying diseases, comorbidities, medications, and lifestyle factors. The overall prevalence of constipation in the Western world has been estimated to be around 15% (Lembo and Camilleri 2003). Opioid-induced constipation rates cited in the literature range from 15% (Moore and McQuay 2005) and 34% (McNicol et al. 2013) at the low end to 85% (Abramowitz et al. 2013) or 95% (Labianca et al. 2012) at the high end. Opioid-induced constipation may be more likely with certain agents or routes of administration than others; for example, oral morphine is more highly associated with constipation than is transdermal fentanyl (Hadley et al. 2013). While tolerance sometimes develops with respect to certain opioid side effects, this is not true of constipation: its prevalence increases with duration of opioid therapy (Tuteja et al. 2010).

The distress constipation can cause patients can be severe enough to discontinue opioid therapy. Opioid-induced constipation has also been associated with increased utilization of healthcare resources and higher costs (Candrilli et al. 2009). While there are strategies to mitigate this side effect, they are not effective in all patients (Sharma and Jamal 2013).

18.2.2 NAUSEA AND VOMITING

For many opioid patients—but not all—nausea and vomiting are transient symptoms that resolve in a few days or weeks (Coluzzi and Pappagallo 2005). Opioid-induced nausea and vomiting (OINV) may be due to enhanced vestibular sensitivity, direct effects on the chemoreceptor trigger zone, and delayed gastric emptying (Porreca and Ossipov 2009). For example, low doses of opioids can activate the mu-opioid receptors in the body's chemoreceptor trigger zone, which regulates vomiting (Smith and Laufer 2014). The chemoreceptor trigger zone is located in the brain's fourth ventricle and is considered a peripheral rather than central site, in that it lacks a complete blood-brain barrier (Smith and Laufer 2014). The neurotransmitters thought to be associated with OINV are serotonin and dopamine. OINV appears to be closely related to chemotherapy-induced nausea and vomiting and postoperative nausea and vomiting.

18.2.3 DYSPHAGIA

Opioids may induce non-peristaltic contraction and incomplete relaxation of the lower esophageal sphincter, which may manifest as dysphagia

(Kraichely et al. 2010; Brock et al. 2012). In a study of eight healthy subjects, morphine reduced reflux by increasing residual lower esophageal sphincter (LES) pressure during transient LES relaxation (Penagini and Bianchi 1997). Impaired LES relaxation may result in dysphagia, suggesting that the esophagus is susceptible to opioid-induced effects (Kraichely et al. 2010).

18.2.4 XEROSTOMIA

Long-term use of opioids may suppress saliva production and result in dry mouth (Moore and McQuay 2005), which in turn may result in discomfort, caries, and periodontal disease. In a survey of 56 cancer pain patients treated with morphine, 95% reported dry mouth, with the investigators unable to establish a dose-response pattern (Glare et al. 2006). Xerostomia might also be associated with dysphagia, problems chewing, and difficulty speaking. Xerostomia may preclude the use of certain drugs, such as oral transmucosal fentanyl and other transmucosal sprays (Davies and Vriens 2005).

18.3 ENDOCRINOPATHY

The long-term use of opioids has been associated with changes in the endocrine system (Rhodin et al. 2010), which are likely to be under-reported in the literature (Brennan 2013) and therefore not fully addressed by guidelines. Opioids affect the hypothalamic–pituitary–gonadal axis (Brennan 2013), which regulates gonadotropin-releasing hormone (GnRH), which in turn targets the pituitary gland and stimulates it to produce luteinizing hormone (LH) and follicle-stimulating hormone (FSH), both of which enter the systemic circulation to stimulate reproductive organs (Katz and Mazer 2009). Preclinical reports show that opioids inhibit this entire system by binding to opioid receptors in the hypothalamus and, in that way, reduce the secretion of GnRH (Katz and Mazer 2009). Opioids may also bind at other points in the axis, for example in the pituitary (this limits LH product, which can result in amenorrhea in women) (Colameco and Coren 2009). In a study of 47 women, aged 30–75 years, taking sustained-action oral or transdermal opioids for managing chronic non-cancer pain matched with 68 controls taking no opioids, LH and FSH values were 30% and 70% lower in opioid-consuming premenopausal and postmenopausal women than controls, respectively (Daniell 2008).

Opioid therapy can also affect the hypothalamic–pituitary–adrenal axis (Brennan 2013). The hypothalamus releases corticotropin-releasing hormone (CRH), targeting the pituitary gland, which releases adrenocorticotropic hormone (ACTH) into the systemic circulation; this induces the adrenals to produce both cortisol and dehydroepiandrosterone (DHEA) (Colameco and Coren 2009). DHEA is a precursor to testosterone and estradiol in men and women, respectively (Katz and Mazer 2009). Opioids inhibit the production of CRH and may also reduce the pituitary gland's ability to respond to CRH (Palm et al. 1997); opioids are also thought to interfere with cortisol and DHEA production, independently from central nervous system downregulation (Daniell 2006).

18.3.1 HYPOGONADISM

A hypogonadism in males (HIM) study found a relationship between hypogonadism and multiple comorbid conditions (n = 2,162), with an overall prevalence of hypogonadism in men ≥ 45 years of 38.7% and higher incidences in specific populations, such as men with diabetes (50.0%), obese men (52.4%), hypertensives (42.4%), and those with asthma or other pulmonary disorders (43.5%) (Mulligan et al. 2006). A retrospective cohort study (n = 81) of men suffering from chronic pain and being treated with chronic opioid therapy found 53% of all men taking opioids daily had hypogonadism (defined as morning total serum testosterone < 250 ng/dL) and that those taking long-acting opioids were more likely to suffer hypogonadism than those taking short-acting opioids (74% vs. 34%, respectively) (Rubinstein et al. 2013).

18.3.2 BONE MINERAL DENSITY

The use of opioids increases the risk of bone fracture (Shorr et al. 1991; Kim et al. 2006; Vestergaard et al. 2006; Miller et al. 2011), but it has been unclear whether this is due to falls caused by opioid-induced dizziness or to opioid-induced hypogonadism leading to osteoporosis. In a study of 20 consecutive male patients treated with intrathecal opioids for managing chronic pain, osteoporosis (defined as a T-score ≤ 2.5 SD) occurred in 21.4% and osteopenia (T-score between −1.0 and −2.5 SD) occurred in 50% (Duarte et al. 2013). That is, 35.7% of the patients in this study were at or above the intervention threshold for hip fracture. This particular study suggested a relationship between hypogonadism and bone mineral density, in that 85% of patients had biochemical hypogonadism (Duarte et al. 2013).

18.3.3 DIABETES

It has been speculated that opioid-induced endocrinopathy might elevate a patient's risk for type 2 diabetes. However, a case-control study of 1.7 million patients taking opioids for non-cancer pain matched type 2 diabetic patients (n = 50,468) with non-diabetic controls taking no opioids (n = 100,415) and found that there was no increased risk of type 2 diabetes among those on long-term opioid therapy for chronic pain (Li et al. 2013). The findings were similar with different opioid types, cumulative opioid use, and timing of opioid dosing.

18.4 RESPIRATORY EFFECTS

Normal respiration involves the interaction between the body's central respiratory pacemakers and chemoreceptors and mechanoreceptors in the central and peripheral nervous systems. When the central respiratory pacemaker is interrupted, irregular breathing patterns can result; humans are most susceptible to such disordered breathing during sleep. Central and peripheral chemoreceptors rely on reflex and tonic inputs to detect changes in carbon dioxide and oxygen ratios and can adjust tidal volume and respiratory rate (Yue and Guilleminault 2010). Peripheral chemore-ceptors respond to arterial oxygen changes and signal the central pattern-generators to regulate breathing. Of course, the patterns, rhythms, and mechanics of breathing change with posture, activity, disease, and sleep cycles; for these reasons, rapid-eye-movement (REM) sleep may be accom-panied by irregular respiration and pauses (Yue and Guilleminault 2010).

Central respiratory pacemakers have been identified in animals in the pontomedullary reticular formation and contain neurons that create rhyth-mic activity that persists even when the downstream activity of these neurons is interrupted chemically (Yue and Guilleminault 2010). The retrotrapezoid/ parafacial respiratory nucleus (in the ventrolateral medulla oblongata) is thought to direct the rhythms of respiration. It is thought that the former sys-tem controls inspiration and the latter expiration, but there are also neuronal systems involved in the post-inhalation phase of the respiratory cycle.

Opioids decrease central respiratory pattern generation and, in that way, may reduce both respiratory rate and tidal volume (Yue and Guilleminault 2010). At lower doses, opioids reduce tidal volume in a dose-proportional way and thus depress respiration. At higher doses, further changes in decreased respiratory rate and impaired respiratory rhythm generation occur. The fact that high doses of opioids virtually eliminate the normal respiratory response to hypoxia suggests that opioids affect the hypoxemic respiratory drive via the peripheral chemoreceptors (Yue and Guilleminault 2010).

Resistance in the airways leads to a concomitant increase in respiratory effort, but opioids have been shown to decrease this compensatory increase in effort. Opioids are associated with a tendency for obstruction of the upper airway level, rigidity of accessory muscles of respiration (intercostal muscles and abdominal muscles), and decreased muscle activity during sleep (Yue and Guilleminault 2010).

Respiratory depression associated with opioid use is thought to diminish with opioid tolerance, at least during wakeful states. However, sleep-disordered breathing has been observed in chronic opioid therapy patients, and central sleep apnea may occur in as many as 30% of patients (Teichtahl and Wang 2007). Awake respiration is affected by mu-opioid receptors, and opioids that stimulate these receptors have been associated with dose-dependent respiratory depression (Teichtahl and Wang 2007). In the awake patient, acute opioid use may result in increased respiratory pauses, delayed expiration, and irregular breathing patterns.

The respiratory adverse events associated with opioids are potentially life threatening. Even among patients using patient-controlled morphine for control of postsurgical pain, abnormal breathing patterns were very common and 50% exhibited at least partial obstruction of breathing (Drummond et al. 2013).

18.4.1 ATAXIA

Ataxic breathing or respiration with irregular pauses and gasping has long been observed in neurological conditions (Biot 1876). Patients on long-term opioid therapy may exhibit the same erratic variability in respiratory rate and effort of breath associated with disordered breathing (Yue and Guilleminault 2010). In a retrospective cohort study (n = 60), ataxic breathing during non-REM sleep occurred significantly more often in opioid patients than controls (70.0% vs. 5.0%, p < 0.001), and the results were dose proportional (92% of those taking morphine equivalents of ≥ 200 mg morphine equivalents daily versus 61% in those taking less) (Walker and Zacny 2002).

18.4.2 CENTRAL SLEEP APNEA

Central sleep apnea has been consistently associated with long-term opioid use (Wang et al. 2005; Alattar and Scharf 2009). The mechanisms involved are not clear, but a study in methadone patients suggests a strong dose-response effect (Wang et al. 2005).

18.5 NEUROINHIBITION AND EXCITATION

Neuroinhibitory and excitatory effects include those that affect low-level consciousness (somnolence, fatigue, disordered sleep), the thinking process (cognitive impairment, agitation, hallucinations, delirium), and those that affect neurons themselves (hyperalgesia, tolerance, and possibly myoclonus). In addition, opioids may also have neurotoxic effects.

Neuroexcitatory effects associated with opioids are less common than some of the other opioid adverse events, but they can occur and may be particularly troublesome in certain patients. In a retrospective review of 54 hospice patients with chronic kidney disease who received hydromorphone therapy for pain control, 20% exhibited tremor, 20% myoclonus, 48% agitation, and 39% cognitive dysfunction. No patient in this study had seizures (Paramanandam et al. 2011). Neuroexcitatory adverse events likely relate to accumulation of opioid metabolites in the system, which elevates the risk of these effects for patients on long-term opioid therapy and those with renal dysfunction (Paramanandam et al. 2011). Neuroexcitatory effects do not occur with all opioids; for example, their incidence is lower with methadone than hydromorphone, since methadone has no active metabolites. However, there are case studies in the literature that report myoclonus in palliative patients taking high doses of parenteral methadone for cancer pain control (Ito and Liao 2008).

18.5.1 SOMNOLENCE

Somnolence is the state of being near sleep, while fatigue, discussed below, refers to the more global condition of feeling tired, weak, and exhausted. Thus, a patient may have fatigue, but not be somnolent. However, in clinical practice, patients and caregivers may blur the distinctions between somnolence and fatigue.

Somnolence is more likely to occur at the initiation of therapy or with dose escalations, as tolerance to the sedative effects of opioids may diminish such effects over time (McNicol et al. 2003). The anticholinergic effects of opioids have been implicated in causing drowsiness and somnolence (Sloan et al. 2005). These effects may relate to the effects opioids can have on non-opioid receptors (Vella-Brincat and Macleod 2007). There is variation among opioids in these effects; for instance, morphine is thought to be more likely than fentanyl to cause somnolence (Ahmedzai 1997).

18.5.2 FATIGUE

Fatigue is a vague, diffuse, and highly subjective symptom that may relate to the underlying disease process, comorbid conditions, medication regimen, or lifestyle. A hallmark of centrally mediated painful syndromes, such as chronic non-malignant pain, includes hyperalgesia and fatigue (Phillips and Clauw 2011). Furthermore, fatigue may be the result of another adverse event, such as sleep apnea or disordered breathing. Opioid-induced fatigue is the result of central nervous system inhibition caused by stimulation of opioid receptors (Zlott and Byrne 2010).

18.5.3 DISORDERED SLEEP ARCHITECTURE

The sleep–wake cycle is regulated by the interplay of various neurotransmitters, including noradrenaline, serotonin, acetylcholine, dopamine, histamine, gamma-aminobutyric acid, and melatonin (Bourne and Mills 2004). Many things can alter this balance, including opioid analgesics and disease. Thus it is not always clear whether sleep disturbances in a chronic pain patient are a result of the underlying disease or the opioid therapy.

Recent studies in animals and humans have shown that acute administration of opioids may decrease total sleep, and decrease REM sleep while increasing stage 2 sleep (Shaw et al. 2005). A study of short-term opioid use to control postsurgical pain was associated with an absence of REM sleep in patients followed by a rebound later (Knill et al. 1990). Opioids have also been associated with increased periods of wakefulness during sleep time (Staedt et al. 1996), although it is sometimes observed anecdotally among clinicians that opioids improve sleep. Few high-quality clinical studies have been carried out in this area, and further research is needed to better understand the effect of opioids on sleep.

18.5.4 COGNITIVE IMPAIRMENT

Cognitive impairment is a continuum of symptoms spanning forgetfulness, decreased attention spans, and foggy thinking at one extreme and disorientation, restlessness, and confusion about time at the other. Agitation, hallucinations, and delirium may be considered forms of cognitive dysfunction. Cognitive impairment may occur for one or several reasons, including the underlying disease processes, comorbid conditions, dementia, other mental health disorders, and drug therapy. Impaired cognition is a frequent byproduct of serious illness or impending death. It has been estimated that cognitive impairment occurs in up to 90% of all patients near death, even those not taking opioids (Lawlor 2002).

Studies in healthy volunteers have reported the paradoxical finding that oral morphine improves cognition and reaction times (Lawlor 2002; Ersek et al. 2004) and that, overall, oral opioids do not have a profound impact on cognition. Meperidine was associated with the most severe forms of cognitive dysfunction (Lawlor 2002). There is some evidence that impairment is more likely during opioid initiation or dose escalation and less likely once a stable dose regimen is achieved (Clark and Lynch 2003; Ersek et al. 2004).

18.5.5 Agitation

Agitation has been associated with high doses and longer-term therapy. In a study of 48 palliative care patients treated with parenteral hydromorphone, agitation was significantly associated with maximum dose of the drug and duration of treatment (Thwaites et al. 2004). The metabolite hydromorphone-3-glucoronide has been theorized as being involved in such neuroexcitatory symptoms (Thwaites et al. 2004). In a study of renally impaired palliative patients administered hydromorphone, the rate of agitation was 48% in a dose-dependent and duration-dependent fashion (Paramanandam et al. 2011).

18.5.6 Hallucinations

Cholinergic projections originating in the basal forebrain and extending into the thalamus and cortex appear to play a key role in the maintenance of consciousness and conscious awareness (Slatkin and Rhiner 2004). Opioids can inhibit cholinergic pathways, thus disrupting signals to the basal forebrain, and/or induce an acetylcholine deficit. The result can be cognitive dysfunction, delirium, or hallucinations. For reasons that are not entirely clear, hallucinations occur more frequently in the postoperative than the long-term care setting (Wheeler et al. 2002). Possibly related to opioid-induced hallucinations are nightmares and vivid dreams, but such events may relate more to the stress of the illness (Dunn and Milch 2002) than to opioid effects.

18.5.7 Delirium

Delirium is a paroxysmal disturbance in consciousness and cognition, which may be related to any number of causes, none of which are mutually exclusive: opioids, severe illness, impending death, stress of diagnosis, other drugs, or any of several mental health disorders. Delirium includes cognitive impairment, and it has been speculated that opioid-induced cognitive impairment may actually be a subclinical manifestation of delirium (Lawlor 2002).

Delirium occurs when the cholinergic and dopaminergic systems within the central nervous system become unbalanced. Many opioids, such as morphine, inhibit central cholinergic activity in numerous cortical and subcortical portions of the brain, as a result of the accumulation in the brain of metabolites. In the case of morphine, these would be morphine-3-glucuronide (M3G) and morphine-6-glucuronide (M6G) (Vella-Brincat and Macleod 2007). This cholinergic inhibition may suffice to induce delirium. Since M6G is excreted mainly in the urine, renally compromised patients taking morphine may be at elevated risk for delirium. Further support for this theory comes from the fact that the cholinergic system is involved in consciousness and acetylcholine modulates cortical arousal (Slatkin and Rhiner 2004).

18.5.8 HYPERALGESIA

Although not a toxicological effect, it is worthwhile to discuss here the opioid-induced hyperalgesia (OIH), because it can be confused with the development of tolerance. OIH is the paradoxical phenomenon in which a patient on opioid therapy suffers increased pain. OIH has been documented in animals, but remains a more controversial phenomenon in humans (Raffa and Pergolizzi 2013). OIH can be clinically difficult to distinguish from opioid tolerance, a phenomenon in which the patient requires increased doses of the opioid analgesic to achieve or maintain the same level of analgesia (Bell and Salmon 2009).

Morphine in particular has been associated with OIH, presumably a result of its metabolites (Anderson et al. 2003); low-dose fentanyl (without metabolites) in a transdermal system has been less associated with OIH (Kornick et al. 2003). However, complete absence of this effect in all patients is not possible. A case report in the literature suggests that high-dose transdermal fentanyl may be associated with OIH and neurotoxic symptoms, presumably because fentanyl can be accumulated in adipose tissue and muscle, leading to prolonged effects (Okon and George 2008). A patient with OIH should be tapered so that the opioid dose is reduced or opioids are discontinued altogether. By contrast, a patient with opioid tolerance (who may present with similar symptoms and complaints) may appropriately require a higher dose of opioid analgesic.

18.5.9 TOLERANCE

Tolerance is not a toxic effect, but tolerance to the therapeutic and adverse effects of an opioid can develop at different rates. Tolerance occurs when a patient requires increased doses of an opioid agent to maintain the same level of analgesia. In the public's mind, tolerance is easily confused with addiction and, indeed, demands for higher doses of opioids are considered *de facto* aberrant drug behaviors (Passik and Kirsh 2004). However, demands for higher doses may relate to inadequate analgesia as a result of physiological tolerance, which is both prevalent and not thoroughly understood.

Tolerance may develop with respect to analgesia (resulting in more pain), but also with respect to other opioid-associated effects (resulting in fewer or milder adverse events) (Faisinger and Bruera 1995). Because tolerance to adverse effects develops individually and at different rates, it is sometimes termed *selective tolerance* (Collett 1998). For example, tolerance to nausea often develops rapidly, whereas the opposite is true for constipation. There is broad inter-patient variability in terms of tolerance development to adverse events. Tolerance can be innate (genetically predetermined and observed with the first dose) or acquired (Collett 1998). Acquired tolerance can be pharmacokinetic or pharmacodynamic (Vella-Brincat and Macleod 2007). Pharmacokinetic tolerance occurs when changes in drug metabolism or disposition as a result of repeated dosing cause changes in serum drug concentrations. Pharmacodynamic tolerance occurs when the drug induces changes to the body over time and these changes decrease the subject's responsiveness to the drug (Collett 1998).

Tolerance may be related to hyperalgesia (Vella-Brincat and Macleod 2007). This theory holds that opioids antagonize N-methyl-D-aspartate (NMDA) receptors, which in turn activate the descending pain pathways for serotonin (5-HT) and noradrenalin. This then inhibits nociception and relieves pain. By the same token, agonism of NMDA receptors can cause pain and has been implicated in the development of tolerance (Meldrum 2003).

Tolerance is a multifactorial process with changes over time to the opioid receptors, the activation of NMDA receptors, and the movements of calcium and magnesium ions (Mao et al. 2002; Glare et al. 2004). When tolerance is developing, NMDA systems increase in activity and function as an anti-opioid system (Glare et al. 2004). There is some evidence of cross-talk between opioid receptors and NMDA receptors (Mercadante et al. 2003); these two distinct receptor systems can exist in the same neuron in the central nervous system (Mao et al. 2002). In a preclinical study, it was shown that chronic opioid use led to neuronal excito-toxicity by way of apoptosis regulated by spinal glutamate transporters (glutamate is an NMDA receptor agonist) (Hall and Sykes 2004). It has also been postulated that tolerance occurs because prolonged activation of opioid receptors leads to tolerance at the cellular level (Mercadante et al. 2003). Persistent changes in the neural circuitry can also affect pain modulation and be involved in tolerance.

18.5.10 MYOCLONUS

Myoclonus is a centrally mediated process resulting in uncontrollable twitching or jerking of the extremities. It involves large-amplitude rhythmic movements that have been shown in preclinical studies to be associated with the brainstem, cerebellum, and cortex and rely on the neurotransmitter γ-aminobutyric acid (GABA) (Lalonde and Strazielle 2012). GABA may be susceptible to interactions with opioids and other substances, such as acetylcholine (Lalonde and Strazielle 2012). The 3-glucuronide metabolites of morphine and hydromorphone have been postulated to cause myoclonus (Smith 2000), but this has been questioned in a more recent study (McCann et al. 2010). Other theories include dopamine antagonism in the basal ganglia (Han et al. 2002) or inhibition of glycine in dorsal horn neurons (Mercadante 1998).

18.5.11 Neurotoxicity

With more patients surviving once-devastating illnesses and long-term opioid therapy becoming more prevalent, opioid-induced neurotoxicity is being reported (Glare et al. 2004). Neurotoxicity refers to damage to nerve cells, which may result in a variety of mechanisms. Neurotoxic effects involve endocytosis of opioid receptors and the activation of NMDA receptors, which are activated by glutamate (Mao et al. 2002). Ketamine is an NMDA antagonist, and its blocking of NMDA receptors has been effective in preventing opioid-induced neurotoxicity (Mao et al. 2002). Modulating glutamate transport may also be helpful. Neurotoxic effects can vary by opioid. For example, methadone has NMDA antagonistic properties and may be less associated with neurotoxic effects than morphine (Tarumi et al. 2002).

18.6 OTHER EFFECTS

The following opioid-associated adverse events are clinically important and sufficiently widespread to merit inclusion here but do not fit neatly into the above categories.

18.6.1 Headache

Opioids and headaches have a paradoxical relationship. On the one hand, headache is a frequently reported adverse event associated with opioid therapy. On the other hand, opioid therapy is sometimes used to treat headaches of other etiologies. The rate of headache reported in patients in opioid studies is 22.7% (Nimmaanrat et al. 2007) to 33.1% (Anastassopoulos et al. 2011). It is difficult to assess the rate of true opioid-caused headaches, in that headaches are prevalent in the general population. In the United States, migraine prevalence is 17.1% and 5.6% for women and men, respectively (Stewart et al. 2013).

The role of opioids in the treatment of migraine, severe headache, and chronic headache remains controversial (Levin 2014). Medication-overuse headache (MOH) is a recently identified syndrome in which excessive use of analgesics (typically acetaminophen products or NSAIDs) is viewed as the cause rather than the consequence of headache. MOH may actually be multiple conditions, as chronic opioid use may exacerbate headaches by activating toll-like receptor-4 on glial cells, causing a pro-inflammatory cascade that results in heightened pain (Johnson et al. 2013). Thus there may be an opioid-overuse headache similar to MOH, which is likely related to opioid-induced hyperalgesia, a paradoxical condition in which use of opioids lowers pain thresholds (Simonnet and Rivat 2003). Both opioid-induced headache and opioid-induced hyperalgesia have been attributed to central sensitization and glial activation (facilitating pain) caused by repeated activation of nociceptive pathways (Raffa and Pergolizzi 2012).

18.6.2 PRURITUS

Pruritus can be a systemic condition rather than a cutaneous disorder (Krajnik and Zylicz 2001), and it may be mediated by endogenous opioids or opioid drug therapy. Opioid-induced pruritus is likely due to histamine release in the periphery (Swegle and Logemann 2006). Opioid-receptor antagonists, such as naltrexone, can suppress pruritus associated with endogenous opioid production (Bottcher and Wildt 2013). This includes the use of topical antagonists (Bigliardi et al. 2007).

The exact mechanisms underlying pruritus are not clear. It has been speculated that there is an "itch center" in the central nervous system, but it has not been anatomically identified (Kumar and Singh 2013). Activation of the medullary dorsal horn could antagonize inhibitory transmitters, which implicates serotonin in pruritus (Kumar and Singh 2013). For example, it may be that serotonergic pathway modulation causes pruritus (Kumar and Singh 2013) and/or influences the subject's perception of pruritus (Szarvas et al. 2003). Morphine is thought to directly release histamines from mast cells, which could account for pruritus and explain why pruritus is more common with some opioids than others (Waxler et al. 2005). One theory is that pain and pruritus are transmitted by the same group of sensory neurons (small, unmyelinated C-fibers) and the release of prostaglandins enhances C-fiber transmission to the central nervous system, which, in turn causes or potentiates pruritus (Kumar and Singh 2013).

Thus it may be that neuraxial opioid-induced pruritus is mediated by a different mechanism than pruritus induced via mu-opioid receptors. This would explain why antihistamines can be more effective in treating pruritus associated with oral opioids than with intrathecal opioids (Ko et al. 2004).

The incidence of pruritus is particularly high in postpartum patients after neuraxial opioid administration, with an estimated incidence of 60–100% (Shah et al. 2000; Yeh et al. 2000; Szarvas et al. 2003), which might be due to an interaction between estrogen and opioid receptors, since the rate of pruritus after intrathecal opioid administration in other surgeries is much lower (for instance, around 30–60% for orthopedic surgery) (Colbert et al. 1999; Dimitriou and Voyagis 1999). Pruritus appears to be dose-dependent, and its occurrence and prevalence can vary by opioid product (Reich and Szepietowski 2010). For instance, morphine is more associated with pruritus than is oxymorphone (Reich and Szepietowski 2010).

18.6.3 URINARY RETENTION

A distended bladder can cause significant morbidity, including hypotension, hypertension, cardiac arrhythmias, and vomiting (Baldini et al. 2009). Opioids may decrease bladder functions, such as the sensation of fullness and urge to void, resulting in urinary symptoms (Wheeler et al. 2002).

18.6.4 DEPRESSION

Cross-sectional studies have demonstrated an association of opioid use with depression in patients with chronic non-cancer pain (Reid et al. 2002; Sullivan et al. 2006; Grattan et al. 2012), but it is not clear whether depressed mood is the result of opioid use, a consequence of chronic pain, or independent of both. To be sure, depression is not uncommon among those taking opioids. In one long-term retrospective study, depressed individuals were more often prescribed opioids and more likely to take opioids long term than non-depressed persons (Braden et al. 2009). Dysphoria is a known side effect of opioid therapy (reported by about 1–5% of patients) (Drugs. com 2013) and may relate to depression or be a subclinical manifestation of it. Euphoria, another well-known opioid side effect, can occur even in the presence of depression; euphoria is generally transient.

In a retrospective database study (n = 49,770) from the Veterans Affairs system, propensity scoring was used for prescription of an opioid analgesic and a diagnosis of depression (Scherrer et al. 2013). The risk of depression increased significantly with the duration of the opioid prescription (hazard ratio [HR] 1.25, 95% confidence interval [CI] 1.05 to 1.46 for 90 to 180 days, and HR 1.51, 95% CI, 1.31 to 1.75 for >180 days) (Scherrer et al. 2013).

18.7 CONCLUSION AND PERSPECTIVE

Just as there is variability in patient response to the therapeutic effects of opioids, there is a great degree of variability among patients, even seemingly similar patients, regarding the adverse and toxic effects of opioids. And the extent of occurrence of adverse effects can vary with the type of opioid and route of administration. For instance, nausea rates are lower with transdermal fentanyl than they are with oral morphine (Okamoto et al. 2010), and transdermal opioid products overall are associated with a lower rate of constipation than are oral morphine controlled-release products

(Tassinari et al. 2008). Other factors can play a role, including dose, illness, other medical treatments or interventions, comorbidities, and product formulation.

Risk stratification for opioid-related adverse effects can be done only in the broadest sense. The most clinically meaningful strategies in opioid therapy are those that limit the use of opioids to the lowest possible dose for the shortest period of time. But the clinical reality of chronic pain is that many patients need opioid therapy for an extended period of time. In such cases, non-opioid pain relievers, such as acetaminophen, may be combined with a smaller dose of opioid and offer an opioid-sparing effect. Since opioid-associated adverse effects are often dose dependent, maintaining a low dose can be the cornerstone of adverse effect management.

Patients should be informed about the risks of opioid-associated adverse effects and offered multimodal pain therapy, including non-pharmacological options. They can be important adjunctive aids in reducing pain and thus may have an opioid-sparing effect. Such non-pharmacological strategies include lifestyle modifications, dietary changes, physical therapy, massage, occupational therapy, exercise, relaxation techniques, cognitive therapies, and complementary and alternative medicine, such as acupuncture.

Opioid-associated adverse effects can be treatment limiting, so careful clinical management is advisable. When possible, prophylactic strategies can be used to try to ameliorate anticipated adverse effects (such as emesis and constipation). In some cases, rotation to a different regimen (agent, formulation, or route of administration) may be helpful.

REFERENCES

Abramowitz, L., Beziaud, N., Labreze, L., et al. 2013. Prevalence and impact of constipation and bowel dysfunction induced by strong opioids: A cross-sectional survey of 520 patients with cancer pain. DYONISOS study. *J Med Econ* 16: 1423–33.

Ahmedzai, S. 1997. New approaches to pain control in patients with cancer. *Eur J Cancer Care (Engl)* 33 Suppl: S8–S14.

Alattar, M. and Scharf, S. 2009. Opioid-associated central sleep apnea: A case series. *Sleep Breath* 13: 201–206.

Anastassopoulos, K.P., Chow, W., Ackerman, S.J., Tapia, C., Benson, C. and Kim, M.S. 2011. Oxycodone-related side effects: Impact on degree of bother, adherence, pain relief, satisfaction, and quality of life. *J Opioid Manag* 7: 203–15.

Anderson, G., Christrup, L. and Sjogren, P. 2003. Relationships among morphine metabolism, pain and side effects during long-term treatment: An update. *J Pain Symptom Manage* 25: 74–91.

Baldini, G., Bagry, H., Aprikian, A. and Carli, F. 2009. Postoperative urinary retention: Anesthetic and perioperative considerations. *Anesthesiology* 110: 1139–57.

Bell, K. and Salmon, A. 2009. Pain, physical dependence and pseudoaddiction: Redefining addiction for 'nice' people? *Int J Drug Policy* 20: 170–78.

Bigliardi, P.L., Stammer, H., Jost, G., Rufli, T., Buchner, S. and Bigliardi-Qi, M. 2007. Treatment of pruritus with topically applied opiate receptor antagonist. *J Am Acad Dermatol* 56: 979–88.

Biot, M. 1876. [Contribution to a study on the Cheyne-Stokes respiratory phenomenon]. *Lyon Med* 23: 517–28, 61–67.

Bottcher, B. and Wildt, L. 2013. Treatment of refractory vulvovaginal pruritus with naltrexone, a specific opiate antagonist. *Eur J Obstet Gynecol Reprod Biol* Dec 15.

Bourne, R. and Mills, G. 2004. Sleep disruption in critically ill patients: Pharmacological considerations. *Anaesthesia* 59: 374–84.

Braden, J.B., Sullivan, M.D., Ray, G.T., et al. 2009. Trends in long-term opioid therapy for noncancer pain among persons with a history of depression. *Gen Hosp Psychiatry* 31: 564–70.

Brennan, M.J. 2013. The effect of opioid therapy on endocrine function. *Am J Med* 126: S12–18.

Brock, C., Olesen, S.S., Olesen, A.E., Frokjaer, J.B., Andresen, T. and Drewes, A.M. 2012. Opioid-induced bowel dysfunction: Pathophysiology and management. *Drugs* 72: 1847–65.

Candrilli, S., Davis, K. and Iyer, S. 2009. Impact of constipation on opioid use patterns, health care resource utilization, and costs in cancer patients on opioid therapy. *J Palliat Care Pharmacother* 23: 231–41.

Clark, A. and Lynch, M. 2003. Therapy of chronic non-malignant pain with opioids (letter). *Can J Anaesthes* 50: 92.

Colameco, S. and Coren, J. 2009. Opioid-induced endocrinopathy. *J Am Osteopath Assoc* 109: 20–25.

Colbert, S., O'Hanlon, D.M., Galvin, S., Chambers, F. and Moriarty, D.C. 1999. The effect of rectal diclofenac on pruritus in patients receiving intrathecal morphine. *Anaesthesia* 54: 948–52.

Collett, B.-J. 1998. Opioid tolerance: The clinical perspective. *Br J Anaesth* 81: 58–68.

Coluzzi, F. and Pappagallo, M. 2005. Opioid therapy for chronic noncancer pain: Practice guidelines for initiation and maintenance of therapy. *Minerva Anesthesiol* 71: 425–33.

Daniell, H. 2006. DHEAS deficiency during consumption of sustained-action prescribed opioids: Evidence for opioid-induced inhibition of adrenal androgen production. *J Pain Palliat Care Pharmacother* 7: 901–907.

Daniell, H. 2008. Opioid endocrinopathy in women consuming prescribed sustained-action opioids for control of nonmalignant pain. *J Pain* 9: 28–36.

Davies, A.N. and Vriens, J. 2005. Oral transmucosal fentanyl citrate and xerostomia. *J Pain Symptom Manage* 30: 496–97.

Dimitriou, V. and Voyagis, G.S. 1999. Opioid-induced pruritus: Repeated vs single dose ondansetron administration in preventing pruritus after intrathecal morphine. *Br J Anaesth* 83: 822–23.

Drugs.com. 2013. Oxycodone side effects *Oxycodone*. Drugs.com.

Drummond, G.B., Bates, A., Mann, J. and Arvind, D.K. 2013. Characterization of breathing patterns during patient-controlled opioid analgesia. *Br J Anaesth* 111: 971–78.

Duarte, R.V., Raphael, J.H., Southall, J.L., Labib, M.H., Whallett, A.J. and Ashford, R.L. 2013. Hypogonadism and low bone mineral density in patients on long-term intrathecal opioid delivery therapy. *BMJ Open* 3.

Dunn, G. and Milch, R. 2002. Is this a bad day, or one of the last days? How to recognize and respond to approaching demis. *J Am Coll Surg* 195: 879–87.

Ersek, M., Cherrier, M., Overman, S. and Irving, G. 2004. The cognitive effects of opioids. *Pain Manage Nurs* 5: 75–93.

Faisinger, R. and Bruera, E. 1995. Is this opioid analgesic tolerance? *J Pain Symptom Manage* 10: 573–77.

Glare, P., Aggarwal, G. and Clark, K. 2004. Ongoing controversies in the pharmacological management of cancer pain. *Intern Med* 34: 45–49.

Glare, P., Walsh, D. and Sheehan, D. 2006. The adverse effects of morphine: A prospective survey of common symptoms during repeated dosing for chronic cancer pain. *Am J Hosp Palliat Care* 23: 229–35.

Grattan, A., Sullivan, M.D., Saunders, K.W., Campbell, C.I. and Von Korff, M.R. 2012. Depression and prescription opioid misuse among chronic opioid therapy recipients with no history of substance abuse. *Ann Fam Med* 10: 304–11.

Hadley, G., Derry, S., Moore, R.A. and Wiffen, P.J. 2013. Transdermal fentanyl for cancer pain. *Cochrane Database Syst Rev* 10: CD010270.

Hall, E. and Sykes, N. 2004. Analgesia for patients with advanced disease. *Postgrad Med J* 80: 148–54.

Han, P., Arnold, R., Bond, G., Janson, D. and Abu-Elmagd, K. 2002. Myoclonus secondary to withdrawal from transdermal fentanyl: Case report and literature review. *J Pain Symptom Manage* 23: 66–72.

Ito, S. and Liao, S. 2008. Myoclonus associated with high-dose parenteral methadone. *J Palliat Med* 11: 838–41.

Johnson, J.L., Hutchinson, M.R., Williams, D.B. and Rolan, P. 2013. Medication-overuse headache and opioid-induced hyperalgesia: A review of mechanisms, a neuroimmune hypothesis and a novel approach to treatment. *Cephalalgia* 33: 52–64.

Katz, N. and Mazer, N. 2009. The impact of opioids on the endocrine system. *Clin J Pain* 25: 170–5.

Kim, T., Alford, D., Malabanan, A., Holick, M. and Samet, J. 2006. Low bone density in patients receiving methadone maintenance treatment. *Drug Alcohol Depend* 85: 258–62.

Knill, R.L., Moote, C.A., Skinner, M.I. and Rose, E.A. 1990. Anesthesia with abdominal surgery leads to intense REM sleep during the first postoperative week. *Anesthesiology* 73: 52–61.

Ko, M., Song, M., Edwards, T., Lee, H.K. and Naughton, N. 2004. The role of central mu opioid receptors in opioid-induced itch in primates. *J Pharmacol Exp Ter* 310: 169–76.

Kornick, C.A., Santiago-Palma, J., Moryl, N., Payne, R. and Obbens, E.A. 2003. Benefit-risk assessment of transdermal fentanyl for the treatment of chronic pain. *Drug Saf* 26: 951–73.

Kraichely, R., Arora, A. and Murray, J. 2010. Opiate-induced oesophageal dysmotility. *Aliment Pharmacol Ther* 31: 601–606.

Krajnik, M. and Zylicz, Z. 2001. Understanding pruritus in systemic disease. *J Pain Symptom Manage* 21: 151–68.

Kumar, K. and Singh, S.I. 2013. Neuraxial opioid-induced pruritus: An update. *J Anaesthesiol Clin Pharmacol* 29: 303–307.

Labianca, R., Sarzi-Puttini, P., Zuccaro, S.M., Cherubino, P., Vellucci, R. and Fornasari, D. 2012. Adverse effects associated with non-opioid and opioid treatment in patients with chronic pain. *Clin Drug Investig* 32 Suppl 1: 53–63.

Lalonde, R. and Strazielle, C. 2012. Brain regions and genes affecting myoclonus in animals. *Neurosci Res* 74: 69–79.

Lawlor, P. 2002. The panorama of opioid-related cognitive dysfunction in patients with cancer. *Cancer Causes Control* 94: 1836–53.

Lembo, A. and Camilleri, M. 2003. Chronic constipation. *N Engl J Med* 349: 1360–68.

Levin, M. 2014. Opioids in headache. *Headache* 54: 12–21.

Li, L., Setoguchi, S., Cabral, H. and Jick, S. 2013. Opioid use for noncancer pain and risk of myocardial infarction amongst adults. *J Intern Med* 273: 511–26.

Mao, J., Sung, B., Ji, R.R. and Lim, G. 2002. Neuronal apoptosis associated with morphine tolerance: Evidence for an opioid–induced neurotoxic mechanism. *J Neurosci* 22: 7650–61.

McCann, S., Yaksh, T.L. and von Gunten, C.F. 2010. Correlation between myoclonus and the 3-glucuronide metabolites in patients treated with morphine or hydromorphone: A pilot study. *J Opioid Manag* 6: 87–94.

McNicol, E., Horowicz-Mehler, N., Fisk, R., et al. 2003. Management of opioid side effects in cancer-related and chronic noncancer pain: A systematic review. *J Pain* 4: 231–56.

McNicol, E.D., Midbari, A. and Eisenberg, E. 2013. Opioids for neuropathic pain. *Cochrane Database Syst Rev* 8: CD006146.

Meldrum, M. 2003. A capsule history of pain management. *JAMA* 290: 2470–75.

Mercadante, S. 1998. Pathophysiology and treatment of opioid-related myoclonus in cancer patients. *Pain Manag Nurs* 74: 5–9.

Mercadante, S., Ferrera, P., Villari, P. and Arcuri, E. 2003. Hyperalgesia: An emerging iatrogenic syndrome. *J Pain Symptom Manage* 26: 769–75.

Miller, M., Sturmer, T., Azrael, D., Levin, R. and Solomon, D.H. 2011. Opioid analgesics and the risk of fractures in older adults with arthritis. *J Am Geriatr Soc* 59: 430–38.

Moore, R.A. and McQuay, H.J. 2005. Prevalence of opioid adverse events in chronic non-malignant pain: Systematic review of randomised trials of oral opioids. *Arthritis Res Ther* 7: R1046–51.

Mulligan, T., Frick, M., Zuraw, Q., Stemhagen, A. and McWhirter, C. 2006. Prevalence of hypogonadism in males aged at least 45 years: The HIM study. *Int J Clin Pract* 60: 762–69.

Nimmaanrat, S., Wasinwong, W., Uakritdathikarn, T. and Cheewadhanaraks, S. 2007. The analgesic efficacy of tramadol in ambulatory gynecological laparoscopic procedures: A randomized controlled trial. *Minerva Anestesiol* 73: 623–28.

Okamoto, Y., Tsuneto, S., Tsugane, M., Takagi, T. and Uejima, E. 2010. A retrospective chart review of opioid-induced nausea and somnolence on commencement for cancer pain treatment. *J Opioid Manag* 6: 431–34.

Okon, T.R. and George, M.L. 2008. Fentanyl-induced neurotoxicity and paradoxic pain. *J Pain Symptom Manage* 35: 327–33.

Palm, S., Moenig, H. and Maier, C. 1997. Effects of oral treatment with sustained release morphine tablets on hypothalamic–pituitary-adrenal axis. *Methods Find Exp Clin Pharmacol* 19: 269–73.

Paramanandam, G., Prommer, E. and Schwenke, D.C. 2011. Adverse effects in hospice patients with chronic kidney disease receiving hydromorphone. *J Palliat Med* 14: 1029–33.

Passik, S.D. and Kirsh, K.L. 2004. Assessing aberrant drug-taking behaviors in the patient with chronic pain. *Curr Pain Headache Rep* 8: 289–94.

Penagini, R. and Bianchi, P.A. 1997. Effect of morphine on gastroesophageal reflux and transient lower esophageal sphincter relaxation. *Gastroenterology* 113: 409–14.

Phillips, K. and Clauw, D.J. 2011. Central pain mechanisms in chronic pain states: Maybe it is all in their head. *Best Pract Res Clin Rheumatol* 25: 141–54.

Porreca, F. and Ossipov, M.H. 2009. Nausea and vomiting side effects with opioid analgesics during treatment of chronic pain: Mechanisms, implications, and management options. *Pain Med* 10: 654–62.

Raffa, R.B. and Pergolizzi, J.V., Jr. 2012. Multi-mechanistic analgesia for opioid-induced hyperalgesia. *J Clin Pharm Ther* 37: 125–27.

Raffa, R.B. and Pergolizzi, J.V., Jr. 2013. Opioid-induced hyperalgesia: Is it clinically relevant for the treatment of pain patients? *Pain Manag Nurs* 14: e67–83.

Reich, A. and Szepietowski, J.C. 2010. Opioid-induced pruritus: An update. *Clin Exp Dermatol* 35: 2–6.

Reid, M., Engles-Horton, L., Weber, M., Kerns, R., Rogers, E. and O'Connor, P. 2002. Use of opioid medications for chronic noncancer pain syndromes in primary care. *J Gen Intern Med* 17: 173–79.

Rhodin, A., Stridsberg, M. and Gordh, T. 2010. Opioid endocrinopathy: A clinical problem in patients with chronic pain and long-term oral opioid treatment. *Clin J Pain* 26: 374–80.

Ripamonti, C., Zecca, E. and Bruera, E. 1997. An update on the clinical use of methadone for cancer pain. *Pain* 70: 109–15.

Rubinstein, A.L., Carpenter, D.M. and Minkoff, J.R. 2013. Hypogonadism in men with chronic pain linked to the use of long-acting rather than short-acting opioids. *Clin J Pain* 29: 840–45.

Scherrer, J.F., Svrakic, D.M., Freedland, K.E., et al. 2013. Prescription opioid analgesics increase the risk of depression. *J Gen Intern Med* 29(3): 491–99.

Shah, M.K., Sia, A.T. and Chong, J.L. 2000. The effect of the addition of ropiva-
caine or bupivacaine upon pruritus induced by intrathecal fentanyl in labour.
Anaesthesia 55: 1008–13.

Sharma, A. and Jamal, M.M. 2013. Opioid induced bowel disease: A twenty-first
century physicians' dilemma. Considering pathophysiology and treatment
strategies. *Curr Gastroenterol Rep* 15: 334.

Sharma, S. 2002. Sphincter of Oddi dysfunction in patients addicted to opium: An
unrecognized entity. *Gastrointest Endosc* 55: 427–30.

Shaw, I.R., Lavigne, G., Mayer, P. and Choiniere, M. 2005. Acute intravenous
administration of morphine perturbs sleep architecture in healthy pain-free
young adults: A preliminary study. *Sleep* 28: 677–82.

Shorr, R., Griffin, M., Daugherty, J. and Ray, W. 1991. Opioid analgesics and the
risk of hip fracture in the elderly: Codeine and propoxyphene. *J Gerontol*
47: M111–M115.

Simonnet, G. and Rivat, C. 2003. Opioid-induced hyperalgesia: Abnormal or normal pain. *Neuroreport* 14: 1–7.

Slatkin, N. and Rhiner, M.I. 2004. Treatment of opioid-induced delirium with
acetylcholinesterase inhibitors: A case report. *J Pain Symptom Manage* 27:
268–73.

Sloan, P., Slatkin, N. and Ahdieh, H. 2005. Effectiveness and safety of oral
extended-release oxymorphone for the treatment of cancer pain: A pilot
study. *Support Care Cancer* 13: 57–65.

Smith, H.S. and Laufer, A. 2014. Opioid induced nausea and vomiting. *Eur J
Pharmacol* 722: 67–78.

Smith, M.T. 2000. Neuroexcitatory effects of morphine and hydromorphone:
Evidence implicating the 3-glucuronide metabolites. *Clin Exp Pharmacol
Physiol* 27: 524–28.

Staedt, J., Wassmuth, F., Stoppe, G., Hajak, G., Rodenbeck, A., Poser, W. and
Ruther, E. 1996. Effects of chronic treatment with methadone and naltrexone on sleep in addicts. *Eur Arch Psychiatry Clin Neurosci* 246: 305–309.

Stewart, W.F., Roy, J. and Lipton, R.B. 2013. Migraine prevalence, socioeconomic
status, and social causation. *Neurology* 81: 948–55.

Sullivan, M.D., Edlund, M.J., Zhang, L., Unutzer, J. and Wells, K.B. 2006.
Association between mental health disorders, problem drug use, and regular
prescription opioid use. *Arch Intern Med* 166: 2087–93.

Swegle, J. and Logemann, C. 2006. Management of common opioid-induced
adverse effects. *Am Fam Physicians* 74: 1347–54.

Szarvas, S., Harmon, D. and Murphy, D. 2003. Neuraxial opioid-induced pruritus:
A review. *J Clin Anesth* 15: 234–39.

Tarumi, Y., Pereira, J. and Watanabe, S. 2002. Methadone and fluconazole:
Respiratory depression by drug interaction. *J Pain Symptom Manage* 23:
148–53.

Tassinari, D., Sartori, S., Tamburini, E., Scarpi, E., Raffaeli, W., Tombesi, P. and
Maltoni, M. 2008. Adverse effects of transdermal opiates treating moderate-
severe cancer pain in comparison to long-acting morphine: A meta-analysis
and systematic review of the literature. *J Palliat Med* 11: 492–501.

Teichtahl, H. and Wang, D. 2007. Sleep-disordered breathing with chronic opioid use. *Expert Opin Drug Saf* 6: 641–49.

Thwaites, D., McCann, S. and Broderick, P. 2004. Hydromorphone neuroexcitation. *J Palliat Med* 7: 545–50.

Tuteja, A.K., Biskupiak, J., Stoddard, G.J. and Lipman, A.G. 2010. Opioid-induced bowel disorders and narcotic bowel syndrome in patients with chronic non-cancer pain. *Neurogastroenterol Motil* 22: 424–30, e96.

Vella-Brincat, J. and Macleod, A. 2007. Adverse effects of opioids on the central nervous systems of palliative care patients. *H Pain Palliative Care Pharmacother* 21: 15–25.

Vestergaard, P., Rejnmark, L. and Mosekilde, L. 2006. Fracture risk associated with the use of morphine and opiates. *J Inter Med* 260: 76–87.

Walker, D.J. and Zacny, J.P. 2002. Analysis of the reinforcing and subjective effects of different doses of nitrous oxide using a free-choice procedure. *Drug Alcohol Depend* 66: 93–103.

Wang, C.Z. and Yuan, C.S. 2013. Pharmacologic treatment of opioid-induced constipation. *Expert Opin Investig Drugs* 22: 1225–27.

Wang, D., Teichtahl, H., Drummer, O., Goodman, C., Cherry, G., Cunnington, D. and Kronborg, I. 2005. Central sleep apnea in stable methadone maintenance treatment patients. *Chest* 128: 1348–56.

Waxler, B., Dadabhoy, Z., Stojilkovic, L. and Rabito, S. 2005. Primer of postoperative pruritus for anesthesiologists. *Anesthesiology* 103: 168–78.

Wheeler, M., Oderda, G.M., Ashburn, M.A. and Lipman, A.G. 2002. Adverse events associated with postoperative opioid analgesia: A systematic review. *J Pain* 3: 159–80.

Wirz, S., Wittmann, M., Schenk, M., et al. 2009. Gastrointestinal symptoms under opioid therapy: A prospective comparison of oral sustained-release hydromorphone, transdermal fentanyl, and transdermal buprenorphine. *Eur J Pain* 13: 737–43.

Yeh, H.M., Chen, L.K., Lin, C.J., et al. 2000. Prophylactic intravenous ondansetron reduces the incidence of intrathecal morphine-induced pruritus in patients undergoing cesarean delivery. *Anesth Analg* 91: 172–75.

Yue, H.J. and Guilleminault, C. 2010. Opioid medication and sleep-disordered breathing. *Med Clin North Am* 94: 435–46.

Zlott, D.A. and Byrne, M. 2010. Mechanisms by which pharmacologic agents may contribute to fatigue. *PM R* 2: 451–55.

19 Toxicology of Mitragynine and Analogs

Surash Ramanathan and Sharif M. Mansor

CONTENTS

19.1 INTRODUCTION

Mitragyna speciosa, also known as ketum in Malaysia and kratom in Thailand, is an indigenous plant of southern Thailand and northern Peninsular Malaysia. At lower doses, the extract helps manual laborers to enhance their work productivity under hot tropical weather, while at higher doses it produces opioid-like effects. Ketum leaves have long been used in this part of the world as an herbal remedy to treat various medical conditions, such as to reduce fever, for its analgesic, antidiarrheal, antidepressant, and antidiabetic effects, and also to suppress opiate withdrawal symptoms (Jansen and Prast 1988; Grewal 1932a; Burkill 1935; Kumarnsit et al. 2006; Farah Idayu et al. 2011).

Mitragynine is the principal component found abundantly in this plant, along with other minor indole alkaloids that are structurally related to mitragynine (12–66%), such as paynantheine (9%), speciogynine (7%), 7-hydroxymitragynine (2%), speciociliatine (1%), mitraphylline (<1%), and rhynchophylline (<1%). The narcotic, stimulant, and other dose-dependent effects produced by the plant (Sabetghadam et al. 2013a; Shellard 1989; Grewal 1932b) have been attributed primarily to mitragynine (Matsumoto et al. 1996a,b). Kratom leaves were already used in the nineteenth century to treat opium withdrawal in opium users in Malaya (now known as Malaysia) (Hooper 1907). Studies directed at central nervous system receptors (Babu et al. 2008; Taufik Hidayat et al. 2010) showed that mitragynine might exert some pharmacological effects that mitigate opioid withdrawal symptoms and blunt cravings (Boyer et al. 2008; Khor et al. 2011).

Kratom is a cheap and affordable herbal preparation. It is thus not surprising that it is reported to be widely used and abused, for example in the northern states of Malaysia, probably due to the country's non-rigid regulation (Vicknasingam et al. 2010). In Malaysia, kratom is placed under the first and third schedules of the Poisons Act of 1952. Kratom is banned in Thailand; in Australia and Myanmar it is under the control of narcotic law, and in Sweden, Poland, and Denmark it is a controlled drug. However, kratom is currently not a controlled substance in the United Kingdom, United States, or Germany (Hassan et al. 2013). Owing to this complex legal status, kratom is widely available and can be purchased in various forms and doses via Internet vendors (Ahmad and Aziz 2013; Troy 2013). Many users tend to assume that kratom is safer than illegal substances, so demand for kratom continues to rise. This may explain the propagation of recreational kratom use, which is being observed not only in Malaysia but also in other parts of the world, including the United States and Europe (Boyer et al. 2008, Kapp et al. 2011; Nelsen et al. 2010). In Malaysia, kratom is consumed by drug addicts as an inexpensive alternative to more expensive opiates, and also to manage withdrawal symptoms. Kratom juice is consumed in unadulterated and adulterated form. In Malaysia, users who seek better euphoric effects will spike their kratom drinks with cough medication or rat poison (active ingredient warfarin), or even synthetic pyrethroid (found in mosquito coils), to add zest to their drinks. To date, there have been no reports of fatal overdose of kratom *per se*. If there are such occurrences, they are probably the result of kratom products contaminated with synthetic adulterants. This issue will be discussed in the section on human toxicity.

At present, there are not many studies on toxicity of mitragynine or crude extracts of *M. speciosa* in the literature. The toxicity of minor alkaloids of *M. speciosa* has yet to be reported. Study of the chronic toxicity of mitragynine is urgently needed in order to understand its long-term safety, since kratom is usually consumed on a long-term basis. This chapter will discuss the toxicity of *M. speciosa*'s major constituent, mitragynine, and its crude extracts in experimental models (including cytotoxicity) and in humans.

19.2 CYTOTOXICITY

Studies of the cytotoxicity of *M. speciosa* extracts and mitragynine are limited. There are a few reported studies describing cytotoxicity employing cell lines, brine shrimp, and gene mutation assay. When human cell lines (HepG2, HEK 293, MCL-5, cHol, and SH-SY5Y) were tested for cytotoxicity, the extract inhibited cell proliferation in a dose-dependent manner (>100 µg/mL), with significant cell death at 1000 µg/mL. There is no clear indication of genotoxicity (gene mutation assay using mouse lymphoma cell), since DNA damage was not observed. Toxicity was observed only at a higher dose (Saidin et al. 2008). Mutagenicity tests carried out in *Salmonella typhimurium* TA 98 and TA 100 bacterial strains showed strong antimutagenicity for *M. speciosa* in the presence of the metabolic activator S9 system (Ghazali et al. 2011). In a brine shrimp study, mitragynine, alkaloid extract, and aqueous extracts were found to be toxic at 44, 62 and 98 µL/mL, respectively (Moklas et al. 2008). Even though, compared to animal models, the conventional in vitro cytotoxicity assessment is less specific and its ability to measure various mechanisms of toxicity factors is limited, it is still a useful screening tool for evaluation of new compounds.

19.3   ANIMAL MODELS

19.3.1   ACUTE TOXICITY STUDIES

There are a number of acute toxicity studies of mitragynine and *M. speciosa* crude extracts, namely methanolic and alkaloid extracts, in the literature (Table 19.1). Sabetghadam et al. (2013a) reported an LD_{50} of 477 mg/kg for mitragynine and 591 mg/kg for alkaloid extract in mice. However, the Therapeutic Index for alkaloid extract and mitragynine was estimated as 3:1 and 20:1, respectively, suggesting that mitragynine is relatively safer compared to the alkaloid extract. Reanmongkol et al. (2007) reported an LD_{50} for methanolic and alkaloid extracts of 4.9 g/kg and 173.2 mg/kg, respectively. Studies conducted by Macko et al. (1972) reported no toxicity up to 920 mg/kg in mice or 806 mg/kg in rats. Conversely, Janchawee et al. (2007) reported fatal effects of an oral dose of 200 mg/kg mitragynine in rats. Death was also observed after administration of 200 mg/kg alkaloid extract of *M. speciosa* to rats (Azizi et al. 2010). In general, alkaloid extract was found to be more toxic than methanol extract. The antinociceptive effect of crude extract of *M. speciosa* is more potent than mitragynine (Watanabe et al. 1992, 1997).

The acute toxicity data of the alkaloid extracts vary significantly between studies conducted in Thailand and those in Malaysia. The mitragynine content in the total alkaloid extracts of *M. speciosa* is also reported to strongly vary by geographical origin. For example, a study in Thailand reported a mitragynine content of 66% in alkaloid extract (Takayama 2004), whereas a Malaysian study reported only 12–25% (Parthasarathy

TABLE 19.1
Animal Toxicity Studies of *Mitragyna speciosa* and Its Principal Alkaloid Mitragynine

No.	Authors	Drug	Dosage	Toxicity
1	Sabetghadam et al. (2013a)	Alkaloid extract and mitragynine	LD_{50} for the alkaloid extract (591 mg/kg) and mitragynine (477 mg/kg) in mice (oral)	The LD_{50} value for mitragynine was lower than that of the alkaloid extract; the margin of safety for mitragynine is wider. This implies that there is a lesser possibility of toxicity effect with mitragynine, when compared to alkaloid extract.
2	Sabetghadam et al. (2013b)	Mitragynine	Three doses of mitragynine (1, 10, 100 mg/kg, for 28 days in rats (oral)	Relatively safe at lower sub-chronic doses (1–10 mg/kg) but showed toxicity (mainly biochemical and hematological) at a higher dose (sub-chronic 28 days: 100 mg/kg), which corresponds to the histopathological changes of the liver and kidney. No mortality was observed for all treatment groups.
3	Janchawee et al. (2007)	Mitragynine	Single dose of 200 mg/kg mitragynine (oral)	Death occurred in the rats.
4	Azizi et al. (2010)	Total alkaloid extract of *M. speciosa*	Single dose 200 mg/kg (oral)	Death occurred in the rats.
5	Reanmongkol et al. (2007)	Methanol and alkaloid extract of *M. speciosa* leaf	LD_{50} for the methanol extract (4.90 g/kg) and alkaloid extract (173.20 mg/kg) in mice (oral)	In acute toxicity test, the signs of toxicity such as lethargy, tremor, fatigue, paralysis, loss of righting reflex, apnea, tonic–clonic convulsion and death were reported.

6	Macko et al. (1972)	Mitragynine (i) hydrochloride salt (ii) ethanedisulfonate salt	(i) Acute study (oral) (ii) Subacute toxicity study	Single dose of 806 mg/kg showed no toxicity in rats. At a dose of 8 mg/kg/day for 5 days, 25% of the animals had diarrhea; no other side effects were observed. No side effects were observed in dogs. (16 mg/kg/day of mitragynine for 5 days). In mice, no evidence of toxicity for dose as high as 920 mg/kg.
			(a) Rats – 5 or 50 mg/kg per day for 6 weeks (oral)	No side effects were observed in either dosing regimens. A significant decrease in liver and kidney weights was observed at high dose for both male and female rats.
			(b) Dogs – 5 mg/kg per day for 3 weeks (oral)	No signs of toxicity were observed.
			(c) Dogs – 20 mg/kg per day for 3 weeks (oral)	At a 20 mg/kg/day dose, no side effects were observed. At a dose of 40 mg/kg/day from day 22 to day 50 leukopenia, granulocytopenia, lymphocytopenia, monocytosis and atypical and immature lymphocytes were reported. These changes were reversed after the drug was withdrawn.
7	Kamal et al. (2012)	Standardized *Mitragyna speciosa* leaf aqueous extract	Rats – 175, 500 and 2000 mg/kg (oral)	The extract had no significant changes in the body weight, food or water consumption. Hemoglobin, albumin, calcium and cholesterol were significantly decreased. Steatosis and centrilobular necrosis were observed on several parts of the liver. No death was observed in all groups.
8	Harizal et al. (2010)	Standardized methanol extract of rats *Mitragyna speciosa*	Rats – 100, 500 and 1000 mg/kg doses (oral)	Significant elevation of ALT, AST, albumin, triglycerides, cholesterol and albumin (p>0.05), at the three dose levels. Nephrotoxicity was seen only at a dose of 1000 mg/kg.

et al. 2013; Takayama 2004). This may partly explain the disparity in the lethal doses reported in studies employing alkaloid extracts derived from different sources. Apart from this, the type of vehicles used to dissolve the drug or the extracts and the particular route of administration would have also contributed to the varying lethal dose. The contribution of other alkaloids in the extract to overall toxicity is not well documented. Data from our laboratory analysis of methanolic and water extract of *M. speciosa* of Malaysian origin indicated a much lower content of mitragynine in methanolic extract (4–5%) compared with aqueous extract (<1%). In a study by Malaysian investigators of mitragynine acute toxicity (14 days) in rats after oral administration of standardized methanolic extract of *M. speciosa*, no sign of toxicity (body weight, food, water consumption, absolute and relative organ weight, and hematological parameters) was observed at the three doses studied (100, 500, and 1000 mg/kg). Only at the highest dose were acute severe hepatotoxicity and mild nephrotoxicity evident, but no damage was observed in axons or dendrites of hippocampal neurons (Harizal et al. 2010). In another study of aqueous extract of *M. speciosa* in rats, no death was observed up to 2000 mg/kg (Kamal et al. 2012). Although the overall findings capture a considerable acute-toxicity profile of mitragynine and its crude extracts in animals, information about long-term safety is lacking.

19.3.2 Sub-Chronic Toxicity Studies

One of the earlier sub-acute toxicity studies was conducted by Macko et al. (1972) in rats and dogs. No side effects were observed in rats after administration of 5 or 50 mg/kg/day of mitragynine for five days per week for six weeks. Similarly no adverse effects were observed at a dose of 20 mg/kg/day six days per week for three weeks. However, when the dose was increased to 40 mg/kg/day six days per week, hematological findings were observed (such as leukopenia, granulocytopenia, lymphocytosis, monocytosis, and immature lymphocytes) from day 22. The changes were reversed after the drug was withdrawn from days 50 to 92. A group of dogs that received oral doses of 5 mg/kg/day of mitragynine for three weeks did not show any signs of toxicity.

The most recent study (Sabetghadam et al. 2013b) evaluated the effects of sub-chronic exposure of male and female rats to mitragynine. Mitragynine was administered at an oral dose of 1, 10, or 100 mg/kg for 28 days. Food intake and relative body weight were measured during the experiment. After completion of drug treatment, biochemical, hematological and histological analyses were performed. No mortality was observed in any of the treatment groups. Toxic effects were not observed in animals given the lower or intermediate doses (1 and 10 mg/kg). However, at the highest dose (100 mg/kg), a decrease in food intake was observed in the female, but not male, rats, and the relative body weight of the female rats decreased significantly. The relative liver weight of both male and female rats increased during treatment with this dose of mitragynine. There were changes in the biochemical and hematological parameters corresponding to the histopathological changes. Serum lactate dehydrogenase levels were elevated, which warrants further study. But there were no significant effects on other biochemical parameters, such as cholesterol, triglycerides, total protein, creatinine, albumin, alkaline phosphates, or glucose levels. However, there were large increases in aspartate aminotransferase (AST), alanine aminotransferase (ALT), and urea levels in the female rats, which are suggestive of hepatocellular damage, but no liver necrosis was observed in any of the treated rats (Tennekoon et al. 1994). Other indices, such as total protein, serum albumin, and triglycerides, were not altered, suggesting that there were no adverse effects on the normal hepatic synthetic or excretory functions in the rats, despite hepatic cell injury (Harizal et al. 2010). With regards to hemopoietic profile, red blood cell count and white blood cell count were significantly reduced at the higher dose, with notable reduction in hemoglobin and hematocrit levels. This is indicative that sub-chronic high-dose mitragynine administration has the potential to produce anemia. Another undesirable effect on the hemopoietic system encountered during sub-chronic administration of mitragynine (1, 10, or 100 mg/kg/for 28 days) was a notable reduction in platelet count (observed in both male and female rats). Detailed histopathological analysis also revealed brain, kidney, and liver toxicity in animals administered a 100-mg/kg dose. Brain abnormalities such as local vacuolation and necrotic and degenerating neurons were noted in female and male rats (Weber et al. 2006). In kidney, swollen glomerulus capsules and the presence of red blood cells between lumens were observed in female rats, perhaps an indication of an early stage of renal toxicity. Liver toxicity (10 and 100 mg/kg; sub-chronic dosing) was evident as moderate destruction of polygonal hepatic lobules, hepatocyte hypertrophy, dilation of sinusoids, and hemorrhage, but no centri lobular necrosis or inflammatory cell infiltration was observed at any of the doses. Serum cholesterol and glucose levels did not change significantly in male or female rats, but a dose-dependent decreasing trend in glucose was observed after prolonged administration of mitragynine. This could possibly be due to the anorexic effect of mitragynine, which is indirectly related to stimulation of peripheral opioid receptors rather than central regulation of blood glucose levels (Tsuchiya et al. 2002; Kumarnsit et al. 2006).

Overall, mitragynine is relatively safe at lower sub-chronic doses (1–10 mg/kg), but exhibits toxicity at 100 mg/kg (sub-chronic, 28 days).

When human equivalent doses were considered as a means to interpret whether or not the animal data are practically relevant for human interpretation, the sub-chronic dose of 1–10 mg/kg used in rats was found to be within the 0.1 to 1.7 mg/kg human equivalent. Analysis of mitragynine content in kratom juice from various geographical locations in northern Peninsular Malaysia, where kratom is regularly consumed, indicates a daily consumption of 0.3 to 5.1 mg/kg, with no serious side effects despite prolonged use (Vicknasingam et al. 2010).

Future chronic toxicity studies in suitable animal models, employing larger sample size, are warranted in order to ascertain the long-term effects of mitragynine. It is important that such evaluations employ dose levels that are relevant to humans by taking into consideration the range of amounts typically consumed by kratom users.

19.4 HUMAN TOXICITY

The first study of mitragynine in humans was reported by Grewal (1932b). Five volunteers, including the author, consumed 50 mg mitragynine acetate in 20 mL of distilled water in four repeated doses, or powdered leaves (0.65 to 1.3 g). Cocaine-like effects were observed in all the subjects. Some developed nausea, slight tremors of the hand and tongue, vomiting, giddiness, feeling of haziness, slight nystagmus, or slight flushing of the face. The effects reported after consumption of powdered leaves are similar to those produced by mitragynine, except that the feeling of lightness of the body and muscle tenseness were well marked. To date, no official clinical trial has been reported for *M. speciosa* extracts or for mitragynine. However, many studies on the social behavior and pattern of kratom usage by humans have been reported by several investigators (e.g., Suwanlert 1975; Assanangkornchai et al. 2007; Vicknasingam et al. 2010; Ahmad and Aziz 2012).

Kratom is widely sold as a decoction drink in villages in the northern states of Malaysia (Kedah and Perlis), which border Thailand (Chan et al. 2005). Kratom (ketum) consumption has always been widespread in these areas, and this has offered a unique opportunity for investigators to evaluate the effects of kratom consumption in human subjects. Vicknasingam and colleagues (2010) studied self-claimed kratom users. They described kratom as an affordable, easily available herbal drug with no serious side effects despite prolonged use. Its primary use was to manage withdrawal symptoms, since it was cheaper than other substances. The observed side effects were mild, such as loss of weight, dehydration, and constipation, and no unwanted medical problems were recorded. Ahmad and Aziz (2012) reported the various uses of kratom: as a tea extract for recreational and social uses; for strength and physical endurance; and for improved sexual performance. They too reported the use of kratom for pain relief and weaning off of opiate addiction, but for whatever reason, a large number of kratom users had difficulty stopping kratom use after a prolonged period of usage. In most cases, typical withdrawal symptoms were experienced by these subjects on cessation of use, such as excessive tearing, malaise, and jerky movement of limbs, but these were not life threatening. Some users claimed that the ketum withdrawal symptoms were less severe than those of heroin or cannabis. Obviously, such assertions must be objectively evaluated with a larger sample size. Other adverse effects observed in both short- and long-term users were light-headedness, vomiting, dehydration, constipation, lethargy, and hyper-pigmentation (Grewal 1932b; Vicknasingam et al. 2010; Ahmad and Aziz 2012). Similar adverse effects were reported in Thai studies (Suwanlert 1975; Assanangkornchai et al. 2007).

Usage of kratom is no longer limited to the country of its origin. Over the past years, there has been a wider distribution of kratom use, evident by the number of Internet sales and Internet forums on the use and management of kratom (Schmidt et al. 2011; Boyer et al. 2007; Babu et al. 2008). In most situations, kratom was ingested intentionally to attain the desirable effects of euphoria, to ameliorate opiate withdrawal, or to treat chronic pain. Most users prefer to self-treat to avoid the risk of being labeled drug-dependent. Despite informal and unsupervised widespread use of kratom, there are no reports of mortalities after ingestion of kratom alone, even after chronic and high-dosage consumption, but serious adverse reactions have been reported in several cases.

A typical case of adverse reaction related to Internet access to kratom was reported by Roche et al. (2008), where the user did not report taking it together with other drugs or substance of abuse. The patient experienced foaming at the mouth and seizure-like movements and later developed fever, aspiration, and pneumonia, and presented an episode of hypotension in response to intravenous fluids (Roche et al. 2008). Another case characterized by seizure and coma was reported by Nelsen et al. (2010). The patient had a history of social alcohol and tobacco use and a medical history of managing chronic pain and depression with amitriptyline, oxycodone, and kratom. He consumed a kratom/datura tea one-half hour before the incident and claimed to have been using kratom regularly to self-treat chronic pain. The analysis of his urine sample indicated the presence of mitragynine (167 ± 15 ng/mL) (Nelsen et al. 2010). Since no human mitragynine pharmacokinetic data are available, and lacking the ability to rule out other possible contributory causes of toxicity, a direct inference cannot be made to mitragynine toxicity. But consumption of kratom along with other drugs might have aggravated severe adverse reactions.

One published case reports kratom dependence of a man who had a history of alcohol dependence and anxiety disorder (McWhirter and Morris 2010). The patient self-administrated kratom initially to ameliorate the effects of opioid withdrawal and later increased the consumption to attain the desired effects of euphoria. He subsequently developed tolerance. He was admitted to a hospital for detoxification and showed withdrawal symptoms consisting of anxiety, restlessness, tremor, sweating, and craving for the substance.

In addition to adverse reactions to mitragynine, two cases of possible drug-induced organ toxicity following high-dose kratom consumption have been reported. One patient had a history of alcohol abuse and kratom addiction. After kratom usage for 4 months, the patient complained of chronic abdominal pain and developed severe primary hypothyroidism (Sheleg and Collins 2011). There is no underlying evidence to relate kratom use with thyroid gland malfunction, but a high dosage of mitragynine possibly has such an effect and this warrants further experimental investigations. In this regard, it is interesting to note that morphine has been reported to suppress thyroid function in animals (Meites et al. 1979; Rauhala et al. 1998). The other patient had drug-induced intrahepatic cholestasis after an ingestion of an overdose of kratom powder accumulated over a period of two weeks. Mitragynine presence was confirmed following analysis of both urine and serum samples, but the presence of other synthetic adulterants, amphetamines, benzodiazepines, or opioids was not detected (Kapp et al. 2011).

Fatal cases of consuming *adulterated* kratom products have been reported (Kroonstad et al. 2011). Like kratom, "Krypton" is available via the Internet and is often advertised as a harmless product. It appears to contain powdered kratom leaves along with another mu-opioid receptor agonist, O-desmethyltramadol. To date, nine cases of death from Krypton intoxication have been reported where both mitragynine and O-desmethyltramadol were detected in the post-mortem blood samples. The blood concentrations of mitragynine and of O-desmethyltramadol ranged from 0.02 to 0.18 μg/g and 0.4 to 4.3 μg/g, respectively. This incident highlights the danger of adding O-desmethyltramadol to powdered leaves of kratom, which could have contributed to unintentional death.

A similar fatal poisoning was reported involving the co-ingestion of propylhexedrine and mitragynine (Holler et al. 2011). Both propylhexedrine (1.7 mg/L) and mitragynine (0.39 mg/L) were found in the deceased's blood. Urine analysis indicated the presence of both drugs along with acetaminophen, morphine, and promethazine. There was an indication of propylhexedrine distribution into the liver, kidney, and spleen, and it was in a higher order of magnitude than mitragynine. Even though the cause of death was ruled propylhexedrine toxicity, the contribution of mitragynine to the accidental death remains unclear.

From the limited data available in the literature, cases involving death through consumption of kratom products have been considered unintentional or accidental, or due to adulterated products. The contribution of other alkaloids of kratom in the above cases is unknown, because there are no reference data on blood concentrations available to date. In addition, there are no data available to show the necessary amount of kratom consumed, the duration of consumption, or the influence of other substances for severe toxicity to be evident in the user.

19.5   FUTURE PERSPECTIVES

Although several studies on the toxicity of mitragynine have been reported since 1972, full knowledge about the toxicity of mitragynine and its analogs is still incomplete. Currently, assessments of mitragynine (kratom) toxicity are confined to acute and sub-chronic safety studies. In most instances, these studies have not given sufficient consideration to the physicochemical properties of mitragynine and factors that can affect its absorption in the gastrointestinal tract. Kratom extract and mitragynine are usually administered either orally or i.p. (intraperitoneally) in toxicity studies. The compounds or crude extracts are dissolved in various co-solvents (e.g., 1% acetic acid) or surfactants (such as Tween® 20, Tween® 80, propylene glycol, or 1% Cremophor) to facilitate solubility. The presence of co-solvents or surfactants might enhance the in vitro solubility for the poorly soluble drug, but its solubility in vivo might be totally different. Thus, this can result in non-predictable pharmacological responses. A typical example occurred in a mice study. Mitragynine antinociceptive effect was found to be dose-dependent up to 35 mg/kg, but when the dose was approximately doubled, the response was reduced instead of increased (Shamima et al. 2012). In addition to this, varied pharmacokinetic responses after oral administration of mitragynine in rats have also been reported in the literature. In these studies, investigators employed various vehicles (surfactants) for drug administration, and the doses employed were in the range of 20 to 50 mg/kg, far lower than the dose used in toxicity studies (Janchawee et al. 2007; de Moraes et al. 2009; Parthasarathy et al. 2010). However, no explanations were given by the investigators for the unexpected responses. It can be hypothesized that the vehicle used to solubilize the drug may apparently show a good in vivo solubility, but the hydrophobic nature of mitragynine could lead to drug precipitation in the gut lumen, particularly when the drug is administered at higher doses. This may ultimately result in poor drug absorption. In addition, excess dosage of the drug, either accidental or intentional, could also result in an exaggerated response to the drug.

Taking both possibilities into account, this may give an explanation for the wide-ranged toxicity reports on kratom extracts and mitragynine. Thus uncertainty about the poor absorption level of mitragynine could lead investigators to under-report its risk assessment. Work carried out in our laboratory has indicated that mitragynine is a poorly water-soluble drug (< 100 μg/mL), is a weak base, and has an acid-degradable nature. Its poor aqueous solubility and a log p value of < 2 indicate both water and lipid limited solubility. In view of this, a concerted effort towards designing a more appropriate toxicity study that eliminates the problem associated with mitragynine's poor absorption is needed to establish a more realistic safety assessment before one can make a correct inference about animal studies for human safety risk prediction. On the other hand, drug toxicity is one aspect of the risk–benefit balance, and it is viewed differently depending upon its intended use. In general, drug toxicity should be acceptable when there is little harm caused relative to its clinical benefits. Indeed, this consideration should mainly be viewed after a comprehensive animal toxicity study has been carried out for kratom or mitragynine. Animal toxicity studies must be conducted at a dose level whereby a meaningful translation of animal data for human interpretation could be derived. It is highly imperative that in future, more toxicological studies following GLP requirements be conducted before any clinical study is carried out on humans.

ACKNOWLEDGEMENTS

Financial support was received from Research University (RUT) Grant Scheme (100/CDADAH/855005) and from the Higher Education Centres of Excellence (HICoE) special funding. The authors gratefully acknowledge the contribution of Ms Vemala Devi, Asokan Muniandy, Ms Nur Aziah Hanapi, and Ms Saadiah Omar for the preparation of this chapter.

REFERENCES

Ahmad, K. and Aziz, Z. 2012. *Mitragyna speciosa* use in the northern states of Malaysia: A cross-sectional study. *Journal of Ethnopharmacology* 141: 446–50.

Assanangkornchai, S., Muekthong, A., Sam-angsri, N. and Pattanasattayawong, U. 2007. The use of *Mitragyna speciosa* (Krathom), an addictive plant, in Thailand. *Substance, Use & Misuse* 42: 2145–57.

Azizi, J., Ismail, S., Mordi, M.N., Ramanathan, S., Said, M.I.M. and Mansor, S. M. 2010. In vitro and in vivo effects of three different *Mitragyna speciosa* Korth leaf extracts on phase II drug metabolizing enzymes – glutathione transferases (GSTs). *Molecules* 15: 432–41.

Babu, K.M., McCurdy, C.R., and Boyer, E.W. 2008. Opioid receptors and legal highs: *Salvia divinorum* and Kratom. *Clinical Toxicology* 46: 146–152.

Boyer, E.W., Babu, K.M., Macalino, G.E. and Compton, W. 2007. Self-treatment of opioid withdrawal with dietary supplement, Kratom. *The American Journal on Addiction* 16: 352–56.

Boyer, E.W., Babu, K.M., Adkins, J.E., McCurdy, C.R. and Halpern, H.N. 2008. Self-treatment of opioid withdrawal using Kratom (*Mitragyna speciosa* Korth.). *Addiction* 103: 1048–50.

Burkill, I.H. 1935. *A Dictionary of the Economic Products of the Malay Peninsula*, Vol. II, pp. 1480–83.

Chan, K.B., Pakiam, C. and Rahim, R. A. 2005. Psychoactive plant abuse: The identification of mitragynine in ketum and in ketum preparations. *Bulletin on Narcotics* LVII, 249–56.

de Moraes, N.V., Moretti, R.A. C., Furr III, E.B., McCurdy, C.R. and Lanchote, V.L. 2009. Determination of mitragynine in rat plasma by LC–MS/MS: Application to pharmacokinetics. *Journal of Chromatography B* 877: 2593–97.

Farah Idayu, N., Taufik Hidayat, M., Moklasm, M.A.M., Sharida, F., Nurul Raudzah, A.R., Shamima, A.R. and Apryani, E. 2011. Antidepressant-like effect of mitragynine isolated from *Mitragyna speciosa* Korth in mice model of depression. *Phytomedicine* 18(5): 402–407.

Ghazali, A.R., Abdullah, R., Ramli, N., Rajab, N.F., Ahmad-Kamal, M.S. and Yahya, N.A. 2011. Mutagenic and antimutagenic activities of *Mitragyna speciosa* Korth extract using Ames test. *Journal of Medicinal Plants Research 5* 8: 1345–48.

Grewal K.S. 1932a. Observation on the pharmacology of mitragynine. *Journal of Pharmacology and Experimental Therapeutics* 46: 251–71.

Grewal, K.S. 1932b. The effect of mitragynine on man. *British Journal of Medical Psychology* 12: 41–58.

Harizal, S.N., Mansor, S.M., Hasnan, J., Tharakan, J.K. and Abdullah, J. 2010. Acute toxicity study of the standardized methanolic extract of *Mitragyna speciosa* Korth in rodent. *Journal of Ethnopharmacology* 2: 404–409.

Hassan, Z., Mustapha, M., Vineswaran, N., Yusoff, N.H.M., Suhaimi, F.W., Rajakumar, V., Vicknasingam, B.K., et al. 2013. From Kratom to mitragynine and its derivatives: Physiological and behavioural effects related to use, abuse and addiction. *Neuroscience and Biohavioural Reviews* 37: 138–51.

Holler, J.M., Vorce, S.P., McDonough-Bender, P.C., Magluilo Jr., J., Solomon, C.J. and Levine, B. 2011. A drug toxicity death involving propylhexedrine and mitragynine. *Journal of Analytical Toxicology* 35: 54–59.

Hooper, D. 1907. The anti-opium leaf. *Pharmaceutical Journal* 78: 453.

Janchawee, B., Keawpradub, N., Chittrakarn, S., Prasettho, S., Wararatananurak, P. and Sawangjareon, K. 2007. A high-performance liquid chromatographic method for determination of mitragynine in serum and its application to a pharmacokinetic study in rats. *Biomedical Chromatography* 21: 176–83.

Jansen, K.L.R. and Prast, C.J. 1988. Psychoactive properties of mitragynine (kratom). *Journal of Psychoactive Drugs* 20(4): 455–57.

Kamal, M.S.A., Ghazali, A.R., Yahya, N.A. Wasiman, M.J. and Ismail, Z. 2012. Acute toxicity study of standardized *M. speciosa* Korth aqueous extract in Sprague Dawley rats. *Journal of Plant Studies* 1(2): 120–28.

Kapp, F.G., Maurer, H.H., Auwärter, V., Winkelmann, M. and Hermanns-Clausen, M. 2011. Intrahepatic cholestasis following abuse of powdered Kratom (*Mitragyna speciosa*). *Journal of Medical Toxicology* 7(3): 227–31.

Khor, B.S., Jamil, M.F., Adenan, M.I. and Shu-Chien, A.C. 2011. Mitragynine attenuates withdrawal syndrome in morphine-withdrawn zebrafish. *PLoS INE* 6(12), e28340.

Kroonstad, R., Roman, M., Thelander, G. and Eriksson, A. 2011. Unintentional fatal intoxications with mitragynine and O-desmethyltramadol from the herbal blend Krypton. *Journal of Analytical Toxicology* 35(4): 242–47.

Kumarnsit, E., Keawpradub, N. and Nuankaew, W. 2006. Acute and long-term effects of alkaloid extract of *Mitragyna speciosa* on food and water intake and body weight in rats. *Fitoterapia* 77: 339–45.

Macko, E., Weisbach, J.A., Douglas, B. 1972. Some observations on the pharmacology of mitragynine. *Archives internationales de pharmacodynamie et de thérapie* 198(1): 145–61.

Matsumoto, K., Mizowaki, M., Suchitra, T., Murakami, Y., Takayama, H., Sakai, S., Aimi, N. and Watanabe, H. 1996a. Central antinociceptive effects of mitragynine in mice: Contribution of descending noradrenergic and serotonergic systems. *European Journal of Pharmacology* 317: 75–81.

Matsumoto, K., Mizowaki, M., Suchitra, T., Takayama, H., Sakai, S., Aimi, N. and Watanabe, H. 1996b. Antinociceptive action of mitragynine in mice: Evidence for the involvement of supraspinal opioid receptors. *Life Sciences* 59: 1149–55.

McWhirter, L. and Morris, S. 2010. A case report of inpatient detoxification after kratom (*Mitragyna speciosa*) dependence. *European Addiction Research* 16(4): 229–31.

Meites, J., Bruni, J.F., Van Vugt, D.A. and Smith, A.F. 1979. Relation of endogenous opioid peptides and morphine to neuroendocrine functions. *Life Sciences* 24(15): 1325–36.

Moklas, M.A.M., Nurul Raudzah A.R., Taufik Hidayat, M., Sharida, F., Farah Idayu, N., Zulkhairi, A. and Shamima, A.R. 2008. A preliminary toxicity study of mitragynine, an alkaloid from *Mitragyna Speciosa* Korth and its effects on locomotor activity in rats. *Advances in Medical and Dental Sciences* 2(3): 56-60.

Nelsen, J.L., Lapoint, J., Hodgman, M.J. and Aldous, K.M. 2010. Seizure and coma following Kratom (*Mitragynina speciosa* Korth) exposure. *Journal of Medical Toxicology* 6(4): 424–26.

Parthasarathy, S., Ramanathan, S., Ismail, S., Adenan, M.I., Mansor, S.M. and Murugaiyah, V. 2010. Determination of mitragynine in plasma with solid-phase extraction and rapid HPLC–UV analysis, and its application to a pharmacokinetic study in rat. *Analytical and Bioanalytical Chemistry* 397: 2023–30.

Rauhala, P., Männistö, P.T. and Tuominen, R.K. 1998. Effect of chronic morphine treatment on thyrotropin and prolactin levels and acute hormone responses in the rat. *Journal of Pharmacology and Experimental Therapeutics* 246(2): 649–54.

Reanmongkol, W., Keawpradub, N. and Sawangjaroen, K. 2007. Effects of the extracts from *Mitragyna speciosa* Korth. Leaves on analgesic and behavioral activities in experimental animals. *Song Journal of Science and Technology* 29(1): 39–48.

Roche, K.M., Hart, K., Sangalli, B., Lefberg, J. and Bayer, M. 2008. Kratom: A case of a legal high. *Clinical Toxicology* 46(7): 598.

Sabetghadam, A., Navaratnam, V. and Mansor, S. M. 2013a. Dose–response relationship, acute toxicity, and therapeutic index between the alkaloid extract of *Mitragyna speciosa* and its main active compound mitragynine in mice. *Drug Development Research* 74: 23–30.

Sabetghadam, A., Ramanathan, S., Sasidharan, S. and Mansor, S.M. 2013b. Subchronic exposure to mitragynine, the principal alkaloid of *Mitragyna speciosa*, in rats. *Journal of Ethnopharmacology* 146: 815–23.

Saidin, N.A., Randall, T., Takayama, H., Holmes, E. and Gooderham, N.J. 2008. Malaysian Kratom, a phyto-pharmaceutical of abuse: Studies on the mechanism of cytotoxicity. *Toxicology* 253(P29): 19–20.

Schmidt, M.M., Sharma, A., Schifano, F. and Feinmann, C. 2011. Legal highs on the net-evaluation of UK-based Websites, products and product information. *Forensic Science International* 206: 92–97.

Shamima, A.R., Fakurazi, S., Hidayat, M.T., Hairuszah, T., Moklas, M.A.M. and Arulselvan, P. 2012. Antinociceptive action of isolated mitragynine from *Mitragyna speciosa* through activation of opioid receptor system. *Int J Mol Sci* 13: 11427–42.

Sheleg, S.V. and Collins, G. B., 2011. A coincidence of addiction to Kratom and severe primary hypothyroidism. *Journal of Addiction Medicine* 5(4): 300–301.

Shellard, E.J. 1989. Ethnopharmacology of kratom and the *Mitragyna* alkaloids. *Journal of Ethnopharmacology* 25(1): 123–24.

Suwanlert, S., 1975.A study of Kratom eaters in Thailand. *Bulletin on Narcotics* 27: 21–27.

Takayama, H. 2004. Chemistry and pharmacology of analgesic indole alkaloids from the rubiaceous plant, Mitragyna speciosa. *Chemical and Pharmaceutical Bulletin* 52(8): 916–28.

Taufik Hidayat, M., Apryani, E., Nabishah, B.M., Moklas, M.A.A., Sharida, F. and Farhan, M.A. 2010. Determination of Mitragynine bound opioid receptors. *Advances in Medical and Dental Sciences* 3(3): 65–70.

Tennekoon, K.H., Jeevathayaparan, S., Angunawala, P., Karunanayake, E.H. and Jayasinghe, K.S.A. 1994. Effect of *Momordica charantia* on key hepatic enzymes. *Journal of Ethnopharmacology* 44(2): 93–97.

Troy, J. D. 2013. New 'legal' highs: Kratom and methoxetamine. *Current Psychiatry* 12(4).

Tsuchiya, S., Miyashita, S., Yamamoto, M., Horie, S., Sakai, S., Aimi, N., Takayama, H. and Watanabe, K. 2002. Effect of mitragynine, derived from Thai folk medicine, on gastric acid secretion through opioid receptor in anesthetized rats. *European Journal of Pharmacology* 443: 185–88.

Vicknasingam, B., Narayanan, S., Beng, G. T. and Mansor, S. M. 2010. The informal use of ketum (*Mitragyna speciosa*) for opioid withdrawal in the northern states of peninsular Malaysia and implications for drug substitution therapy. *International Journal of Drug Policy* 21: 283–88.

Watanabe, K., Yano, S., Horie, S., Sakai, S., Takayama, H. and Ponglux, D. 1992. Pharmacological profiles of 'kratom' (*Mitragyna speciosa*), a Thai medicinal plant, with special reference to its analgesic activity. In: Tongroach,

P., Watanabe, H., Ponglux, D., Suvanakoot, U., and Ruangrungsi, N. (eds). *Advances in Research on Pharmacologically Active Substances from Natural Products*. 125–32. Chiang Mai: Chiang Mai Univ Bull.

Watanabe, K., Yano, S., Horie, S. and Yamamoto, L.T. 1997. Inhibitory effect of mitragynine, an alkaloid with analgesic effect from Thai medicinal plat *Mitragyna speciosa*, on electrically stimulated contraction of isolated guinea-pig ileum through the opioid receptor. *Life Sciences* 60(12): 933–42.

Weber, M., Modemann, S., Schipper, P., Trauer, H., Franke, H. and Illes, P. 2006. Increased polysialic acid neural cell adhesion molecule expression in human hippocampus of heroin addicts. *Neuroscience* 138: 1215–23.

20 The Use of Animal Models to Measure Tolerance, Dependence, and Abuse Potential

Scott M. Rawls

CONTENTS

20.1 INTRODUCTION

Mitragynine is the major alkaloid in a psychoactive plant preparation known as kratom, among other names. Evidence from human users suggests that mitragynine possesses abuse liability, produces tolerance, and can cause withdrawal symptoms following discontinuation after extended use. Preclinical evidence suggests that mitragynine may possess significant risk of abuse liability as well as a spectrum of therapeutic effects, including antinociception, muscle relaxation, and anti-inflammatory effects. Its mechanism remains poorly elucidated, but it does act within the central nervous system, and is thought to involve opioid receptor activation, modulation of calcium channel activity, and impacts on monoamine systems (Matsumoto et al. 2005, 2008). The goal of this review is to first present a general comparison of mitragynine with other prevalent natural products and designer drugs and, second, to discuss animal models that can be used to develop the neuropharmacological profile of mitragynine, with a focus on abuse and dependence liability.

The addiction potential of designer drugs and naturally occurring opioids has emerged as an international health concern for clinicians and government officials. Perhaps the greatest attention has been directed toward new classes of designer synthetic drugs synthesized to mimic the effects of established drugs of abuse, with a 20-fold increase in reported human exposures from 2010 to 2011 (Deluca et al. 2009; James et al. 2011). Among these new classes of drugs are the synthetic cathinones, a group of β-ketone amphetamine compounds derived from cathinone, the active stimulant in the khat plant (*Catha edulis*) (Carroll et al. 2012). Chemical alterations, and functional group substitutions, to the core structure of the parent cathinone compound have yielded a large number of new synthetic cathinone psychostimulants, the most commonly abused being MDPV (3,4-methylenedioxypyrovalerone) and methylone (3,4-methylenedioxy-N-methylcathinone) and mephedrone (4-methylmethcathinone) in the United Kingdom. In user reports, "bath salts" are described as having similar psychostimulant effects to those found with cocaine, MDMA, and methamphetamine (Brandt et al. 2010; Brunt et al. 2011; Deluca et al. 2009; Schifano et al. 2011). These negative clinical presentations led the United States government to categorize MEPH, MDPV and methylone as Schedule I drugs in October 2011, eventually leading to a permanent Schedule I distinction for MEPH and MDPV in July 2012, and methylone in 2013 (Wood 2013).

Among naturally occurring opioids, salvinorin (Sal) A, a kappa opioid receptor agonist with a short duration of action in vivo, has received a great deal of publicity. Chewing or smoking the leaves of *Salvia divinorum* or drinking juice of crushed leaves causes hallucinations in humans (Valdés 1994; Siebert 1994). Sal A is a potent and selective agonist for the kappa opioid receptor that binds with a high affinity similar to U50,488H, a selective KOPR agonist, but does not show significant affinities to mu and delta opioid receptors or the nociceptin/orphanin FQ receptor (Roth et al. 2002; Wang et al. 2005). Sal A also promotes internalization and down-regulation of the human kappa opioid receptors in cultured cells (Wang et al. 2005).

20.2 BODY TEMPERATURE: INVOLVEMENT OF OPIOIDS

Despite the significant influence of mu, kappa, and delta opioid receptors on thermoregulation, effects of kratom alkaloids on body temperature have not been investigated. Based on the rich history of using body temperature assays to define the pharmacological profile of novel opioids, it can be predicted that much could be learned about the profile of mitragynine and its derivatives by defining their effects on body temperature. Much of the early work investigating a role for opioid receptors in thermoregulation was accomplished with morphine. The administration of morphine to rats at doses of 4–15 mg/kg morphine produces robust hyperthermia, but progressively higher doses induce hypothermia (Geller et al. 1983). Experiments using selective opioid receptor agonists and antagonists reveal that mu opioid receptor activation is responsible for the hyperthermic response to morphine, whereas kappa opioid receptor activation mediates the hypothermic effect of morphine (Spencer et al. 1988; Handler et al. 1992; Adler and Geller 1987, 1993; Adler 1983; Geller et al. 1986).

Early work suggested only a minor role for delta opioid receptors in thermoregulation, but studies using selective, non-peptide delta opioid agonists now suggest a more significant role (Broccardo and Imprato 1992; Handler et al. 1992, 1994; Pohorecky et al. 1999; Baker and Meert 2002; Salmi et al. 2003; Rawls et al. 2005; Rawls and Cowan 2006). Low doses of deltorphin-II, a delta-2 opioid receptor agonist, injected centrally produce hypothermia, whereas higher doses produce a biphasic response of hypothermia followed by hyperthermia (Salmi et al. 2003). The hypothermic effect of deltorphin-II is antagonized by the delta opioid receptor antagonist naltrindole, thus indicating that delta opioid receptor activation was the mechanism.

In contrast to the hypothermic effect produced by kappa and delta opioid receptor activation, mu opioid receptor activation results in hyperthermia (Geller et al. 1982, 1983; Spencer et al. 1988; Adler and Geller 1993; Rawls et al. 2003). Morphine administered systemically at doses of 5–16 mg/kg produces hyperthermia in rats, whereas higher doses produce a biphasic response of a brief hyperthermia followed by a pronounced, enduring hypothermia (Geller et al. 1982, 1983; Adler and Geller 1993). Pretreatment of rats with naltrexone completely blocks the hyperthermic response to morphine, indicating that low doses of morphine produce hyperthermia through activation of mu opioid receptors (Geller et al. 1983; Chen et al. 1996; Rawls et al. 2003; Spencer et al. 1988). The importance of mu opioid receptors in the production of hyperthermia is further confirmed by experiments showing that the mu opioid agonists DAMGO ([D-Ala2, N-MePhe4, Gly-ol]-enkephalin) and PLO17 ([N-MePhe3,D-Pro4]morphiceptin), injected into the ventricles or hypothalamus, induce hyperthermia that is blocked by mu opioid receptor antagonists CTAP and beta-funaltrexamine (Adler and Geller 1993; Handler et al. 1992). Results from knockout mice reveal that genetic deletion of the mu opioid receptor does not impact basal body temperature or circadian body temperature rhythm, but the hyperthermia produced by a low dose of morphine (1 mg/kg) is enhanced in wild-type mice relative to mu opioid receptor knockout mice (Benamar et al. 2007).

Predictions about the effects of mitragynine on body temperature can be inferred from its opioid receptor profile, which indicates that it displays a high affinity for mu opioid receptors (Yamamoto et al. 1999). Relative to morphine, the affinity of mitragynine to kappa and delta opioid receptors is substantially lower. The antinociceptive effects of mitragynine are mediated by mu and delta opioid receptors located in supraspinal regions (Matsumoto et al. 1996; Tohda et al. 1997; Thongpradichote et al. 1998). On the basis of mitragynine's strong mu opioid receptor profile, morphine-like effects on body temperature would be predicted, with hyperthermia manifesting at low doses resulting primarily from activation of mu opioid receptors. At higher doses of mitragynine, hyperthermia and hypothermia might be predicted, with the decline in body temperature being mediated by activation of delta and kappa opioid receptors. It is also possible that metabolites of mitragynine, such as pseudoindoxyl, that display potent agonist properties at opioid receptors will contribute to its effects on body temperature.

20.3 ABUSE LIABILITY TESTING

The addictive properties of mitragynine have not been investigated extensively in animals or humans, but mitragynine has a pharmacological profile that is suggestive of an addictive substance. In addition to producing morphine-like effects through mu opioid receptor activation, mitragynine displays stimulant-like effects, likely due to enhancement of monoamine levels, which may underlie its acute rewarding effects (Farah Idayu et al. 2011). Mu opioid receptors also mediate antinociceptive effects of mitragynine reported in a number of standard preclinical assays, including the acetic acid writing and tail-flick tests, as naloxone completely blocks the antinociception (Takayama et al. 2002; Horie et al. 2005). The antinociceptive effect of mitragynine is about 10-fold more potent than morphine; however, it is about 50-fold less potent than 7-hydroxymitragynine (7-HMG), a minor constituent of kratom extracts (Matsumoto et al. 1996; 2005; Watanabe et al. 1997), outcomes that are presumably related to the greater lipophilicity and brain penetrability of 7-HMG (Matsumoto et al. 2006).

Abuse liability for mitragynine is suggested not only by its pharmacological profile but also by anecdotal evidence from humans. With regard to the latter, human users have reported feeling energetic and euphoric minutes after consumption of mitragynine, and these types of effects are often harbingers of addictive consumption of drugs (Müller and Schumann 2011). Further, withdrawal symptoms, including anxiety and craving, have been reported following discontinuation of chronic mitragynine consumption (Suwanlert 1975).

While the therapeutic potential of mitragynine has been demonstrated in animal models of pain, as discussed above, the potential for mitragynine to cause addiction has not been extensively investigated in preclinical studies. A number of animal models, such as conditioned place preference (CPP), self-administration, and drug discrimination, are useful in the preclinical evaluation of abuse and dependence liability. Abuse and dependence are not the same. Abuse involves the consumption of drugs for nonmedical purposes, normally because the drug produces positive reinforcing effects, whereas dependence refers to the need to continue taking a drug to avoid withdrawal effects on drug discontinuation and is not necessarily associated with any reinforcing properties of the drug. The difference between abuse and dependence can be demonstrated pharmacologically. Some substances such as fluoxetine (Prozac), the world's best-selling antidepressant, causes mild withdrawal effects upon discontinuation of consumption but does not possess positive reinforcing properties (Zajecka et al. 1997). On the other hand, substances such as morphine, heroin, cocaine, and amphetamines produce both abuse and dependence. The distinction between abuse and dependence is important, and different in vivo assays are utilized to determine this distinction.

For drug abuse evaluation of novel substances, such as mitragynine, a dual approach is often pursued. The first step is simple in that it compares the novel test substance with a prototypical drug that has an established profile. This part of the approach can determine whether the drug enters the central nervous system, interacts with receptors, enzymes, or transporters with which prototypical drugs interact, and produces an effect that is suggestive of addiction. The study can be discontinued if results from these assays are not indicative of abuse or dependence liability. However, if these assays reveal the potential for abuse liability, as in the case of mitragynine, then a second level of evaluations are necessary to determine whether the test substance produces reinforcing effects and how closely the test substance resembles more established drugs of abuse. To assess the positive reinforcing properties of a substance, two assays, CPP and self-administration, are normally employed. To assess mechanism of action and how closely the test substance resembles a prototypic drug, drug-discrimination assays are used. Dependence is evaluated by the occurrence of withdrawal symptoms upon discontinuation of exposure to the novel compound. The following sections discuss these assays in further detail, with a focus on how they can be used to define the neuropharmacological profile of mitragynine.

20.3.1 CONDITIONED PLACE PREFERENCE (CPP)

The CPP assay is the simplest and most rapid procedure for investigating rewarding effects of a substance. In the CPP procedure, animals are repeatedly exposed to a distinct environment in the presence of a positively reinforcing substance. When an animal is later given the choice between the environment that has been paired with the reinforcing drug and an environment that has not been paired with the drug, it will show a preference for the previously drug-paired environment (Tzschentke 1998; Soderman and Unterwald 2008; Watterson et al. 2013). The paradigm can vary, but normally consists of three phases, the pre-test, conditioning, and post-test, and can utilize a biased or unbiased design. A pre-test is conducted in the absence of drug to determine the amount of time spent in each environment. With a biased design, the animal being tested shows a natural preference for a specific environment even in the absence of a drug. During conditioning the positive reinforcer is paired with the non-preferred environment. The conditioning period can vary with the animal and the reinforcer. Following conditioning, a post-test is conducted in the absence of drug in which the animal is given a choice to spend time in the previously drug-paired or the environment that was not previously associated with the drug. A preference score, defined as the difference between the post-test and pre-test times, is then calculated to determine whether the rewarding effects of the drug shifted the preference of the animal more toward the least desirable environment. It is well established that a number of species, including rats, mice, and invertebrates, display CPP to established drugs of abuse (Collier and Echevarria 2013; Tallarida et al. 2013). Further, the CPP assay is sensitive to a wide range of substances. Strong reinforcers such as cocaine produce robust CPP under a variety of conditions, whereas CPP produced by weaker reinforcers, such as ecstasy and benzodiazepines, is more dependent on experimental design, including the length of the conditioning phase and the dose (Sanchis-Segura and Spanagel 2006).

The specific ability of mitragynine itself to produce CPP has not yet been investigated; however, kratom extracts, which contain a number of other alkaloids and their derivatives, have been investigated in a place preference assay (Matsumoto et al. 2008). The mitragynine derivative 7-HMG, discussed above, produced a preference shift relative to vehicle in mice, while a number of other derivatives of mitragynine were ineffective (Matsumoto et al. 2008).

20.3.2 Self-Administration

While the CPP results cited above suggest that mitragynine is likely to possess positive reinforcing effects, it is important to test this prediction in animal models of self-administration. Self-administration assays are used to investigate the intrinsic rewarding effects of a substance and are generally regarded as the most effective preclinical "models" of addiction. One reason for the status of the self-administration model in the field of addiction is that it is a contingent model, which means that the animal, rather than the experimenter, delivers the drug, usually by pressing a lever (Oleson and Roberts 2008). Contingency means that the rewarding effect of the drug is dependent on the correct response of the animal (e.g., pressing the correct lever). Further, because the animal is administering the drug, it can titrate drug-taking behavior to match reward and avoid aversive side effects that may be associated with higher doses. Because of these combined factors, the self-administration model more closely models human drug-taking behavior compared to non-contingent models such as CPP.

A defining feature of addictive behavior is the increase in time and energy spent in gaining access to drugs, which is equated to how motivated the subject is willing to work to procure the reward. Self-administration assays, through the use of progressive-ratio (PR) schedules of reinforcement, can assess the motivation of animals to respond for a specific drug (Griffiths et al. 1978; Griffiths and Balster 1979). Under PR schedules, the receipt of a reward is made progressively more difficult for the animal to obtain. For example, in the early stages of a self-administration experiment, an animal may receive a reward following one correct lever press. However, over time, the animal may be required to provide 10 correct lever presses for a single reward, and so on. Eventually, the animal reaches a "breaking point," which is defined as the point at which the animal is no longer willing to work to achieve the reward. In general, the higher the breaking point, the harder the animal is willing to work for the reward, and vice versa.

Another defining feature of addiction is the craving that a subject experiences upon discontinuation of chronic consumption of a drug, and it is this craving that is often responsible for the high relapse rates among drug addicts. The reinstatement model of drug-seeking behavior is a widely used animal model to assess craving and relapse (Steketee and Kalivas 2011). It is important to note that standard self-administration assays may fail to incorporate another key component of the addiction process, which is an escalation or heightened drive toward drug-seeking or drug-taking behavior that develops over time and is coupled with greater negative health and social consequences in humans. Recognition of the importance of modeling this transition from recreational drug use to drug addiction and understanding its neural correlates has resulted in the development of additional models of sensitized self-administration behavior (Ahmed and Koob 1998; Roberts et al. 2007).

It is important to note that self-administration assays do not reveal mechanism. To better assess similarities between a test drug, such as mitragynine, and a prototypical drug of abuse, drug discrimination assays are preferable options for linking abuse liability with a specific mechanism of action. A drawback of the drug discrimination assay is that it may have limited applicability to a test substance possessing novel, or multiple, mechanisms of action. One approach for circumventing this potential confound is to use the test substance as the training drug, which then enables mechanistic comparisons between the test drug and prototypic drugs of abuse with established mechanisms of action. In effect, if an animal is unable to discriminate between a test drug and an established drug (i.e., the drugs fully substitute for each other), their mechanisms are assumed to be similar. However, in the case in which an animal is able to discriminate, or partially discriminate, the test drug is assumed to work through a mechanism of action that is not entirely the same as the established drug. In the case of mitragynine, because it acts through mu opioid and monoaminergic mechanisms, one might predict that it would fully substitute for opioid-based drugs such as heroin and morphine while partially substituting for psychostimulants such as cocaine and amphetamine.

20.4 DRUG DEPENDENCE

Drug tolerance occurs when the effect of a drug progressively declines over time, thus requiring that the dose of the drug be escalated to achieve the desirable effect. A common example of this is the tolerance that develops to the analgesic effects of morphine during chronic exposure. In effect the analgesic efficacy of morphine decreases during repeated exposure and requires that greater doses of morphine be administered to counter the loss in efficacy. As the doses are escalated to maintain analgesic efficacy, dependence can develop to the drug. Drug dependence is different from drug taking in that dependence is characterized by the manifestation of withdrawal symptoms following discontinuation of consumption. Yet dependence is associated with features of drug taking, including frequency and duration of consumption, pharmacokinetics of the drugs, and the degree of dose escalation that is required to maintain the desired effect, and offset tolerance, over a period of time. Drug abuse can cause dependence, as is the case with opioid drugs, nicotine, alcohol, and psychostimulants. Further, in order to avoid the aversive symptoms of withdrawal that occur following drug discontinuation, subjects often relapse and continue drug use (West and Gossop 1994). Dependence, however, can also be produced by drugs that do not possess positive reinforcing effects, such as drugs that block 5-HT reuptake (Zajecka et al. 1997).

For mitragynine derivatives, animal models have been used to assess tolerance, dependence, and withdrawal. Repeated exposure to 7-HMG for 5 days results in the development of tolerance in mice, as determined by the loss of analgesic efficacy (Matsumoto et al. 2005, 2008). Furthermore, the tolerance to 7-HMG is mediated by mu opioid receptors, and 7-HMG displays bidirectional cross-tolerance to morphine. Finally, naloxone precipitated withdrawal signs following repeated exposure to 7-HMG, indicating that this mitragynine derivative, akin to morphine, is capable of producing physical dependence (Matsumoto et al. 2005).

20.5   CONCLUSION

Major strides in the development of the neuropharmacological profile of mitragynine have been made over the past decade, particularly regarding its potential therapeutic uses as an analgesic, anti-inflammatory agent, and muscle relaxant. Although it is apparent from the ability of mitragynine to enhance opioid and monoamine transmission that it may have some therapeutic utility, its relative liability for abuse and dependence remains unclear. What is known is that kratom extracts can produce rewarding effects in non-contingent animal models of addiction and produce tolerance and physical dependence following repeated exposure. The next step is to use self-administration assays to assess the positive reinforcing and drug-seeking effects of mitragynine and drug discrimination assays to assess mechanistic similarities between mitragynine and prototypic drugs of abuse.

REFERENCES

Adler, M.W. 1982. Multiple opiate receptors and their different ligand profiles. *Ann N Y Acad Sci* 398: 340–51.

Adler, M.W. and Geller, E.B. 1987. Hypothermia and poikilothermia induced by a kappa-agonist opioid and a neuroleptic. *Eur J Pharmacol* 140: 233–37.

Adler, M.W., Geller, E.B., Rosow, C.E., and Cochin, J. 1988. The opioid system and temperature regulation. *Annu Rev Pharmacol Toxicol* 28: 429–49.

Adler, M.W. and Geller, E.B. 1993. Physiological functions of opioids: Temperature regulation. In: *Handbook of Experimental Pharmacology*, Herz, A., Akil, H., and Simon, E.J., eds. Vol. 104/2: *Opioids II*. 205–38. Berlin, Heidelberg: Springer-Verlag.

Ahmed, S.H. and Koob, G.F. 1998. Transition from moderate to excessive drug intake: Change in hedonic set point. *Science* 282: 298–300.

Baker, A.K. and Meert, T.F. 2002. Functional effects of systemically administered agonists and antagonists of mu, delta, and kappa opioid receptor subtypes on body temperature in mice. *J Pharmacol Exp Ther* 302: 1253–64.

Benamar, K., Yondorf, M., Barreto, V.T., Geller, E.B., and Adler, M.W. 2007. Deletion of mu-opioid receptor in mice alters the development of acute neuroinflammation. *J Pharmacol Exp Ther* 323: 990–94.

Brandt, S.D., Sumnall, H.R., Measham, F., and Cole, J. 2010. Analyses of second-generation 'legal highs' in the UK: Initial findings. *Drug Test Anal* 2: 377–82.

Broccardo, M., Improta, G., and Tabacco, A. 1998. Central effect of SNC 80, a selective and systemically active delta-opioid receptor agonist, on gastrointestinal propulsion in the mouse. *Eur J Pharmacol* 342: 247–51.

Brunt, T.M., Poortman, A., Niesink, R.J., and van den Brink, W. 2011. Instability of the ecstasy market and a new kid on the block: Mephedrone. *J Psychopharmacol* 25: 1543–47.

Carroll, F.I., Lewin, A.H., Mascarella, S.W., Seltzman, H.H., and Reddy, P.A. 2012. Designer drugs: A medicinal chemistry perspective. *Ann N Y Acad Sci* 1248: 18–38.

Chen, X.H., Geller, E.B., DeRiel, J.K, Liu-Chen, L.Y., and Adler, M.W. 1996. Antisense confirmation of mu- and kappa-opioid receptor mediation of morphine's effects on body temperature in rats. *Drug Alcohol Depend* 43: 119–24.

Collier, A.D. and Echevarria, D.J. 2013. The utility of the zebrafish model in conditioned place preference to assess the rewarding effects of drugs. *Behav Pharmacol* 24: 375–83.

Deluca, P., Schifano, F., Davey, Z., et al. 2010. Mephedrone Report: Psychonaut Web Mapping research report. http: //www.psychonautproject.eu/documents/reports/Mephedrone.pdf (last accessed Feb 22, 2014).

Geller, E.B., Hawk, C., Keinath, S.H., Tallarida, R.J., and Adler, M.W. 1983. Subclasses of opioids based on body temperature change in rats: Acute subcutaneous administration. *J Pharmacol Exp Ther* 225: 391–98.

Geller, E.B., Hawk, C., Tallarida, R.J., and Adler, M.W. 1982. Postulated thermoregulatory roles for different opiate receptors in rats. *Life Sci* 31: 2241–44.

Geller, E.B., Rowan, C.H., and Adler, M.W. 1986. Body temperature effects of opioids in rats: Intracerebroventricular administration. *Pharmacol Biochem Behav* 24: 1761–55.

Griffiths, R.R. and Balster, R.L. 1979. Opioids: Similarity between evaluations of subjective effects and animal self-administration results. *Clin Pharmacol Ther* 25: 611–17.

Griffiths, R.R., Bigelow, G.E., and Liebson, I. 1978. Experimental drug self-administration: Generality across species and type of drug. *NIDA Res Monogr* 20: 24–43.

Handler, C.M., Geller, E.B., and Adler, M.W. 1992. Effect of mu-, kappa-, and delta-selective opioid agonists on thermoregulation in the rat. *Pharmacol Biochem Behav* 43: 1209–16.

Handler, C.M., Piliero, T.C., Geller, E.B., and Adler, M.W. 1994. Effect of ambient temperature on the ability of mu-, kappa- and delta-selective opioid agonists to modulate thermoregulatory mechanisms in the rat. *J Pharmacol Exp Ther* 268: 847–55.

Horie, S., Koyama, F., Takayama, H., et al. 2005. Indole alkaloids of a Thai medicinal herb, *Mitragyna speciosa*, that has opioid agonistic effect in guinea-pig ileum. *Planta Med* 71: 231–36.

Idayu, N.F., Hidayat, M.T., Moklas, M.A., et al. 2011. Antidepressant-like effect of mitragynine isolated from *Mitragyna speciosa* Korth in mice model of depression. *Phytomedicine* 18: 402–407.

James, D., Adams, R.D., Spear,s R., et al. 2011. Clinical characteristics of mephedrone toxicity reported to the U.K. National Poisons Information Service. *Emerg Med J* 28: 686–89.

Matsumoto, K., Hatori, Y., Murayama, T., et al. 2006. Involvement of mu-opioid receptors in antinociception and inhibition of gastrointestinal transit induced by 7-hydroxymitragynine, isolated from Thai herbal medicine *Mitragyna speciosa*. *Eur J Pharmacol* 549: 63–70.

Matsumoto, K., Horie, S., Takayama, H., et al. 2005. Antinociception, tolerance and withdrawal symptoms induced by 7-hydroxymitragynine, an alkaloid from the Thai medicinal herb *Mitragyna speciosa*. *Life Sci* 78: 2–7.

Matsumoto, K., Mizowaki, M., Suchitra, T., et al. 1996. Antinociceptive action of mitragynine in mice: Evidence for the involvement of supraspinal opioid receptors. *Life Sci* 59: 1149–55.

Matsumoto, K., Takayama, H., Narita, M., et al. 2008. MGM-9 [(E)-methyl 2-(3-ethyl-7a,12a-(epoxyethanoxy)-9-fluoro-1,2,3,4,6,7,12,12b-octahydro-8-methoxyindolo[2,3-a]quinolizin-2-yl)-3-methoxyacrylate], a derivative of the indole alkaloid mitragynine: A novel dual-acting mu- and kappa-opioid agonist with potent antinociceptive and weak rewarding effects in mice. *Neuropharmacology* 55: 154–65.

Oleson, E.B. and Roberts, D.C. 2012. Cocaine self-administration in rats: Threshold procedures. *Methods Mol Biol* 829: 303–19.

Pohorecky, L.A., Skiandos, A., Zhang, X., Rice, K.C., and Benjamin, D. 1999. Effect of chronic social stress on delta-opioid receptor function in the rat. *J Pharmacol Exp Ther* 290: 196–206.

Rawls, S.M., Adler, M.W., Gaughan, J.P., Baron, A., Geller, E.B., and Cowan, A. 2003. NMDA receptors modulate morphine-induced hyperthermia. *Brain Res* 984: 76–83.

Rawls, S.M. and Cowan, A. 2006. Modulation of delta opioid-evoked hypothermia in rats by WAY 100635 and fluoxetine. *Neurosci Lett* 398: 319–24.

Roberts, D.C., Morgan, D., and Liu, Y. 2007. How to make a rat addicted to cocaine. *Prog Neuropsychopharmacol Biol Psychiatry* 31: 1614–24.

Roth, B.L., Baner, K., Westkaemper, R., et al. 2002. Salvinorin A: A potent naturally occurring nonnitrogenous kappa opioid selective agonist. *Proc Natl Acad Sci USA* 99: 11934–39.

Salmi, P., Kela, J., Arvidsson, U., and Wahlestedt, C. 2003. Functional interactions between delta- and mu-opioid receptors in rat thermoregulation. *Eur J Pharmacol* 458: 101–106.

Sanchis-Segura, C. and Spanagel, R. 2006. Behavioural assessment of drug reinforcement and addictive features in rodents: An overview. *Addict Biol* 11: 2–38.

Schifano, F., Albanese, A., Fergus, S., et al. 2011. Mephedrone (4-methylmethcathinone; 'meow meow'): chemical, pharmacological and clinical issues. *Psychopharmacology (Berl)* 214: 593–602.

Siebert, D.J. 1994. Salvia divinorum and salvinorin A: New pharmacologic findings. *J Ethnopharmacol* 43: 53–56.

Soderman, A.R. and Unterwald, E.M. 2008. Cocaine reward and hyperactivity in the rat: Sites of mu opioid receptor modulation. *Neuroscience* 154: 1506–16.

Spencer, R.L., Hruby, V.J., and Burks, T.F. 1988. Body temperature response profiles for selective mu, delta and kappa opioid agonists in restrained and unrestrained rats. *J Pharmacol Exp Ther* 246: 92–101.

Steketee, J.D. and Kalivas, P.W. 2011. Drug wanting: Behavioral sensitization and relapse to drug-seeking behavior. *Pharmacol Rev* 63: 348–65.

Suwanlert, S. 1975. A study of kratom eaters in Thailand. *Bull Narc* 27: 21–27.

Takayama H, Ishikawa H, Kurihara M, et al. 2002. Studies on the synthesis and opioid agonistic activities of mitragynine-related indole alkaloids: Discovery of opioid agonists structurally different from other opioid ligands. *J Med Chem* 45: 1949–56.

Tallarida, C.S., Egan, E., Alejo, G.D., Raffa, R., Tallarida, R.J., and Rawls, S.M. 2014. Levamisole and cocaine synergism: A prevalent adulterant enhances cocaine's action in vivo. *Neuropharmacology* 79C: 590–95.

Thongpradichote, S., Matsumoto, K., Tohda, M., et al. 1998. Identification of opioid receptor subtypes in antinociceptive actions of supraspinally-administered mitragynine in mice. *Life Sci* 62: 1371–78.

Tohda, M., Thongpradichote, S., Matsumoto, K., et al. 1997. Effects of mitragynine on cAMP formation mediated by delta-opiate receptors in NG108-15 cells. *Biol Pharm Bull* 20: 338–40.

Tzschentke, T.M. 1998. Measuring reward with the conditioned place preference paradigm: A comprehensive review of drug effects, recent progress and new issues. *Prog Neurobiol* 56: 613–72.

Valdés, L.J. 1994. *Salvia divinorum* and the unique diterpene hallucinogen, salvinorin (divinorin) A. *J Psychoactive Drugs* 26: 277–83.

Wang, Y., Tang, K., Inan, S., et al. 2005. Comparison of pharmacological activities of three distinct kappa ligands (salvinorin A, TRK-820 and 3FLB) on kappa opioid receptors in vitro and their antipruritic and antinociceptive activities in vivo. *J Pharmacol Exp Ther* 312: 220–30.

Watanabe, K., Yano, S., Horie, S., and Yamamoto, L.T. 1997. Inhibitory effect of mitragynine, an alkaloid with analgesic effect from Thai medicinal plant Mitragyna speciosa, on electrically stimulated contraction of isolated guinea-pig ileum through the opioid receptor. *Life Sci* 60: 933–42.

Watterson, L.R., Watterson, E., and Olive, M.F. 2013. Abuse liability of novel 'legal high' designer stimulants: Evidence from animal models. *Behav Pharmacol* 24: 341–55.

Wood, K.E. 2013. Exposure to bath salts and synthetic tetrahydrocannabinol from 2009 to 2012 in the United States. *J Pediatr* 163: 213–16.

Yamamoto, L.T., Horie, S., Takayama, H., et al. 1999. Opioid receptor agonistic characteristics of mitragynine pseudoindoxyl in comparison with mitragynine derived from Thai medicinal plant *Mitragyna speciosa*. *Gen Pharmacol* 33: 73–81.

Zajecka, J., Tracy, K.A., and Mitchell, S. 1997. Discontinuation symptoms after treatment with serotonin reuptake inhibitors: A literature review. *J Clin Psychiatry* 58: 291–97.

21 Kratom
The Epidemiology, Use and Abuse, Addiction Potential, and Legal Status

Zoriah Aziz

CONTENTS

21.1 INTRODUCTION

Kratom (*Mitragyna speciosa* Korth.) is a tree native to areas of Southeast Asia, namely Thailand, Malaysia, Myanmar, and Indonesia. The plant is also known as *ithang, krathom, kakuam,* and *thom* (Thailand), *biak* or *ketum* (Malaysia), and *mambog* (Philippines). Its leaves have historically been used for recreational and medicinal purposes.

Considerable scientific interest has been generated around kratom for two main reasons. First, chronic kratom consumption can lead to addiction. Second, its psychoactive effects are described as unusual because of its dual properties acting both as a stimulant and narcotic. Several reports claim that it has stimulant effects at lower doses and opioid effects at higher doses. Yet the pharmacological bases underlying these two oppos-ing effects (stimulant versus narcotic) are still unclear.

This chapter reviews the epidemiology, types of preparation, use and abuse, addiction potential, and legal status of kratom. The physical and mental effects will only be described briefly where relevant, as these effects are covered elsewhere in this book.

21.2 EPIDEMIOLOGY OF KRATOM

Compared to khat, a plant that also has psychoactive effects, little is known about the extent of kratom use outside countries in Southeast Asia. With wider use of the Internet in the 1990s, interest in kratom has increased and kratom is now available in many countries worldwide. Over the past decade, sales of kratom increased considerably, as evidenced by the number of advertisements for its sale on the Internet (Schmidt et al. 2011). Kratom is often marketed as giving a "legal high," implying that it gives similar effects to illegal drugs such as cannabis or opium, but that it is a legal substance to possess or to use. An online survey conducted in 2011 by the European Monitoring Centre for Drugs and Drug Addiction showed kratom to be one of the new psychoactive substances that is most widely offered, with 20% of online retailers (128 out of 631) shipping it to the EU (European Monitoring Centre for Drugs and Drug Addiction 2012). A survey on natural psychoac-tive substances in 2012 by the United Nation Office of Drugs and Crime also reported that kratom was among the top plant-based substances used, along with khat and *Salvia divinorum* (United Nations Office on Drugs and Crime 2013). Although the prevalence of kratom use is poorly described in countries outside of Southeast Asia, it appears from the Internet that the use is increasing in at least two populations: individuals who self-treat their opi-oid withdrawal, and individuals who self-manage their chronic pain without physician supervision (Boyer et al. 2007). More recently, a large number of websites on the Internet are promoting the use of kratom as a recreational drug, with detailed instructions on how to prepare and use it to achieve the euphoric effects. Given its increasing popularity on the Internet, it is possible that the use of kratom could be fast spreading worldwide.

In Southeast Asia, its use is most prevalent in Thailand, particularly the southern region (Ministry of Justice 2010). The 2007 National Household Survey of Substance Use in Thailand indicated that kratom was the major substance used, surpassing the use of cannabis, inhalants, methamphet-amine, and opium (Assanangkornchai et al. 2008). The highest rate of kra-tom use was in the southern region of Thailand, with about 9.3% of its adult population being lifetime users, even though its use is illegal. The users are mainly villagers and agricultural workers. The same survey conducted in

2008 showed an increase in the number of people who used kratom, from 1.02 million in 2007 to 1.08 million in 2008 (Tanguay 2011). Overall, in Thailand kratom remains one of the three most commonly used illegal substances along with cannabis and methamphetamine (United Nations Office on Drugs and Crime 2009).

In Malaysia, kratom is widely available as drinks and teas in the northern states that share a common border with Thailand (Chan et al. 2005). While there is limited clinical data on human use of kratom in Malaysia, a cross-sectional national survey conducted among youths in Malaysia found that of all drugs abused, 0.5% involved the use of kratom, mainly abused in the northern states of Peninsular Malaysia (Nazar Mohamed et al. 2008). A more recent survey of 530 rural users in the northern states of Malaysia found that about 88% of the respondents use kratom on a daily basis (Ahmad and Aziz 2012). The majority of the respondents obtained kratom from trees grown in the village, and about 12% of respondents bought it from their village shops. Among the reasons given for using kratom were stamina and endurance, social and recreational use, to aid in sleeping, and to reduce opiate withdrawal symptoms. Therefore it appears that, similar to Thailand, in Malaysia kratom remains a substance associated with the rural areas.

In Southeast Asia, particularly in Thailand and Malaysia, kratom has always been associated with males of the older age groups from the rural areas, whereas substances such as marijuana and methamphetamines are more prevalent among teenagers and young adults. However, recent trends in Thailand and Malaysia indicate that more and more adolescents and younger adults are now turning to kratom mainly as a replacement for other, unobtainable, recreational substances (Assanangkornchai et al. 2007; Nazar Mohamed et al. 2008). Compounding the widespread use among adolescents is the use of kratom with a cocktail of other additive or toxic substances—for example, the mixing of kratom with cough syrup containing dextromethorphan or with mosquito coils containing pyrethrum (Tanguay 2011). In Sweden, powdered kratom has been mixed with O-desmethyltramadol (the active metabolite of tramadol), which has high affinity for the mu-opioid receptor (Kronstrand et al. 2011). This mixture, known as Krypton, has been suspected to be the cause of nine deaths.

21.3 PREPARATIONS OF KRATOM

Wray (1907) first reported two methods of preparing the leaves for consumption. In the first method, the leaves were dried in the sun. The dried crisp leaves were then rubbed between the palms of the hands to form a powder. The powder was then mixed with water or coffee for drinking. The second method involved boiling the sun-dried leaves in water.

The mixture was then filtered and the filtrate evaporated. The thick syrup obtained was mixed with hot water for drinking or smoked like opium using a special pipe. It was recorded in the same report that the effects of the leaves, whether taken internally or smoked, resemble those of opium.

Typically, in Thailand and Malaysia the leaves are either chewed when freshly picked or brewed and drunk as tea. Kratom, either fresh or dried, is seldom smoked because the effects were reported to be mild. Many Internet vendors are selling kratom in various preparations, including leaves (whole or crushed), powder, extract, powder in capsules, and resin extract. The most common method of consumption outside Southeast Asia is by boiling dried, powdered kratom, straining, and drinking the tea. Some Internet vendors also offer seeds and seedlings for kratom cultivation.

21.4   USE AND MISUSE OF KRATOM

21.4.1   HISTORICAL USE

Kratom use was first reported in the 1930s, and since then several uses have been described. It is used mainly for its psychoactive effects: a central stimulant to increase work productivity, substitute for opium, and treatment of opioid withdrawal symptoms. Among its medicinal uses are as a wound poultice, antipyretic, cough suppressant, antidiarrheal, intestinal deworming agent, and antidiabetic, as well as to wean addicts off heroin (Burkill and Haniff 1930; Jansen and Prast 1988; Vicknasingam et al. 2010; Ahmad and Aziz 2012). Outside Southeast Asia, anecdotal reports claimed that kratom preparations are used mainly for the self-treatment of chronic pain and managing opioid withdrawal syndromes.

Its use as a stimulant was probably first documented by Grewal (1932). He examined the effects of the powdered leaves and its main alkaloid, mitragynine, in Thai laborers and farmers. Both forms were found to increase energy and the tolerance to heat, thus increasing work productivity. He observed that in this aspect, the leaves of kratom were a psychostimulant and the effect resembled cocaine. He further correlated this with the widespread use of the leaves at that time in Thailand as a stimulant to increase work efficiency and to tolerate heat. The pharmacological mechanism responsible for these stimulant effects is still unclear. While there are few clinical data to provide evidence on the psychomotor stimulant effects of kratom, numerous studies have documented its use in enhancing energy levels in order to increase work productivity (Burkill and Haniff 1930; Suwanlert 1975; Jansen and Prast 1988; Assanangkornchai et al. 2007; Ahmad and Aziz 2012). Many of the current users in Thailand and Malaysia who use kratom for its stimulant effects consider consuming kratom similar to drinking coffee.

As for its use to suppress opiate withdrawal symptoms, Burkill and Haniff (1930) were the first to record the use of kratom in Malaya (present-day Peninsular Malaysia) for that purpose. There are numerous anecdotal reports on the Internet of its use to manage opioid dependence (Boyer et al. 2007). One Malaysian study documented that 88% of 136 active drug users (including users of heroin) used kratom to manage their opioid withdrawal symptoms (Vicknasingam et al. 2010). At least two recent surveys have documented the use of kratom for the treatment of opiate addiction in Malaysia and Thailand (Tanguay 2011; Ahmad and Aziz 2012). However, with the exception of a case report published in 2008 (Boyer et al. 2008), kratom has yet to be proven in controlled clinical trials to be effective in opiate withdrawal.

21.4.2 ANIMAL STUDIES

Evidence from animal studies seems to indicate that kratom acts on mu-, delta- and kappa-opioid receptor subtypes (Matsumoto et al. 1996; Takayama et al. 2002). Since both the analgesic and euphoric actions involve the mu-receptors, this could probably explain the pharmacological basis of using kratom as a substitute for opium. Although kratom has been widely advertised for use as an opium substitute (Babu et al. 2008; Schmidt et al. 2011), few clinical studies have examined its opioid-like effects. The majority of reports are anecdotal and based on the experiences of individual users. Users have described feeling euphoric after 5 to 10 minutes of consuming a moderate to high dose of kratom (Jansen and Prast 1988; Adkins et al. 2011). Of interest is that even though kratom's euphoric effects were described as mild compared to opium and heroin (Jansen and Prast 1988; Assanangkornchai et al. 2007), drug users still preferred kratom. The preference is probably due to its lower cost and greater availability.

Even though pharmacological studies in animals (Matsumoto et al. 1996; Takayama 2004) indicate that kratom has analgesic effects and many Internet websites promote the use of kratom for chronic pain, there have yet to be controlled clinical trials on whether kratom is beneficial for relieving pain in humans.

21.4.3 ADDICTION POTENTIAL

Given the widespread use of kratom, health authorities worldwide are concerned about its potential of abuse and addiction. There remains considerable debate about the addiction problems associated with kratom use. Anecdotal descriptions on the Internet of kratom dependency are contradictory, and descriptions in the published literature are limited.

There are many definitions of addiction. The more clinically useful definition is by the National Institute on Drug Abuse (NIDA), part of the National Institutes of Health under the US Department of Health and Human Services. NIDA defines addiction as "a chronic, relapsing brain disease that is characterized by compulsive drug seeking and use, despite harmful consequences. It is considered a brain disease because drugs change the brain; they change its structure and how it works. These brain changes can be long lasting and can lead to many harmful, often self-destructive behaviors."

Animals make a convenient tool for researchers to study addiction because animals and humans share the same reward pathway and thus animals are also susceptible to drug addiction. To demonstrate the addiction potential of a substance in an animal model, a single test is not satisfactory. For example, assessment of tolerance alone is not sufficient, since tolerance is a widespread phenomenon that occurs with both addictive and non-addictive substances. Because of the non-specificity of tolerance development in relation to addiction, tolerance on its own cannot be used as a predictor for addiction potential. A variety of approaches including tests of self-administration, drug discrimination, tolerance, and dependence need to be considered to examine the potential of a substance to cause addiction (Balster 1991).

Drug discrimination procedures have enabled researchers to classify and distinguish closely related psychoactive compounds (Barry 1974). Aziz and Latiff (2006) conducted drug discrimination procedures in rats to identify the psychoactive class of kratom. Rats trained to discriminate between the kratom extract and saline required many training sessions to reach the performance criteria, indicating that kratom produced weak control over differential lever responding compared to the more readily discriminable drugs such as d-amphetamine and pentobarbital. Cocaine (psychostimulant), morphine (opioid), d-amphetamine (psychostimulant), and pentobarbital (sedative hypnotic) did not substitute for the discriminative stimulus effects of either kratom or mitragynine. Taken together, the findings of this study suggested that kratom has weak but unique discriminative stimulus properties that were different from the four classes of psychoactive substances tested.

It is known that a substance with rewarding properties can cause dependence and addiction. At present, scientific examinations of the rewarding properties of kratom and its constituents using the self-administration approach in animals are scarce. Conditioned place preference is not as rigorous as self-administration in determining the rewarding properties of a substance, but it has some predictive value in identifying drugs that might have abuse potential in humans. The rewarding properties of kratom extract and mitragynine have been examined using a conditioned place preference approach (Sufka et al. 2013). The study found that rats showed place preference to mitragynine that was similar to that of $S(+)$-amphetamine, suggesting that mitragynine has rewarding properties, thus addiction potential.

A few exploratory studies have documented chronic use and potential abuse of kratom preparations and more recently of mitragynine itself (Assanangkornchai et al. 2007; Boyer et al. 2008). One of the first case records of addiction to kratom published in medical proceedings described the chronic user as having marked abstinence symptoms on sudden withdrawal of the leaves (Thuan 1957). Otherwise, the user was in good health and appeared normal both mentally and physically. Suwanlert (1975) reported 5 cases of addiction out of 30 Thai chronic users of kratom. The leaves were mostly chewed or taken as a powder. The stimulant effect was reported to occur within 5 to 10 minutes after the leaves were chewed. The withdrawal symptoms reported were similar to those of opioid withdrawal and included irritability, musculo-skeletal aches, jerky limb movements, and rhinorrhea.

Several studies reported on individuals exhibiting significant tolerance to the effects of kratom and showing symptoms of withdrawal (Babu et al. 2008; Boyer et al. 2008). A more recent case report described kratom dependence in a patient who had previously used kratom for three years (McWhirter and Morris 2010). The patient's craving and opioid-like withdrawal symptoms responded to dihydrocodeine (an opioid agonist) and lofexidine (an alpha-adrenergic antagonist), suggesting that kratom dependence syndrome is probably due to short-acting opioid receptor agonist activity. In the West, increasing reports on kratom in the media have suggested that kratom is risky to consumer health, with a high potential for addiction.

Despite the anecdotal reports of kratom use to manage withdrawal from other opioid substances of abuse, a Malaysian survey of kratom users reported that 87% of 530 respondents were unable to stop when they wanted to (Ahmad and Aziz 2012) and a Thai study reported that 61% of regular users were unable to stop (Assanangkornchai et al. 2007). Additionally, the most recent qualitative study (Saingam et al. 2013) revealed that regular users admitted that they were dependent on kratom, thought of it all the time, and needed to use it continuously. Several regular users also tried unsuccessfully to stop taking kratom and experienced withdrawal symptoms as a result. Suffering withdrawal symptoms and difficulty in stopping use indicates that users can possibly become dependent on it.

21.5 LEGAL STATUS

The current legal framework for international drug control is laid out in three international conventions under the auspices of the United Nations. Unlike other psychoactive substances such as heroin, opium, and cocaine, neither the kratom plant nor mitragynine or other alkaloids from the plant are listed in any of the schedules of the United Nations drug conventions. Thus the legal status of kratom plant and its constituents varies around the world. For example, it is illegal only in several states in the United States, a number of countries in Europe, Australia, New Zealand, and several countries in Asia, such as Thailand and Malaysia. It is still legal in most countries.

Governments have applied a variety of methods to attempt to control kratom or its alkaloids within their legislative frameworks. For example, in Thailand, kratom has been illegal since 1943 and was classified as a controlled substance in the Narcotics Act in 1979, along with cannabis and psychotropic mushroom species. Under this Act, the cultivation, possession, import, export, and use of kratom are prohibited and punishable by law. The penalties are imprisonment of no more than two years and a fine of not more than 200,000 Thai Baht (USD 6,168) for the production, import, and export of kratom, while for use the penalties are imprisonment of not more than one month and a fine of not more than 2,000 Thai Baht (USD 62) (Tanguay 2011). In Malaysia, mitragynine, the major alkaloid of kratom, is listed in the schedule of psychotropic substances in the Poisons Act, 1952. Under this Act, a person found to possess or sell kratom leaves or other kratom preparations, such as drinks and teas containing mitragynine, might be fined a maximum penalty of RM 10,000 (USD 3,300), be given a four-year jail sentence, or both (Poisons Act 1952).

In the United States, kratom is not scheduled under the Controlled Substances Act. However, drug information on kratom is provided on the United States Drug Enforcement Agency's (DEA) website, which states that there is no legitimate medical use for kratom in the United States (US Department of Justice 2013). Therefore kratom cannot legally be advertised as a treatment for any medical conditions. Laws controlling kratom are still pending in several states, such as Massachusetts, Arizona, and Vermont (Erowid 2013). Other states, for example Indiana and Tennessee, have made it illegal to possess mitragynine and 7-hydroxymitragynine (an alkaloid of kratom), but not the kratom plant. Yet other states restrict its sale to a certain population group; for example, Louisiana banned the sale of *M. speciosa* (kratom) to minors in May 2012 under Act No. 355 (Louisiana Senate Bill 130 2012).

Australia placed mitragynine in 2003 and the kratom plant in 2004 in Schedule 9 (S9) of the Standard for the Uniform Scheduling of Drugs and

Poisons (SUSDP). Substances and preparations placed under S9 by law may be used only for research purposes. Thus the Australian law strictly prohibits the sale, distribution, use, and manufacture of kratom and mitragynine without a permit.

Under New Zealand law, both kratom and mitragynine are classified under the same Schedule 1 as prescription drugs in the Medicines Amendment Regulations Act of 2009 (Medicines Amendment Regulations 2009). To use kratom, this law requires the user to obtain a prescription, and therefore it is illegal to sell kratom to individuals without a prescription. Because of this, certain people interpret importing and possessing kratom products as technically legal in New Zealand.

Currently, kratom and its alkaloids (mitragynine and 7-hydroxymitragynine) are controlled in a number of EU Member States such as Denmark, Latvia, Lithuania, Poland, Romania, and Sweden (European Monitoring Centre for Drugs and Drug Addiction 2012). In Denmark, kratom and the two alkaloids may be used only for medicinal and scientific purposes (Ministry of Health and Prevention Denmark 2009). Other countries that control kratom and its alkaloids include Myanmar, South Korea, Israel, Sweden, and Germany.

21.6 CONCLUSION

The available information suggests that kratom is being used extensively for its psychoactive effects. Kratom and its alkaloids might be potentially useful for the management of opioid withdrawal symptoms. However, like almost all psychoactive substances, it has the potential to be addictive. Kratom has been reported to produce dependence, especially at the doses and frequency of use in Southeast Asia, where it is readily available and cheap. Further studies to support the efficacy of kratom for managing opioid withdrawal symptoms, safety, and its addiction potential are warranted.

REFERENCES

Adkins, J.E., Boyer, E.W. and McCurdy, C.R. 2011. *Mitragyna speciosa*, a psychoactive tree from Southeast Asia with opioid activity. *Current Topics in Medicinal Chemistry* 11: 1165–75.

Ahmad, K., and Aziz. Z. 2012. *Mitragyna speciosa* use in the northern states of Malaysia: A cross-sectional study. *Journal of Ethnopharmacology* 141: 446–50.

Assanangkornchai, S., Aramrattana, A., Perngparn, U., Kanato, M., Kanika, N. and Na Ayudhya, A.S. 2008. Current situation of substance-related problems in Thailand. *The Journal of the Psychiatric Association of Thailand* 53: S24–S36.

Assanangkornchai, S., Muekthong, A., Sam-Angsri, N. and Pattanasattayawong, U. 2007. The use of *Mitragyna speciosa* ('Krathom'), an addictive plant, in Thailand. *Substance Use and Misuse* 42: 2145–57.

Aziz, Z. and Latiff, A.A. 2006. The discriminative stimulus properties of *Mitragyna speciosa* extract in rats. Abstract. *Revista Cubana de Farmacia* 40: 56.

Babu, K.M., McCurdy, C.R. and Boyer, E.W. 2008. Opioid receptors and legal highs: *Salvia divinorum* and Kratom. *Clinical Toxicology (Philadelphia, Pa)* 46: 146–52.

Balster, R.L. 1991. Drug abuse potential evaluation in animals. *British Journal of Addiction* 86: 1549–58.

Barry, H., 3rd. 1974. Classification of drugs according to their discriminable effects in rats. *Federation Proceedings* 33: 1814–24.

Boyer, E.W., Babu, K.M., Adkins, J.E., McCurdy, C.R. and Halpern, J.H. 2008. Self-treatment of opioid withdrawal using kratom (*Mitragynia speciosa* Korth). *Addiction* 103: 1048–50.

Boyer, E.W., Babu, K.M. and Macalino, G.E. 2007. Self-treatment of opioid withdrawal with a dietary supplement, Kratom. *The American Journal on Addictions* 16: 352–56.

Burkill, I.H., and Haniff, M. 1930. Malay village medicine. *The Gardens' Bulletin, Straits Settlements* 6: 165–207.

Chan, K.B., Pakiam, C. and Rahim, R.A. 2005. Psychoactive plant abuse: The identification of mitragynine in ketum and in ketum preparations. *Bulletin on Narcotics* 57: 249–56.

Erowid. 2013. Kratom: Legal status. Erowid.org. www.erowid.org/plants/kratom /kratom_law.shtml (accessed 10 November 2013).

European Monitoring Centre for Drugs and Drug Addiction. 2012. Kratom (*Mitragyna speciosa*). http://www.emcdda.europa.eu/publications/drug-profiles /kratom#control (accessed 20 December 2013).

Grewal, K.S. 1932. The effect of mitragynine on man. *British Journal of Medical Psychology* 12: 41–58.

Jansen, K.L.R. and Prast, C.J. 1988. Ethnopharmacology of kratom and the mitragyna alkaloids. *Journal of Ethnopharmacology* 23: 115–19.

Kronstrand, R., Roman, M., Thelander, G. and Eriksson, A. 2011. Unintentional fatal intoxications with mitragynine and O-desmethyltramadol from the herbal blend Krypton. *Journal of Analytical Toxicology* 35: 242–47.

Louisiana Senate Bill 130. 2012. Prohibits distribution of products containing *Mitragyna speciosa* in minors. http://legiscan.com/LA/text/SB130/id/651753 (accessed 20 December 2013).

Matsumoto, K., Mizowaki, M., Suchitra, T., Murakami, Y., Takayama, H., Sakai, S.I., Aimi, N., et al. 1996. Central antinociceptive effects of mitragynine in mice: Contribution of descending noradrenergic and serotonergic systems. *European Journal of Pharmacology* 317: 75–81.

McWhirter, L., and Morris, S. 2010. A case report of inpatient detoxification after kratom (*Mitragyna speciosa*) dependence. *European Addiction Research* 16: 229–31.

Medicines Amendment Regulations. 2009. www.legislation.govt.nz/regulation /public/2009/0212/latest/DLM2220001.html (accessed 21 December 2013).

Ministry of Justice, Office of the Narcotics Control Board Thailand. 2010. Thailand Narcotics Control Annual Report 2010. http: //en.oncb.go.th/file /publications.html (accessed 21 December 2013).

Nazar Mohamed, M., Marican, S., Elias, N. and Don, Y. 2008. Pattern of substance abuse and drug misuse among youth in Malaysia. *Jurnal Antidadah Malaysia* 3–4: 1–56.

Poisons Act. 1952. www.pharmacy.gov.my/v2/sites/default/files/document-upload /poisons-act-1952-act-366.pdf (accessed 20 December 2013).

Saingam, D., Assanangkornchai, S., Geater, A.F. and Balthip, Q. 2013. Pattern and consequences of krathom (*Mitragyna speciosa* Korth.) use among male villagers in southern Thailand: A qualitative study. *The International Journal on Drug Policy* 24: 351–58.

Schmidt, M.M., Sharma, A., Schifano, F. and Feinmann, C. 2011. 'Legal highs' on the net: Evaluation of UK-based websites, products and product information. *Forensic Science International* 206: 92–97.

Sufka, K.J., Loria, M.J., Lewellyn, K., Zjawiony, J.K., Ali, Z., Abe, N. and Khan, I.A. 2014. The effect of *Salvia divinorum* and *Mitragyna speciosa* extracts, fraction and major constituents on place aversion and place preference in rats. *Journal of Ethnopharmacology* 151: 361–64.

Suwanlert, S. 1975. A study of kratom eaters in Thailand. *Bulletin on Narcotics* 27: 21–27.

Takayama, H. 2004. Chemistry and pharmacology of analgesic indole alkaloids from the rubiaceous plant, *Mitragyna speciosa*. *Chemical and Pharmaceutical Bulletin (Tokyo)* 52: 916–28.

Takayama, H., Ishikawa, H., Kurihara, M., Kitajima, M., Aimi, N., Ponglux, D., Koyama, F., et al. 2002. Studies on the synthesis and opioid agonistic activities of mitragynine-related indole alkaloids: Discovery of opioid agonists structurally different from other opioid ligands. *Journal of Medicinal Chemistry* 45: 1949–56.

Tanguay, P. 2011. Kratom in Thailand: Decriminalisation and community control? *Legislative Reform of Drug Policies* 13: 1–16.

Thuan, L.C. 1957. Addiction to *Mitragyna speciosa*. *Proceeding of the Alumni Association* 10: 322–24.

United Nations Office on Drugs and Crime. 2009. Patterns and trends of amphetamine-type stimulants and other drugs in East and South-East Asia (and neighbouring regions). www.unodc.org/documents/scientific/2009 _Patterns_and_Trends.pdf (accessed 21 December 2013).

United Nations Office on Drugs and Crime. 2013. The challenge of new psychoactive substances. www.unodc.org/documents/scientific/NPS_2013_SMART .pdf (accessed 20 December 2013).

US Department of Justice, Drug Enforcement Administration, Office of Diversion Control. 2013. Kratom (*Mitragyna speciosa* Korth.) (Street Names: Thang, Kakuam, Thom, Ketum, Biak). www.deadiversion.usdoj.gov/drug_chem_ info/kratom.pdf (accessed 21 December 2013).

Vicknasingam, B., Narayanan, S., Beng, G.T. and Mansor, S.M. 2010. The informal use of ketum (*Mitragyna speciosa*) for opioid withdrawal in the northern states of peninsular Malaysia and implications for drug substitution therapy. *The International Journal on Drug Policy* 21: 283–88.

Wray, L. 1907. 'Biak': An opium substitute. *Journal of the Federated Malay States Museums* 2: 53–56.

Bibliography

Abdullah, J.M. 2011. Interesting Asian plants
Their compounds and effects on electrophysiology and behaviour. *Malays J Med Sci* 18: 1–4.

Adkins, J.E., Boyer, E.W. and McCurdy, C.R. 2011. *Mitragyna speciosa*, a psychoactive tree from Southeast Asia with opioid activity. *Curr Top Med Chem* 11: 1165–75.

Ahmad, K. and Aziz, Z. 2012. *Mitragyna speciosa* use in the northern states of Malaysia: A cross-sectional study. *J Ethnopharmacol* 141: 446–50.

Aji, B.M., Onyeyili, P.A. and Osunkwo, U.A. 2001. The central nervous effects of *Mitragyna africanus* (Willd) stembark extract in rats. *J Ethnopharmacol* 77: 143–49.

Akinsinde, K.A. and Olukoya, D.K. 1995. Vibriocidal activities of some local herbs. *J Diarrhoeal Dis Res* 13: 127–29.

Akpata, T.V. 1987. Effects of sawdust pollution on the germination of fungal spores in Lagos Lagoon. *Environ Pollut* 44: 37–48.

Al-Saffar, Y., Stephanson, N.N. and Beck, O. 2013. Multicomponent LC-MS/MS screening method for detection of new psychoactive drugs, legal highs, in urine-experience from the Swedish population. *J Chromatogr B Analyt Technol Biomed Life Sci* 930: 112–20.

Apryani, E., Hidayat, M.T., Moklas, M.A., Fakurazi, S. and Idayu, N.F. 2010. Effects of mitragynine from *Mitragyna speciosa* Korth leaves on working memory. *J Ethnopharmacol* 129: 357–60.

Arndt, T., Claussen, U., Gussregen, B., Schrofel, S., Sturzer, B., Werle, A. and Wolf, G. 2011. Kratom alkaloids and O-desmethyltramadol in urine of a 'Krypton' herbal mixture consumer. *Forensic Sci Int* 208: 47–52.

Asase, A., Kokubun, T., Grayer, R.J., Kite, G., Simmonds, M.S., Oteng-Yeboah, A.A. and Odamtten, G.T. 2008. Chemical constituents and antimicrobial activity of medicinal plants from Ghana: *Cassia sieberiana, Haemastostaphis barteri, Mitragyna inermis* and *Pseudocedrela kotschyi. Phytother Res* 22: 1013–16.

Asita, A.O., Matsui, M., Nohmi, T., et al. 1991. Mutagenicity of wood smoke condensates in the Salmonella/microsome assay. *Mutat Res* 264: 7–14.

Assanangkornchai, S., Muekthong, A., Sam-Angsri, N. and Pattanasattayawong, U. 2007. The use of *Mitragynine speciosa* ('Krathom'), an addictive plant, in Thailand. *Subst Use Misuse* 42: 2145–57.

Azas, N., Laurencin, N., Delmas, F., Di, G.C., Gasquet, M., Laget, M. and Timon-David, P. 2002. Synergistic in vitro antimalarial activity of plant extracts used as traditional herbal remedies in Mali. *Parasitol Res* 88: 165–71.

Azizi, J., Ismail, S. and Mansor, S.M. 2013. *Mitragyna speciosa* Korth leaves extracts induced the CYP450 catalyzed aminopyrine–N-demethylase (APND) and UDP-glucuronosyl transferase (UGT) activities in male Sprague-Dawley rat livers. *Drug Metabol Drug Interact* 28: 95–105.

Azizi, J., Ismail, S., Mordi, M.N., Ramanathan, S., Said, M.I. and Mansor, S.M. 2010. In vitro and in vivo effects of three different *Mitragyna speciosa* Korth leaf extracts on phase II drug metabolizing enzymes: Glutathione transferases (GSTs). *Molecules* 15: 432–41.

Babu, K.M., McCurdy, C.R. and Boyer, E.W. 2008. Opioid receptors and legal highs: *Salvia divinorum* and Kratom. *Clin Toxicol (Phila)* 46: 146–52.

Backstrom, B.G., Classon, G., Lowenhielm, P. and Thelander, G. 2010. [Krypton—new, deadly Internet drug. Since October 2009 nine young persons have died in Sweden]. *Lakartidningen* 107: 3196–97.

Beckett, A.H., and Casy, A.F. 1954. Synthetic analgesics: Stereochemical considerations. *J Pharm Pharmacol* 6: 986–1001.

Beckett, A.H., Dwuma-Badu, D. and Haddock, R.E. 1969. Some new *mitragyna*-type indoles and oxindoles: The influence of stereochemistry on mass spectra. *Tetrahedron* 25: 5961–69.

Beckett, A.H., Shellard, E.J., Phillipson, J.D. and Lee, C.M. 1965. Alkaloids from *Mitragyna speciosa* (Korth.). *J Pharm Pharmacol* 17: 753–55.

Beckett, A.H., Shellard, E.J., Phillipson, J.D. and Lee, C.M. 1966a. The *Mitragyna* species of Asia. VI. Oxindole alkaloids from the leaves of *Mitragyna speciosa* Korth. *Planta Med* 14: 266–76.

Beckett, A.H., Shellard, E.J., Phillipson, J.D. and Lee, C.M. 1966b. The *Mitragyna* species of Asia. VII. Indole alkaloids from the leaves of *Mitragyna speciosa* Korth. *Planta Med* 14: 277–88.

Beckett, A.H., Shellard, E.J. and Tackie, A.N. 1963a. The *Mitragyna* species of Ghana. The alkaloids of the leaves of *Mitragyna ciliata* Aubr. Et Pellegr. *J Pharm Pharmacol* 15: Suppl 166–69.

Beckett, A.H., Shellard, E.J. and Tackie, A.N. 1963b. The *Mitragyna* species of Ghana. The alkaloids of the leaves of *Mitragyna stipulosa*. *J Pharm Pharmacol* 15: Suppl 158–65.

Beckett, A.H. and Tackie, A.N. 1963. The structures of the alkaloids from *Mitragyna* species of Ghana. *J Pharm Pharmacol* 15: Suppl 267–69.

Belmain, S.R., Neal, G.E., Ray, D.E. and Golob, P. 2001. Insecticidal and vertebrate toxicity associated with ethnobotanicals used as post-harvest protectants in Ghana. *Food Chem Toxicol* 39: 287–91.

Bhardwaj, M., Bharadwaj, L., Trigunayat, K. and Trigunayat, M.M. 2011. Insecticidal and wormicidal plants from Aravalli hill range of India. *J Ethnopharmacol* 136: 103–10.

Bidie, A.D., Koffi, E., N'Guessan, J.D., Djaman, A.J. and Guede-Guina, F. 2008. Influence of *Mitragyna ciliata* (MYTA) on the microsomal activity of ATPase Na+/K+ dependent extract on a rabbit heart. *Afr J Tradit Complement Altern Med* 5: 294–301.

Bouyer, J., Guerrini, L., Cesar, J., de la Rocque, S. and Cuisance, D. 2005. A phyto-sociological analysis of the distribution of riverine tsetse flies in Burkina Faso. *Med Vet Entomol* 19: 372–78.

Boyer, E.W., Babu, K.M., Adkins, J.E., McCurdy, C.R. and Halpern, J.H. 2008. Self-treatment of opioid withdrawal using kratom (*Mitragynia speciosa* Korth). *Addiction* 103: 1048–50.

Boyer, E.W., Babu, K.M. and Macalino, G.E. 2007. Self-treatment of opioid with-drawal with a dietary supplement, Kratom. *Am J Addict* 16: 352–56.

Burillo-Putze, G., Lopez Briz, E., Climent Diaz, B., Munne Mas, P., Nogue Xarau, S., Pinillos, M.A. and Hoffman, R.S. 2013. [Emergent drugs (III): Hallucinogenic plants and mushrooms]. *An Sist Sanit Navar* 36: 505–18.

Cao, X.F., Wang, J.S., Wang, X.B., Luo, J., Wang, H.Y. and Kong, L.Y. 2013. Monoterpene indole alkaloids from the stem bark of *Mitragyna diversifolia* and their acetylcholine esterase inhibitory effects. *Phytochemistry* 96: 389–96.

Chan, K.B., Pakiam, C. and Rahim, R.A. 2005. Psychoactive plant abuse: The identification of mitragynine in ketum and in ketum preparations. *Bull Narc* 57: 249–56.

Charoonratanaa, T., Wungsintaweekul, J., Pathompak, P., Georgiev, M.I., Choi, Y.H. and Verpoorte, R. 2013. Limitation of mitragynine biosynthesis in *Mitragyna speciosa* (Roxb.) Korth. through tryptamine availability. *Zeitschrift für Naturforschung C* 68: 394–405.

Chee, J.W., Amirul, A.A., Majid, M.I. and Mansor, S.M. 2008. Factors influencing the release of *Mitragyna speciosa* crude extracts from biodegradable P(3HB-co-4HB). *Int J Pharm* 361: 1–6.

Cheng, Z.H., Yu, B.Y. and Yang, X.W. 2002a. 27-Nor-triterpenoid glycosides from *Mitragyna inermis*. *Phytochemistry* 61: 379–82.

Cheng, Z.H., Yu, B.Y., Yang, X.W. and Zhang, J. 2002b. [Triterpenoid saponins from bark *Mitragyna inermis*]. *Zhongguo Zhong Yao Za Zhi* 27: 274–77.

Chittrakarn, S., Keawpradub, N., Sawangjaroen, K., Kansenalak, S. and Janchawee, B. 2010. The neuromuscular blockade produced by pure alkaloid, mitragynine and methanol extract of kratom leaves (*Mitragyna speciosa* Korth.). *J Ethnopharmacol* 129: 344–49.

Chittrakarn, S., Penjamras, P. and Keawpradub, N. 2012. Quantitative analysis of mitragynine, codeine, caffeine, chlorpheniramine and phenylephrine in a kratom (*Mitragyna speciosa* Korth.) cocktail using high-performance liquid chromatography. *Forensic Sci Int* 217: 81–86.

Chittrakarn, S., Sawangjaroen, K., Prasettho, S., Janchawee, B. and Keawpradub, N. 2008. Inhibitory effects of kratom leaf extract (*Mitragyna speciosa* Korth.) on the rat gastrointestinal tract. *J Ethnopharmacol* 116: 173–78.

Cornara, L., Borghesi, B., Canali, C., Andrenacci, M., Basso, M., Federici, S. and Labra, M. 2013. Smart drugs: Green shuttle or real drug? *Int J Legal Med* 127: 1109–23.

de Moraes, N.V., Moretti, R.A., Furr, E.B., 3rd, McCurdy, C.R. and Lanchote, V.L. 2009. Determination of mitragynine in rat plasma by LC-MS/MS: Application to pharmacokinetics. *J Chromatogr B Analyt Technol Biomed Life Sci* 877: 2593–97.

Dhawan, B.N., Cesselin, F., Raghubir, R., Reisine, T., Bradley, P.B., Portoghese, P.S., and Hamon, M. 1996. International Union of Pharmacology. XII. Classification of opioid receptors. *Pharmacol Rev* 48: 567–92.

Dongmo, A., Kamanyi, M.A., Tan, P.V., Bopelet, M., Vierling, W. and Wagner, H. 2004. Vasodilating properties of the stem bark extract of *Mitragyna ciliata* in rats and guinea pigs. *Phytother Res* 18: 36–39.

Dongmo, A.B., Kamanyi, A., Dzikouk, G., et al. 2003. Anti-inflammatory and analgesic properties of the stem bark extract of *Mitragyna ciliata* (Rubiaceae) Aubrev. & Pellegr. *J Ethnopharmacol* 84: 17–21.

Dresen, S., Ferreiros, N., Putz, M., Westphal, F., Zimmermann, R. and Auwarter, V. 2010. Monitoring of herbal mixtures potentially containing synthetic cannabinoids as psychoactive compounds. *J Mass Spectrom* 45: 1186–94.

Edwankar, C.R., Edwankar, R.V., Namjoshi, O.A., Rallapalli, S.K., Yang, J. and Cook, J.M. 2009. Recent progress in the total synthesis of indole alkaloids. *Curr Opin Drug Discov Devel* 12: 752–71.

Fakurazi, S., Rahman, S.A., Hidayat, M.T., Ithnin, H., Moklas, M.A. and Arulselvan, P. 2013. The combination of mitragynine and morphine prevents the development of morphine tolerance in mice. *Molecules* 18: 666–81.

Fatima, N., Tapondjou, L.A., Lontsi, D., Sondengam, B.L., Atta Ur, R. and Choudhary, M.I. 2002. Quinovic acid glycosides from *Mitragyna stipulosa*: First examples of natural inhibitors of snake venom phosphodiesterase I. *Nat Prod Lett* 16: 389–93.

Finch, N., Gemenden, C.W., Hsu, I. H.-C. and Taylor, W.I. 1963. Oxidative transformations of indole alkaloids. II. The preparation of oxindoles from *cis*-DE-yohimbinoid alkaloids: The partial synthesis of carapanaubine. *J Am Chem Soc* 85: 1520–23.

Finch, N., Gemenden, C.W., Hsu, I. H.-C., Kerr, A., Sim, G.A. and Taylor, W.I. 1965. Oxidative transformations of indole alkaloids. III. Pseudoindoxyls from yohimbinoid alkaloids and their conversion to 'invert' alkaloids. *J Am Chem Soc* 87: 2229–35.

Fiot, J., Baghdikian, B., Boyer, L., et al. 2005. HPLC quantification of uncarine D and the anti-plasmodial activity of alkaloids from leaves of *Mitragyna inermis* (Willd.) O. Kuntze. *Phytochem Anal* 16: 30–33.

Fiot, J., Sanon, S., Azas, N., et al. 2006. Phytochemical and pharmacological study of roots and leaves of *Guiera senegalensis* J.F. Gmel (Combretaceae). *J Ethnopharmacol* 106: 173–78.

Forrester, M.B. 2013. Kratom exposures reported to Texas poison centers. *J Addict Dis* 32: 396–400.

Gang, T., Liu, T., Zhu, Y. and Liu, Z.Y. 2008. [Molecular identification of medicinal plant genus *Uncaria* in Guizhou]. *Zhong Yao Cai* 31: 825–28.

Gongora-Castillo, E., Childs, K.L., Fedewa, G., et al. 2012. Development of transcriptomic resources for interrogating the biosynthesis of monoterpene indole alkaloids in medicinal plant species. *PLoS One* 7: e52506.

Gueller, G. and Borschberg, H.J. 1992. Synthesis of Aristotelia-type alkaloids. Part X. Biomimetic transformation of synthetic (+)-aristotelline into (–)-alloaristoteline. *Tetrahedron: Asymmetry* 3: 1197–1204.

Hanapi, N.A., Ismail, S. and Mansor, S.M. 2013. Inhibitory effect of mitragynine on human cytochrome P450 enzyme activities. *Pharmacognosy Res* 5: 241–46.

Harizal, S.N., Mansor, S.M., Hasnan, J., Tharakan, J.K. and Abdullah, J. 2010. Acute toxicity study of the standardized methanolic extract of *Mitragyna speciosa* Korth in rodent. *J Ethnopharmacol* 131: 404–9.

Hassan, Z., Muzaimi, M., Navaratnam, V., et al. 2013. From Kratom to mitragynine and its derivatives: Physiological and behavioural effects related to use, abuse, and addiction. *Neurosci Biobehav Rev* 37: 138–51.

Hazim, A.I., Ramanathan, S., Parthasarathy, S., Muzaimi, M. and Mansor, S.M. 2014. Anxiolytic-like effects of mitragynine in the open-field and elevated plus-maze tests in rats. *J Physiol Sci.*

Helander, A., Beck, O., Hagerkvist, R. and Hulten, P. 2013. Identification of novel psychoactive drug use in Sweden based on laboratory analysis: Initial experiences from the STRIDA project. *Scand J Clin Lab Invest* 73: 400–406.

Holler, J.M., Vorce, S.P., McDonough-Bender, P.C., Magluilo, J., Jr., Solomon, C.J. and Levine, B. 2011. A drug toxicity death involving propylhexedrine and mitragynine. *J Anal Toxicol* 35: 54–59.

Horie, S., Koyama, F., Takayama, H., et al. 2005. Indole alkaloids of a Thai medicinal herb, *Mitragyna speciosa*, that has opioid agonist effect in guinea-pig ileum. *Planta Med* 71: 231–36.

Horie, S., Yamamoto, L.T., Futagami, Y., Yano, S., Takayama, H., Sakai, S., Aimi, N., Ponglux, D., Shan, J., Pang, P.K.T., and Watanabe, K. 1995. Analgesic, neuronal Ca^{2+} channel-blocking and smooth muscle relaxant activities of mitragynine, an indole alkaloid, from the Thai folk medicine 'Kratom.' *J Traditional Med* 12: 366–67.

Horie, S., Koyama, F., Takayama, H., Ishikawa, H., Aimi, N., Ponglux, D., Matsumoto, K., and Murayama, T. 2005. Indole alkaloids of a Thai medicinal herb, *Mitragyna speciosa*, that has opioid agonist effect in guinea-pig ileum. *Planta Med* 71: 231–36.

Houghton, P.J., Lala, P.K., Shellard, E.J. and Sarpong, K. 1976. The alkaloids of *Mitragyna stipulosa* (D.C.) O. Kuntze. *J Pharm Pharmacol* 28: 664.

Houghton, P.J., Latiff, A., and Said, I.M. 1991. Alkaloids from *Mitragyna speciosa*. *Phytochemistry* 30: 347–50.

Houghton, P.J. and Shellard, E.J. 1973. Proceedings: The interconversion of open and closed E ring oxindole alkaloids in *Mitragyna* species. *J Pharm Pharmacol* 25 Suppl: 143P–44.

Houghton, P.J. and Shellard, E.J. 1974. The *Mitragyna* species of Asia. 28. The alkaloidal pattern in *Mitragyna rotundifolia* from Burma. *Planta Med* 26: 104–12.

Idayu, N.F., Hidayat, M.T., Moklas, M.A., Sharida, F., Raudzah, A.R., Shamima, A.R. and Apryani, E. 2011. Antidepressant-like effect of mitragynine isolated from *Mitragyna speciosa* Korth in mice model of depression. *Phytomedicine* 18: 402–407.

Ingsathit, A., Woratanarat, P., Anukarahanonta, T., et al. 2009. Prevalence of psychoactive drug use among drivers in Thailand: A roadside survey. *Accid Anal Prev* 41: 474–78.

Ishikawa, H., Takayama, H., and Aimi, N. 2002. Dimerization of indole derivatives with hypervalent iodines(III): A new entry for the concise total synthesis of *rac*- and *meso*-chimonanthines. *Tetrahedron Lett* 43: 5637–39.

Ishikawa, H., Kitajima, M., and Takayama, H. 2004. *m*-Chloroperbenzoic acid oxidation of corynanthe-type indole alkaloid, mitragynine, afforded unusual dimerization products. *Heterocycles* 63: 2597–604.

Jamil, M.F., Subki, M.F., Lan, T.M., Majid, M.I. and Adenan, M.I. 2013. The effect of mitragynine on cAMP formation and mRNA expression of mu-opioid receptors mediated by chronic morphine treatment in SK-N-SH neuroblastoma cell. *J Ethnopharmacol* 148: 135–43.

Janchawee, B., Keawpradub, N., Chittrakarn, S., Prasettho, S., Wararatananurak, P. and Sawangjareon, K. 2007. A high-performance liquid chromatographic method for determination of mitragynine in serum and its application to a pharmacokinetic study in rats. *Biomed Chromatogr* 21: 176–83.

Jansen, K.L. and Prast, C.J. 1988a. Ethnopharmacology of kratom and the *Mitragyna* alkaloids. *J Ethnopharmacol* 23: 115–19.

Jansen, K.L. and Prast, C.J. 1988b. Psychoactive properties of mitragynine (kratom). *J Psychoactive Drugs* 20: 455–57.

Kan, T. and Fukuyama, T. 2004. Ns strategies: A highly versatile synthetic method for amines. *Chem. Commun.* 353–59.

Kang, W.Y., Shi, Y.Y. and Hao, X.J. 2007. [Quinovic acid triterpenoid saponins from bark of *Mitragyna rotundifolia*]. *Zhongguo Zhong Yao Za Zhi* 32: 2015–18.

Kang, W.Y., Zhang, B.R., Xu, Q.T., Li, L. and Hao, X.J. 2006. [Study on the chemical constituents of *Mitragyna rotundifolia*]. *Zhong Yao Cai* 29: 557–60.

Kapp, F.G., Maurer, H.H., Auwarter, V., Winkelmann, M. and Hermanns-Clausen, M. 2011. Intrahepatic cholestasis following abuse of powdered kratom (*Mitragyna speciosa*). *J Med Toxicol* 7: 227–31.

Karou, S.D., Tchacondo, T., Ilboudo, D.P. and Simpore, J. 2011. Sub-Saharan Rubiaceae: A review of their traditional uses, phytochemistry and biological activities. *Pak J Biol Sci* 14: 149–69.

Kerschgens, I.P., Claveau, E., Wanner, M.J., Ingemann, S., van Maarseveen, J.H. and Hiemstra, H. 2012. Total syntheses of mitragynine, paynantheine and speciogynine via an enantioselective thiourea-catalysed Pictet–Spengler reaction. *Chem Commun (Camb)* 48: 12243–45.

Khor, B.S., Jamil, M.F., Adenan, M.I. and Shu-Chien, A.C. 2011. Mitragynine attenuates withdrawal syndrome in morphine-withdrawn zebrafish. *PLoS One* 6: e28340.

Kim, J., Schneekloth, J.S., Jr. and Sorensen, E.J. 2012. A chemical synthesis of 11-methoxy mitragynine pseudoindoxyl featuring the interrupted Ugi reaction. *Chem Sci* 3: 2849–52.

Kitajima, M., Misawa, K., Kogure, N., Said, I.M., Horie, S., Hatori, Y., Murayama, T., and Takayama, H. 2006. A new indole alkaloid, 7-hydroxyspeciociliatine, from the fruits of Malaysian *Mitragyna speciosa* and its opioid agonistic activity. *J Nat Med* 60: 28–35.

Kong, W.M., Chik, Z., Ramachandra, M., Subramaniam, U., Aziddin, R.E. and Mohamed, Z. 2011. Evaluation of the effects of *Mitragyna speciosa* alkaloid extract on cytochrome P450 enzymes using a high throughput assay. *Molecules* 16: 7344–56.

Kowalczuk, A.P., Lozak, A. and Zjawiony, J.K. 2013. Comprehensive methodology for identification of Kratom in police laboratories. *Forensic Sci Int* 233: 238–43.

Kronstrand, R., Roman, M., Thelander, G. and Eriksson, A. 2011. Unintentional fatal intoxications with mitragynine and O-desmethyltramadol from the herbal blend Krypton. *J Anal Toxicol* 35: 242–47.

Kumarnsit, E., Keawpradub, N. and Nuankaew, W. 2006. Acute and long-term effects of alkaloid extract of *Mitragyna speciosa* on food and water intake and body weight in rats. *Fitoterapia* 77: 339–45.

Kumarnsit, E., Keawpradub, N. and Nuankaew, W. 2007a. Effect of *Mitragyna speciosa* aqueous extract on ethanol withdrawal symptoms in mice. *Fitoterapia* 78: 182–85.

Kumarnsit, E., Vongvatcharanon, U., Keawpradub, N. and Intasaro, P. 2007b. Fos-like immunoreactivity in rat dorsal raphe nuclei induced by alkaloid extract of *Mitragyna speciosa*. *Neurosci Lett* 416: 128–32.

Le, D., Goggin, M.M. and Janis, G.C. 2012. Analysis of mitragynine and metabolites in human urine for detecting the use of the psychoactive plant kratom. *J Anal Toxicol* 36: 616–25.

Lee, C.M., Trager, W.F. and Beckett, A.H. 1967. Corynantheidine-type alkaloids. II. Absolute configuration of mitragynine, speciociliatine, mitraciliatine and speciogynine. *Tetrahedron* 23: 375–85.

Leon, F., Habib, E., Adkins, J.E., Furr, E.B., McCurdy, C.R. and Cutler, S.J. 2009. Phytochemical characterization of the leaves of *Mitragyna speciosa* grown in U.S.A. *Nat Prod Commun* 4: 907–10.

Lersten, N.R. and Horner, H.T. 2011. Unique calcium oxalate 'duplex' and 'concretion' idioblasts in leaves of tribe Naucleeae (Rubiaceae). *Am J Bot* 98: 1–11.

Lim, E.L., Seah, T.C., Koe, X.F., et al. 2013. in Vitro evaluation of cytochrome P450 induction and the inhibition potential of mitragynine, a stimulant alkaloid. *Toxicol in vitro* 27: 812–24.

Liu, H., McCurdy, C.R. and Doerksen, R.J. 2010. Computational study on the conformations of mitragynine and mitragynaline. *Theochem* 945: 57–63.

Logan, B.K., Reinhold, L.E., Xu, A. and Diamond, F.X. 2012. Identification of synthetic cannabinoids in herbal incense blends in the United States. *J Forensic Sci* 57: 1168–80.

Lu, S., Tran, B.N., Nelsen, J.L. and Aldous, K.M. 2009. Quantitative analysis of mitragynine in human urine by high performance liquid chromatography–tandem mass spectrometry. *J Chromatogr B Analyt Technol Biomed Life Sci* 877: 2499–505.

Ma, J., Yin, W., Zhou, H. and Cook, J.M. 2007. Total synthesis of the opioid agonistic indole alkaloid mitragynine and the first total syntheses of 9-methoxygeissoschizol and 9-methoxy-N_b-methylgeissoschizol. *Org Lett* 9: 3491–94.

Ma, J., Yin, W., Zhou, H., Liao, X. and Cook, J.M. 2009. General approach to the total synthesis of 9-methoxy-substituted indole alkaloids: Synthesis of mitragynine, as well as 9-methoxygeissoschizol and 9-methoxy-N_b-methyl-geissoschizol. *J Org Chem* 74: 264–73.

Macko, E., Weisbach, J.A. and Douglas, B. 1972. Some observations on the pharmacology of mitragynine. *Arch Int Pharmacodyn Ther* 198: 145–61.

Martey, O.N., Armah, G. and Okine, L.K. 2010. Absence of organ specific toxicity in rats treated with Tonica, an aqueous herbal haematinic preparation. *Afr J Tradit Complement Altern Med* 7: 231–40.

Maruyama, T., Kawamura, M., Kikura-Hanajiri, R., Takayama, H. and Goda, Y. 2009. The botanical origin of kratom (*Mitragyna speciosa*; Rubiaceae) available as abused drugs in the Japanese markets. *J Nat Med* 63: 340–44.

Matsumoto, K., Hatori, Y., Murayama, T., et al. 2006a. Involvement of mu-opioid receptors in antinociception and inhibition of gastrointestinal transit induced by 7-hydroxymitragynine, isolated from Thai herbal medicine *Mitragyna speciosa*. *Eur J Pharmacol* 549: 63–70.

Matsumoto, K., Horie, S., Ishikawa, H., Takayama, H., Aimi, N., Ponglux, D. and Watanabe, K. 2004. Antinociceptive effect of 7-hydroxymitragynine in mice: Discovery of an orally active opioid analgesic from the Thai medicinal herb *Mitragyna speciosa*. *Life Sci* 74: 2143–55.

Matsumoto, K., Horie, S., Takayama, H., et al. 2005a. Antinociception, tolerance and withdrawal symptoms induced by 7-hydroxymitragynine, an alkaloid from the Thai medicinal herb *Mitragyna speciosa*. *Life Sci* 78: 2–7.

Matsumoto, K., Mizowaki, M., Suchitra, T., et al. 1996a. Central antinociceptive effects of mitragynine in mice: Contribution of descending noradrenergic and serotonergic systems. *Eur J Pharmacol* 317: 75–81.

Matsumoto, K., Mizowaki, M., Suchitra, T., Takayama, H., Sakai, S., Aimi, N. and Watanabe, H. 1996b. Antinociceptive action of mitragynine in mice: Evidence for the involvement of supraspinal opioid receptors. *Life Sci* 59: 1149–55.

Matsumoto, K., Mizowaki, M., Takayama, H., Sakai, S., Aimi, N. and Watanabe, H. 1997. Suppressive effect of mitragynine on the 5-methoxy-N,N-dimethyltryptamine-induced head-twitch response in mice. *Pharmacol Biochem Behav* 57: 319–23.

Matsumoto, K., Narita, M., Muramatsu, N., et al. 2014. Orally active opioid mu/delta dual agonist mgm-16, a derivative of the indole alkaloid mitragynine, exhibits potent antiallodynic effect on neuropathic pain in mice. *J Pharmacol Exp Ther* 348: 383–92.

Matsumoto, K., Takayama, H., Ishikawa, H., Aimi, N., Ponglux, D., Watanabe, K. and Horie, S. 2006b. Partial agonistic effect of 9-hydroxycorynantheidine on μ-opioid receptor in the guinea-pig ileum. *Life Sci* 78: 2265–71.

Matsumoto, K., Takayama, H., Narita, M., et al. 2008. MGM-9 [(E)-methyl 2-(3-ethyl-7a,12a-(epoxyethanoxy)-9-fluoro-1,2,3,4,6,7,12,12b-octahydro-8-methoxy indolo[2,3-a]quinolizin-2-yl)-3-methoxyacrylate], a derivative of the indole alkaloid mitragynine: A novel dual-acting mu- and kappa-opioid agonist with potent antinociceptive and weak rewarding effects in mice. *Neuropharmacology* 55: 154–65.

Matsumoto, K., Yamamoto, L.T., Watanabe, K., et al. 2005b. Inhibitory effect of mitragynine, an analgesic alkaloid from Thai herbal medicine, on neurogenic contraction of the vas deferens. *Life Sci* 78: 187–94.

Maurer, H.H. 2010. Chemistry, pharmacology, and metabolism of emerging drugs of abuse. *Ther Drug Monit* 32: 544–49.

McWhirter, L. and Morris, S. 2010. A case report of inpatient detoxification after kratom (*Mitragyna speciosa*) dependence. *Eur Addict Res* 16: 229–31.

Mohamad Zuldin, N.N., Said, I.M., Mohd Noor, N., Zainal, Z., Jin Kiat, C. and Ismail, I. 2013. Induction and analysis of the alkaloid mitragynine content of a *Mitragyna speciosa* suspension culture system upon elicitation and precursor feeding. *Scientific World Journal* 2013: 209434.

Monjanel-Mouterde, S., Traore, F., Gasquet, M., et al. 2006. Lack of toxicity of hydroethanolic extract from *Mitragyna inermis* (Willd.) O. Kuntze by gavage in the rat. *J Ethnopharmacol* 103: 319–26.

Muazu, J. and Kaita, A.H. 2008. A review of traditional plants used in the treatment of epilepsy amongst the Hausa/Fulani tribes of northern Nigeria. *Afr J Tradit Complement Altern Med* 5: 387–90.

Mustofa, Valentin, A., Benoit-Vical, F., Pelissier, Y., Kone-Bamba, D. and Mallie, M. 2000. Antiplasmodial activity of plant extracts used in West African traditional medicine. *J Ethnopharmacol* 73: 145–51.

Neerman, M.F., Frost, R.E. and Deking, J. 2013. A drug fatality involving Kratom. *J Forensic Sci* 58 Suppl 1: S278–79.

Nelsen, J.L., Lapoint, J., Hodgman, M.J. and Aldous, K.M. 2010. Seizure and coma following Kratom (*Mitragynina speciosa* Korth) exposure. *J Med Toxicol* 6: 424–26.

Ogata, J., Uchiyama, N., Kikura-Hanajiri, R. and Goda, Y. 2013. DNA sequence analyses of blended herbal products including synthetic cannabinoids as designer drugs. *Forensic Sci Int* 227: 33–41.

Ongley, P.A. 1953. [The alkaloids of *Mitragyna*]. *Ann Pharm Fr* 11: 594–602.

Orio, L., Alexandru, L., Cravotto, G., Mantegna, S. and Barge, A. 2012. UAE, MAE, SFE-CO2 and classical methods for the extraction of *Mitragyna speciosa* leaves. *Ultrason Sonochem* 19: 591–95.

Ouedraogo, S., Ralay Ranaivo, H., Ndiaye, M., Kabore, Z.I., Guissou, I.P., Bucher, B. and Andriantsitohaina, R. 2004. Cardiovascular properties of aqueous extract from *Mitragyna inermis* (wild). *J Ethnopharmacol* 93: 345–50.

Pandey, R., Singh, S.C. and Gupta, M.M. 2006. Heteroyohimbinoid type oxindole alkaloids from *Mitragyna parvifolia*. *Phytochemistry* 67: 2164–69.

Parthasarathy, S., Bin Azizi, J., Ramanathan, S., Ismail, S., Sasidharan, S., Said, M.I. and Mansor, S.M. 2009. Evaluation of antioxidant and antibacterial activities of aqueous, methanolic and alkaloid extracts from *Mitragyna speciosa* (Rubiaceae family) leaves. *Molecules* 14: 3964–74.

Parthasarathy, S., Ramanathan, S., Ismail, S., Adenan, M.I., Mansor, S.M. and Murugaiyah, V. 2010. Determination of mitragynine in plasma with solid-phase extraction and rapid HPLC-UV analysis, and its application to a pharmacokinetic study in rat. *Anal Bioanal Chem* 397: 2023–30.

Parthasarathy, S., Ramanathan, S., Murugaiyah, V., Hamdan, M.R., Said, M.I., Lai, C.S. and Mansor, S.M. 2013. A simple HPLC-DAD method for the detection and quantification of psychotropic mitragynine in *Mitragyna speciosa* (ketum) and its products for the application in forensic investigation. *Forensic Sci Int* 226: 183–87.

Philipp, A.A., Meyer, M.R., Wissenbach, D.K., Weber, A.A., Zoerntlein, S.W., Zweipfenning, P.G. and Maurer, H.H. 2011a. Monitoring of kratom or Krypton intake in urine using GC-MS in clinical and forensic toxicology. *Anal Bioanal Chem* 400: 127–35.

Philipp, A.A., Wissenbach, D.K., Weber, A.A., Zapp, J. and Maurer, H.H. 2010a. Phase I and II metabolites of speciogynine, a diastereomer of the main Kratom alkaloid mitragynine, identified in rat and human urine by liquid chromatography coupled to low- and high-resolution linear ion trap mass spectrometry. *J Mass Spectrom* 45: 1344–57.

Philipp, A.A., Wissenbach, D.K., Weber, A.A., Zapp, J. and Maurer, H.H. 2011b. Metabolism studies of the Kratom alkaloid speciociliatine, a diastereomer of the main alkaloid mitragynine, in rat and human urine using liquid chromatography-linear ion trap mass spectrometry. *Anal Bioanal Chem* 399: 2747–53.

Philipp, A.A., Wissenbach, D.K., Weber, A.A., Zapp, J. and Maurer, H.H. 2011c. Metabolism studies of the Kratom alkaloids mitraciliatine and isopaynantheine, diastereomers of the main alkaloids mitragynine and paynantheine, in rat and human urine using liquid chromatography-linear ion trap-mass spectrometry. *J Chromatogr B Analyt Technol Biomed Life Sci* 879: 1049–55.

Philipp, A.A., Wissenbach, D.K., Weber, A.A., Zapp, J., Zoerntlein, S.W., Kanogsunthornrat, J. and Maurer, H.H. 2010b. Use of liquid chromatography coupled to low- and high-resolution linear ion trap mass spectrometry for studying the metabolism of paynantheine, an alkaloid of the herbal drug Kratom in rat and human urine. *Anal Bioanal Chem* 396: 2379–91.

Philipp, A.A., Wissenbach, D.K., Zoerntlein, S.W., Klein, O.N., Kanogsunthornrat, J. and Maurer, H.H. 2009. Studies on the metabolism of mitragynine, the main alkaloid of the herbal drug Kratom, in rat and human urine using liquid chromatography-linear ion trap mass spectrometry. *J Mass Spectrom* 44: 1249–61.

Phongprueksapattana, S., Putalun, W., Keawpradub, N. and Wungsintaweekul, J. 2008. *Mitragyna speciosa*: Hairy root culture for triterpenoid production and high yield of mitragynine by regenerated plants. *Zeitschrift für Naturforschung C* 63: 691–98.

Pichainarong, N., Chaveepojnkamjorn, W., Khobjit, P., Veerachai, V. and Sujirarat, D. 2004. Energy drinks consumption in male construction workers, Chonburi province. *J Med Assoc Thai* 87: 1454–58.

Pillay, M.S. 1964. Anatomy of the leaves and young stem of *Mitragyna inermis* (O. Kuntze). *J Pharm Pharmacol* 16: 820–27.

Ponglux, D., Wongseripipatana, S., Takayama, H., et al. 1994. A new indole alkaloid, 7 alpha-hydroxy-7H-mitragynine, from *Mitragyna speciosa* in Thailand. *Planta Med* 60: 580–81.

Portoghese, P.S. 1965. A new concept on the mode of interaction of narcotic analgesics with receptors. *J Med Chem* 8: 609–16.

Posch, T.N., Muller, A., Schulz, W., Putz, M. and Huhn, C. 2012. Implementation of a design of experiments to study the influence of the background electrolyte on separation and detection in non-aqueous capillary electrophoresis-mass spectrometry. *Electrophoresis* 33: 583–98.

Prozialeck, W.C., Jivan, J.K. and Andurkar, S.V. 2012. Pharmacology of kratom: An emerging botanical agent with stimulant, analgesic and opioid-like effects. *J Am Osteopath Assoc* 112: 792–99.

Purintrapiban, J., Keawpradub, N., Kansenalak, S., Chittrakarn, S., Janchawee, B. and Sawangjaroen, K. 2011. Study on glucose transport in muscle cells by extracts from *Mitragyna speciosa* (Korth) and mitragynine. *Nat Prod Res* 25: 1379–87.

Raffa, R.B., Beckett, J.R., Brahmbhatt, V.N., et al. 2013. Orally active opioid compounds from a non-poppy source. *J Med Chem* 56: 4840–48.

Raymond, H. 1950. [The alkaloids of *Mitragyna speciosa* Korthals]. *Ann Pharm Fr* 8: 482–90.

Razafimandimbison, S.G. and Bremer, B. 2002. Phylogeny and classification of Naucleeae s.l. (Rubiaceae) inferred from molecular (ITS, rBCL, and tRNT-F) and morphological data. *Am J Bot* 89: 1027–41.

Razafimandimbison, S.G., Kellogg, E.A. and Bremer, B. 2004. Recent origin and phylogenetic utility of divergent ITS putative pseudogenes: A case study from Naucleeae (Rubiaceae). *Syst Biol* 53: 177–92.

Rosenbaum, C.D., Carreiro, S.P. and Babu, K.M. 2012. Here today, gone tomorrow … and back again? A review of herbal marijuana alternatives (K2, Spice), synthetic cathinones (bath salts), kratom, *Salvia divinorum*, methoxetamine, and piperazines. *J Med Toxicol* 8: 15–32.

Sabetghadam, A., Ramanathan, S. and Mansor, S.M. 2010. The evaluation of antinociceptive activity of alkaloid, methanolic, and aqueous extracts of Malaysian *Mitragyna speciosa* Korth leaves in rats. *Pharmacognosy Res* 2: 181–85.

Sabetghadam, A., Ramanathan, S., Sasidharan, S. and Mansor, S.M. 2013. Subchronic exposure to mitragynine, the principal alkaloid of *Mitragyna speciosa*, in rats. *J Ethnopharmacol* 146: 815–23.

Said, I.M., Chun, N.C. and Houghton, P.J. 1991. Ursolic acid from *Mitragyna speciosa*. *Planta Med* 57: 398.

Saingam, D., Assanangkornchai, S., Geater, A.F. and Balthip, Q. 2013. Pattern and consequences of krathom (*Mitragyna speciosa* Korth.) use among male villagers in southern Thailand: A qualitative study. *Int J Drug Policy* 24: 351–58.

Saxton, J.E. (ed.) 1994. The chemistry of heterocyclic compounds, Part 4, Suppl. Vol. 25. *Indoles: The Monoterpenoid Indole Alkaloids*. Chichester: John Wiley & Sons.

Schmidt, M.M., Sharma, A., Schifano, F. and Feinmann, C. 2011. 'Legal highs' on the net: Evaluation of UK-based websites, products and product information. *Forensic Sci Int* 206: 92–97.

Shaik Mossadeq, W.M., Sulaiman, M.R., Tengku Mohamad, T.A., et al. 2009. Anti-inflammatory and antinociceptive effects of *Mitragyna speciosa* Korth methanolic extract. *Med Princ Pract* 18: 378–84.

Shamima, A.R., Fakurazi, S., Hidayat, M.T., Hairuszah, I., Moklas, M.A. and Arulselvan, P. 2012. Antinociceptive action of isolated mitragynine from *Mitragyna speciosa* through activation of opioid receptor system. *Int J Mol Sci* 13: 11427–42.

Sheleg, S.V. and Collins, G.B. 2011. A coincidence of addiction to 'Kratom' and severe primary hypothyroidism. *J Addict Med* 5: 300–301.

Shellard, E.J. 1971. The genus *Mitragyna*. *Pharm Weekbl* 106: 224–36.

Shellard, E.J. 1974. The alkaloids of *Mitragyna* with special reference to those of *Mitragyna speciosa*, Korth. *Bull Narc* 26: 41–55.

Shellard, E.J. 1983. *Mitragyna*: A note on the alkaloids of African species. *J Ethnopharmacol* 8: 345–47.

Shellard, E.J. 1989. Ethnopharmacology of kratom and the *Mitragyna* alkaloids. *J Ethnopharmacol* 25: 123–24.

Shellard, E.J. and Alam, M.Z. 1968a. The quantitative determination of some mitragyna oxindole alkaloids after separation by thin layer chromatography. 3. Densitometry. *J Chromatogr* 33: 347–69.

Shellard, E.J. and Alam, M.Z. 1968b. The quantitative determination of some *Mitragyna* oxindole alkaloids after separation by thin-layer chromatography. II. Colorimetry, using the Vitali-Morin reaction. *J Chromatogr* 32: 489–501.

Shellard, E.J. and Alam, M.Z. 1968c. The quantitative determination of some *Mitragyna* oxindole alkaloids after separation by thin-layer chromatography. IV. Comparison of ultra-violet spectrophotometry, colorimetry and densitometry as methods for the quantitative determination of oxindole alkaloids in plant material. *J Chromatogr* 35: 72–82.

Shellard, E.J. and Alam, M.Z. 1968d. The quantitative determination of some *Mitragyna* oxindole alkaloids after separation by thin-layer chromatograppy. I. Ultraviolet spectrophotometry. *J Chromatogr* 32: 472–88.

Shellard, E.J. and Alam, M.Z. 1968e. The quantitative determination of some *Mitragyna* oxindole alkaloids. I. Ultra violet spectrophotometry. *Planta Med* 16: 127–36.

Shellard, E.J. and Alam, M.Z. 1968f. The quantitative determination of some *Mitragyna* oxindole alkaloids. II. Colorimetry (using the Vitali-Morin reaction). *Planta Med* 16: 248–55.

Shellard, E.J., Beckett, A.H., Tantivatana, P., Phillipson, J.D. and Lee, C.M. 1966. Alkaloids from *Mitragyna javanica*, Koord. and Valeton and *Mitragyna hirsuta*, Havil. *J Pharm Pharmacol* 18: 553–55.

Shellard, E.J., Beckett, A.H., Tantivatana, P., Phillipson, J.D. and Lee, C.M. 1967a. The *Mitragyna* species of Asia. 8. The alkaloids of the leaves of *Mitragyna javanica* var. *microphylla* Koord and Valeton. *Planta Med* 15: 245–54.

Shellard, E.J. and Houghton, P.J. 1971a. The distribution of alkaloids in *Mitragyna parvifolia* (Roxb.) Korth in young plants grown from Ceylon seed. *J Pharm Pharmacol* 23: 245S.

Shellard, E.J. and Houghton, P.J. 1971b. The *Mitragyna* species of Asia. XIX. The alkaloid pattern in *Mitragyna parvifolia* (Roxb.) Korth. *Planta Med* 20: 82–89.

Shellard, E.J. and Houghton, P.J. 1972a. The conversion of pseudo heteroyohimbine alkaloids to oxindoles. II. In vivo studies in *Mitragyna parvifolia* (Roxb.) Korth. *Planta Med* 21: 16–21.

Shellard, E.J. and Houghton, P.J. 1972b. The *Mitragyna* species of Asia. XX. The alkaloidal pattern in *Mitragyna parvifolia* (Roxb.) Korth. from Burma. *Planta Med* 21: 263–66.

Shellard, E.J. and Houghton, P.J. 1972c. The *Mitragyna* species of Asia. XXI. The distribution of alkaloids in young plants of *Mitragyna parvifolia* grown from seed obtained from Ceylon. *Planta Med* 21: 382–92.

Shellard, E.J. and Houghton, P.J. 1972d. The *Mitragyna* species of Asia. XXII. The distribution of alkaloids in young plants of *Mitragyna parvifolia* grown from seeds obtained from Uttar Pradesh State of India. *Planta Med* 22: 97–102.

Shellard, E.J. and Houghton, P.J. 1973a. The *Mitragyna* species of Asia. XXIV. The isolation of dihydrocorynantheol and corynantheidol from the leaves of *Mitragyna parvifolia* (Roxb.) Korth from Sri Lanka (Ceylon). *Planta Med* 24: 13–17.

Shellard, E.J. and Houghton, P.J. 1973b. The *Mitragyna* species of Asia. XXV. 'In vivo' studies, using 14C-alkaloids, in the alkaloidal pattern in young plants of *Mitragyna parvifolia* grown from seed obtained from Ceylon. *Planta Med* 24: 341–52.

Shellard, E.J. and Houghton, P.J. 1974. The *Mitragyna* species of Asia. XXVI. Further 'in vivo' studies using 14C-alkaloids, in the alkaloidal pattern in young plants of *Mitragyna parvifolia* (Roxb.) Korth grown from seed obtained from Sri Lanka (Ceylon). *Planta Med* 25: 80–87.

Shellard, E.J. and Lala, P.K. 1978. The alkaloids of *Mitragyna rubrostipulata* (Schum.) Havil. *Planta Med* 33: 63–69.

Shellard, E.J. and Lees, M.D. 1965. The *Mitragyna* species of Asia. V. The anatomy of the leaves of *Mitragyna speciosa* Korth. *Planta Med* 13: 280–90.

Shellard, E.J., Phillipson, J.D. and Gupta, D. 1968a. The *Mitragyna* species of Asia. 13. The alkaloids of the leaves of *Mitragyna parvifolia* (Roxb.) Korth. obtained from India. *Planta Med* 16: 436–45.

Shellard, E.J., Phillipson, J.D. and Gupta, D. 1968b. The *Mitragyna* species of Asia. XI. The alkaloids of the leaves of *Mitragyna parvifolia* (R oxb.) Korth. obtained from the Maharashtra State of India. *Planta Med* 16: 20–8.

Shellard, E.J., Phillipson, J.D. and Gupta, D. 1969. The *Mitragyna* species of Asia. XV. The alkaloids from the bark of *Mitragyna parvifolia* (Roxb.) Korth and a possible biogenetic route for the oxindole alkaloids. *Planta Med* 17: 146–63.

Shellard, E.J. and Rungsiyakul, D. 1973. The *Mitragyna* species of Asia. 23. The alkaloids of the leaves of *Mitragyna tubulosa* (Havil). *Planta Med* 23: 221–25.

Shellard, E.J. and Sarpong, K. 1969. The alkaloids of the leaves of *Mitragyna inermis* (Willd.) O. Kuntze. *J Pharm Pharmacol* 21 Suppl: 113S+.

Shellard, E.J. and Sarpong, K. 1970. The alkaloidal pattern in the leaves, stem-bark and root-bark of *Mitragyna* species from Ghana. *J Pharm Pharmacol* Suppl: 34S+.

Shellard, E.J. and Sarpong, K. 1971. Isolation of speciogynine from the leaves of *Mitragyna inermis* (Willd.), O. Kuntze. *J Pharm Pharmacol* 23: 559–60.

Shellard, E.J. and Shadan, P. 1963. The *Mitragyna* species of Ghana. The anatomy of the leaves of *Mitragyna stipulosa* (D.C.) O. Kuntze and *Mitragyna ciliata* Aubr. Et Pellegr. *J Pharm Pharmacol* 15 Suppl: 278–91.

Shellard, E.J., Tantivatana, P. and Beckett, A.H. 1967b. The *Mitragyna* species of Asia. X. The alkaloids of the leaves of *Mitragyna hirsuta* Havil. *Planta Med* 15: 366–70.

Shellard, E.J. and Wade, A. 1967. The morphology and anatomy of the flowers of *Mitragyna ciliata* Aubr. et Pellegr. and *Mitragyna stipulosa* (D.C.) O. Kuntze. *J Pharm Pharmacol* 19: 744–59.

Shellard, E.J. and Wade, A. 1969. The morphology and anatomy of the flowers of *Mitragyna inermis* (Willd.) O. Kuntze. *J Pharm Pharmacol* 21 Suppl: 102S+.

Shukla, A. and Srinivasan, B.P. 2012. 16,17-Dihydro-17b-hydroxy isomitraphylline alkaloid as an inhibitor of DPP-IV, and its effect on incretin hormone and beta-cell proliferation in diabetic rat. *Eur J Pharm Sci* 47: 512–19.

Sinou, V., Fiot, J., Taudon, N., et al. 2010. High-performance liquid chromatographic method for the quantification of *Mitragyna inermis* alkaloids in order to perform pharmacokinetic studies. *J Sep Sci* 33: 1863–69.

Stolt, A.C., Schroder, H., Neurath, H., et al. 2014. Behavioral and neurochemical characterization of kratom (*Mitragyna speciosa*) extract. *Psychopharmacology (Berl)* 231: 13–25.

Sufka, K.J., Loria, M.J., Lewellyn, K., Zjawiony, J.K., Ali, Z., Abe, N. and Khan, I.A. 2014. The effect of *Salvia divinorum* and *Mitragyna speciosa* extracts, fraction and major constituents on place aversion and place preference in rats. *J Ethnopharmacol* 151: 361–64.

Sukrong, S., Zhu, S., Ruangrungsi, N., Phadungcharoen, T., Palanuvej, C. and Komatsu, K. 2007. Molecular analysis of the genus *Mitragyna* existing in Thailand based on rDNA ITS sequences and its application to identify a narcotic species: *Mitragyna speciosa. Biol Pharm Bull* 30: 1284–88.

Sun, X. and Ma, D. 2011. Organocatalytic approach for the syntheses of corynantheidol, dihydrocorynantheol, protoemetinol, protoemetine, and mitragynine. *Chem Asian J* 6: 2158–65.

Suwanlert, S. 1975. A study of kratom eaters in Thailand. *Bull Narc* 27: 21–27.

Sy, G.Y., Sarr, A., Dieye, A.M. and Faye, B. 2004. Myorelaxant and antispasmodic effects of the aqueous extract of *Mitragyna inermis* barks on Wistar rat ileum. *Fitoterapia* 75: 447–50.

Takayama, H. 2004. Chemistry and pharmacology of analgesic indole alkaloids from the rubiaceous plant, *Mitragyna speciosa. Chem Pharm Bull (Tokyo)* 52: 916–28.

Takayama, H., Aimi, N. and Sakai, S. 2000. [Chemical studies on the analgesic indole alkaloids from the traditional medicine (*Mitragyna speciosa*) used for opium substitute]. *Yakugaku Zasshi* 120: 959–67.

Takayama, H., Ishikawa, H., Kitajima, M. and Aimi, N. 2002a. Formation of an unusual dimeric compound by lead tetraacetate oxidation of a corynanthe-type indole alkaloid, mitragynine. *Chem Pharm Bull (Tokyo)* 50: 960–63.

Takayama, H., Ishikawa, H., Kitajima, M., Aimi, N. and Aji, B.M. 2004. A new 9-methoxyyohimbine-type indole alkaloid from *Mitragyna africanus. Chem Pharm Bull (Tokyo)* 52: 359–61.

Takayama, H., Ishikawa, H., Kurihara, M., Kitajima, M., Sakai, S., Aimi, N., Seki, H., Yamaguchi, K., Said, I. M., and Houghton, P. J. 2001. Structure revision of mitragynaline, an indole alkaloid in *Mitragyna speciosa. Tetrahedron Lett* 42: 1741–43.

Takayama, H., Ishikawa, H., Kurihara, M., et al. 2002b. Studies on the synthesis and opioid agonistic activities of mitragynine-related indole alkaloids: Discovery of opioid agonists structurally different from other opioid ligands. *J Med Chem* 45: 1949–56.

Takayama, H., Kitajima, M., and Kogure, N. 2005. Chemistry of Corynanthe-type related indole alkaloids from the *Uncaria*, *Nauclea*, and *Mitragyna* plants. *Curr Org Chem* 9: 1445–64.

Takayama, H., Kurihara, M., Kitajima, M., Said, I.M. and Aimi, N. 1999. Isolation and asymmetric total synthesis of a new *Mitragyna* indole alkaloid, (-)-itralactonine. *J Org Chem* 64: 1772–73.

Takayama, H., Kurihara, M., Kitajima, M., Said, I. M., and Aimi, N. 2000b. Structure elucidation and chiral-total synthesis of a new indole alkaloid, (–)-9-methoxymitralactonine, isolated from *Mitragyna speciosa* in Malaysia. *Tetrahedron* 56: 3145–51.

Takayama, H., Kurihara, M., Subhadhirasakul, S., Kitajima, M., Aimi, N., and Sakai, S. 1996. Stereochemical assignment of pseudoindoxyl alkaloids. *Heterocycles* 42: 87–92.

Takayama, H., Maeda, M., Ohbayashi, S., Kitajima, M., Sakai, S., and Aimi, N. 1995. The first total synthesis of (–)-mitragynine, an analgesic indole alkaloid in *Mitragyna speciosa. Tetrahedron Lett* 36: 9337–40.

Takayama, H., Misawa, K., Okada, N., et al. 2006. New procedure to mask the 2,3-pi bond of the indole nucleus and its application to the preparation of potent opioid receptor agonists with a Corynanthe skeleton. *Org Lett* 8: 5705–08.

Tapondjou, L.A., Lontsi, D., Sondengam, B.L., Choudhary, M.I., Park, H.J., Choi, J. and Lee, K.T. 2002. Structure-activity relationship of triterpenoids isolated from *Mitragyna stipulosa* on cytotoxicity. *Arch Pharm Res* 25: 270–74.

Thongpradichote, S., Matsumoto, K., Tohda, M., Takayama, H., Aimi, N., Sakai, S. and Watanabe, H. 1998. Identification of opioid receptor subtypes in antinociceptive actions of supraspinally-administered mitragynine in mice. *Life Sci* 62: 1371–78.

Tohda, M., Thongpraditchote, S., Matsumoto, K., et al. 1997. Effects of mitragynine on cAMP formation mediated by delta-opiate receptors in NG108-15 cells. *Biol Pharm Bull* 20: 338–40.

Toure, H., Balansard, G., Pauli, A.M. and Scotto, A.M. 1996. Pharmacological investigation of alkaloids from leaves of *Mitragyna inermis* (Rubiaceae). *J Ethnopharmacol* 54: 59–62.

Traore, F., Gasquet, M., Laget, M., et al. 2000. Toxicity and genotoxicity of antimalarial alkaloid rich extracts derived from *Mitragyna inermis* O. Kuntze and *Nauclea latifolia. Phytother Res* 14: 608–11.

Traore-Keita, F., Gasquet, M., Di Giorgio, C., et al. 2000. Antimalarial activity of four plants used in traditional medicine in Mali. *Phytother Res* 14: 45–57.

Tsuchiya, S., Miyashita, S., Yamamoto, M., et al. 2002. Effect of mitragynine, derived from Thai folk medicine, on gastric acid secretion through opioid receptor in anesthetized rats. *Eur J Pharmacol* 443: 185–88.

Ulbricht, C., Costa, D., Dao, J., Isaac, R., LeBlanc, Y.C., Rhoades, J. and Windsor, R.C. 2013. An evidence-based systematic review of kratom (*Mitragyna speciosa*) by the Natural Standard Research Collaboration. *J Diet Suppl* 10: 152–70.

Utar, Z., Majid, M.I., Adenan, M.I., Jamil, M.F. and Lan, T.M. 2011. Mitragynine inhibits the COX-2 mRNA expression and prostaglandin E(2) production induced by lipopolysaccharide in RAW264.7 macrophage cells. *J Ethnopharmacol* 136: 75–82.

Vicknasingam, B., Narayanan, S., Beng, G.T. and Mansor, S.M. 2010. The informal use of ketum (*Mitragyna speciosa*) for opioid withdrawal in the northern states of peninsular Malaysia and implications for drug substitution therapy. *Int J Drug Policy* 21: 283–88.

Vuppala, P.K., Jamalapuram, S., Furr, E.B., McCurdy, C.R. and Avery, B.A. 2013. Development and validation of a UPLC-MS/MS method for the determination of 7-hydroxymitragynine, a mu-opioid agonist, in rat plasma and its application to a pharmacokinetic study. *Biomed Chromatogr* 27(12): 1726–32.

Ward, J., Rosenbaum, C., Hernon, C., McCurdy, C.R. and Boyer, E.W. 2011. Herbal medicines for the management of opioid addiction: Safe and effective alternatives to conventional pharmacotherapy? *CNS Drugs* 25: 999–1007.

Watanabe, K., Yano, S., Horie, S. and Yamamoto, L.T. 1997. Inhibitory effect of mitragynine, an alkaloid with analgesic effect from Thai medicinal plant *Mitragyna speciosa*, on electrically stimulated contraction of isolated guinea-pig ileum through the opioid receptor. *Life Sci* 60: 933–42.

Watanabe, K., Yano, S., Horie, S., Yamamoto, L. T., Takayama, H., Aimi, N., Sakai, S., Ponglux, D., Tongroach, P., Shan, J., and Pang, P.K.T. 1999. Pharmacological properties of some structurally related indole alkaloids contained in the Asian herbal medicines, hirsutine and mitragynine, with special reference to their Ca^{2+} antagonistic and opioid-like effects. In *Pharmacological Research on Traditional Herbal Medicines*, ed. by Watanabe, H., and Shibuya, T., 163–77. Amsterdam: Harwood Academic Publishers.

Wetzel, U., Boldt, A., Lauschke, J., et al. 2005. Expression of connexins 40 and 43 in human left atrium in atrial fibrillation of different aetiologies. *Heart* 91: 166–70.

Winterfeldt, E. 1971. Reaktionen an Indolderivaten, XIII. Chinolon-Derivate durch Autoxydation. *Liebigs Annalen der Chemie* 745: 23–30.

Wungsintaweekul, J., Choo-Malee, J., Charoonratana, T. and Keawpradub, N. 2012. Methyl jasmonate and yeast extract stimulate mitragynine production in *Mitragyna speciosa* (Roxb.) Korth. shoot culture. *Biotechnol Lett* 34: 1945–50.

Xu, J., Dong, J., Ye, X., Ma, X. and Zhang, H. 2009. [Studies on triterpenoid saponins in *Hemsleya chinensis*]. *Zhongguo Zhong Yao Za Zhi* 34: 291–93.

Yamamoto, L.T., Horie, S., Takayama, H., et al. 1999. Opioid receptor agonistic characteristics of mitragynine pseudoindoxyl in comparison with mitragynine derived from Thai medicinal plant *Mitragyna speciosa*. *Gen Pharmacol* 33: 73–81.

Zarembo, J.E., Douglas, B., Valenta, J. and Weisbach, J.A. 1974. Metabolites of mitragynine. *J Pharm Sci* 63: 1407–15.

Index

A

Acetaminophen products, 270
Acetylcholine, 23, 27
Acetylcholinesterase (AChE), 25, 28
AChE, *see* Acetylcholinesterase
ACTH, *see* Adrenocorticotropic hormone
AD, *see* Alzheimer's disease
ADH, *see* Aldehyde dehydrogenases
ADMC of mitragynine and analogs,
    167–175
  absorption, 167–168
    bioavailability, 168
    physiological factors, 168
    xenobiotics, oral ingestion of, 168
  distribution, 168
  metabolism of mitragynine, 170–174
    analogs, 174
    mitragynine, 171–174
  pharmacokinetic profile of
      mitragynine, 170
  xenobiotic metabolism, 168–170
    aldehyde dehydrogenases, 169
    epoxide hydrolases, 169
    flavin-containing
        monooxygenases, 169
    glutathion transferases, 170
    phase I, 169
    phase II, 170
    UDP-glucuronosyl transferases, 170
Adrenocorticotropic hormone (ACTH),
    262
Adulterants, 69
Adulterated kratom products, 290
Adverse effects and toxicity, opioid-
    induced, 257–279
  bowel dysfunction, 259–261
    chemotherapy-induced nausea, 260
    constipation, 260
    dysphagia, 260–261
    motor neurons, 259
    nausea and vomiting, 260
    neurotransmitters, 259

sphincter of Oddi disorder, 259
vasoactive intestinal peptide, 259
xerostomia, 261
endocrinopathy, 261–263
  adrenocorticotropic hormone, 262
  bone mineral density, 262
  corticotropinreleasing hormone, 261
  dehydroepiandrosterone, 262
  follicle-stimulating hormone, 261
  gonadotropin-releasing hormone, 261
  hypogonadism, 262
  luteinizing hormone, 261
kappa-opioid-receptor agonists, 258
mu-opioid-receptor agonists, 258
neuroinhibition and excitation, 264–270
  agitation, 267
  cognitive impairment, 266
  delirium, 267–268
  disordered sleep architecture, 266
  fatigue, 265–266
  hallucinations, 267
  hydromorphone, 265
  hyperalgesia, 268
  methadone, 265
  myoclonus, 269–270
  neurotoxicity, 270
  REM sleep, 266
  selective tolerance, 269
  somnolence, 265
  tolerance, 268–269
neuronal apoptosis, 258
other effects, 270–272
  depression, 272
  headache, 270–271
  itch center, 271
  medication-overuse headache, 270
  pruritus, 271–272
  urinary retention, 272
respiratory effects, 263–264
  airway resistance, 263
  ataxia, 264
  awake respiration, 264
  central respiratory pacemakers, 263

Printed in the United States
by Baker & Taylor Publisher Services